AF726537

IET CONTROL, ROBOTICS AND SENSORS SERIES 114

Signal Processing and Machine Learning for Brain–Machine Interfaces

Other volumes in this series:

Volume 8 **A History of Control Engineering, 1800–1930** S. Bennett
Volume 18 **Applied Control Theory, 2nd Edition** J.R. Leigh
Volume 20 **Design of Modern Control Systems** D.J. Bell, P.A. Cook and N. Munro (Editors)
Volume 28 **Robots and Automated Manufacture** J. Billingsley (Editor)
Volume 33 **Temperature Measurement and Control** J.R. Leigh
Volume 34 **Singular Perturbation Methodology in Control Systems** D.S. Naidu
Volume 35 **Implementation of Self-tuning Controllers** K. Warwick (Editor)
Volume 37 **Industrial Digital Control Systems, 2nd Edition** K. Warwick and D. Rees (Editors)
Volume 39 **Continuous Time Controller Design** R. Balasubramanian
Volume 40 **Deterministic Control of Uncertain Systems** A.S.I. Zinober (Editor)
Volume 41 **Computer Control of Real-time Processes** S. Bennett and G.S. Virk (Editors)
Volume 42 **Digital Signal Processing: Principles, devices and applications** N.B. Jones and J.D.McK. Watson (Editors)
Volume 44 **Knowledge-based Systems for Industrial Control** J. McGhee, M.J. Grimble and A. Mowforth (Editors)
Volume 47 **A History of Control Engineering, 1930–1956** S. Bennett
Volume 49 **Polynomial Methods in Optimal Control and Filtering** K.J. Hunt (Editor)
Volume 50 **Programming Industrial Control Systems Using IEC 1131-3** R.W. Lewis
Volume 51 **Advanced Robotics and Intelligent Machines** J.O. Gray and D.G. Caldwell (Editors)
Volume 52 **Adaptive Prediction and Predictive Control** P.P. Kanjilal
Volume 53 **Neural Network Applications in Control** G.W. Irwin, K. Warwick and K.J. Hunt (Editors)
Volume 54 **Control Engineering Solutions: A practical approach** P. Albertos, R. Strietzel and N. Mort (Editors)
Volume 55 **Genetic Algorithms in Engineering Systems** A.M.S. Zalzala and P.J. Fleming (Editors)
Volume 56 **Symbolic Methods in Control System Analysis and Design** N. Munro (Editor)
Volume 57 **Flight Control Systems** R.W. Pratt (Editor)
Volume 58 **Power-plant control and instrumentation: The control of boilers and HRSG systems** D. Lindsley
Volume 59 **Modelling Control Systems Using IEC 61499** R. Lewis
Volume 60 **People in Control: Human factors in control room design** J. Noyes and M. Bransby (Editors)
Volume 61 **Nonlinear Predictive Control: Theory and practice** B. Kouvaritakis and M. Cannon (Editors)
Volume 62 **Active Sound and Vibration Control** M.O. Tokhi and S.M. Veres
Volume 63 **Stepping Motors, 4th Edition** P.P. Acarnley
Volume 64 **Control Theory, 2nd Edition** J.R. Leigh
Volume 65 **Modelling and Parameter Estimation of Dynamic Systems** J.R. Raol, G. Girija and J. Singh
Volume 66 **Variable Structure Systems: From principles to implementation** A. Sabanovic, L. Fridman and S. Spurgeon (Editors)
Volume 67 **Motion Vision: Design of compact motion sensing solution for autonomous systems** J. Kolodko and L. Vlacic
Volume 68 **Flexible Robot Manipulators: Modelling, simulation and control** M.O. Tokhi and A.K.M. Azad (Editors)
Volume 69 **Advances in Unmanned Marine Vehicles** G. Roberts and R. Sutton (Editors)
Volume 70 **Intelligent Control Systems Using Computational Intelligence Techniques** A. Ruano (Editor)
Volume 71 **Advances in Cognitive Systems** S. Nefti and J. Gray (Editors)
Volume 72 **Control Theory: A guided tour, 3rd Edition** J.R. Leigh
Volume 73 **Adaptive Sampling with Mobile WSN** K. Sreenath, M.F. Mysorewala, D.O. Popa and F.L. Lewis
Volume 74 **Eigenstructure Control Algorithms: Applications to aircraft/rotorcraft handling qualities design** S. Srinathkumar
Volume 75 **Advanced Control for Constrained Processes and Systems** F. Garelli, R.J. Mantz and H. De Battista
Volume 76 **Developments in Control Theory towards Global Control** L. Qiu, J. Chen, T. Iwasaki and H. Fujioka (Editors)
Volume 77 **Further Advances in Unmanned Marine Vehicles** G.N. Roberts and R. Sutton (Editors)
Volume 78 **Frequency-Domain Control Design for High-Performance Systems** J. O'Brien
Volume 80 **Control-oriented Modelling and Identification: Theory and practice** M. Lovera (Editor)
Volume 81 **Optimal Adaptive Control and Differential Games by Reinforcement Learning Principles** D. Vrabie, K. Vamvoudakis and F. Lewis
Volume 83 **Robust and Adaptive Model Predictive Control of Nonlinear Systems** M. Guay, V. Adetola and D. DeHaan
Volume 84 **Nonlinear and Adaptive Control Systems** Z. Ding
Volume 86 **Modeling and Control of Flexible Robot Manipulators, 2nd Edition** M.O. Tokhi and A.K.M. Azad
Volume 88 **Distributed Control and Filtering for Industrial Systems** M. Mahmoud
Volume 89 **Control-based Operating System Design** A. Leva *et al.*
Volume 90 **Application of Dimensional Analysis in Systems Modelling and Control Design** P. Balaguer
Volume 91 **An Introduction to Fractional Control** D. Valério and J. Costa
Volume 92 **Handbook of Vehicle Suspension Control Systems** H. Liu, H. Gao and P. Li
Volume 93 **Design and Development of Multi-Lane Smart Electromechanical Actuators** F.Y. Annaz
Volume 94 **Analysis and Design of Reset Control Systems** Y. Guo, L. Xie and Y. Wang
Volume 95 **Modelling Control Systems Using IEC 61499, 2nd Edition** R. Lewis and A. Zoitl
Volume 96 **Cyber-Physical System Design with Sensor Networking Technologies** S. Zeadally and N. Jabeur (Editors)
Volume 99 **Practical Robotics and Mechatronics: Marine, Space and Medical Applications** I. Yamamoto
Volume 100 **Organic Sensors: Materials and Applications** E. Garcia-Breijo and P Cosseddu (Editors)
Volume 102 **Recent Trends in Sliding Mode Control** L. Fridman J.P. Barbot and F. Plestan (Editors)
Volume 104 **Control of Mechatronic Systems** L. Guvenc, B.A. Guvenc, B. Demirel and M.T Emirler
Volume 105 **Mechatronic Hands: Prosthetic and Robotic Design** P.H. Chappell
Volume 107 **Solved Problems in Dynamical Systems and Control** D. Valério, J. T Machado, A.M. Lopes and A.M. Galhano
Volume 108 **Wearable Exoskeleton Systems: Design, Control and Applications** S. Bai, G.S. Virk and T.G. Sugar
Volume 111 **The Inverted Pendulum in Control Theory and Robotics: From Theory to New Innovations** O. Boubaker and R. Iriarte (Editors)
Volume 112 **RFID Protocol Design, Optimization, and Security for the Internet of Things** A.X. Liu, M. Shahzad, X Liu and K. Li
Volume 113 **Design of Embedded Robust Control Systems Using MATLAB®/Simulink®** P.H. Petkov, T.N. Slavov and J.K. Kralev
Volume 119 **Swarm Intelligence Volumes 1–3** Y. Tan (Editor)
Volume 121 **Integrated Fault Diagnosis and Control Design of Linear Complex Systems** M. Davoodi, N. Meskin and K. Khorasani

Signal Processing and Machine Learning for Brain–Machine Interfaces

Edited by
Toshihisa Tanaka and Mahnaz Arvaneh

The Institution of Engineering and Technology

Published by The Institution of Engineering and Technology, London, United Kingdom

The Institution of Engineering and Technology is registered as a Charity in England & Wales (no. 211014) and Scotland (no. SC038698).

© The Institution of Engineering and Technology 2018

First published 2018

This publication is copyright under the Berne Convention and the Universal Copyright Convention. All rights reserved. Apart from any fair dealing for the purposes of research or private study, or criticism or review, as permitted under the Copyright, Designs and Patents Act 1988, this publication may be reproduced, stored or transmitted, in any form or by any means, only with the prior permission in writing of the publishers, or in the case of reprographic reproduction in accordance with the terms of licences issued by the Copyright Licensing Agency. Enquiries concerning reproduction outside those terms should be sent to the publisher at the undermentioned address:

The Institution of Engineering and Technology
Michael Faraday House
Six Hills Way, Stevenage
Herts, SG1 2AY, United Kingdom

www.theiet.org

While the authors and publisher believe that the information and guidance given in this work are correct, all parties must rely upon their own skill and judgement when making use of them. Neither the authors nor publisher assumes any liability to anyone for any loss or damage caused by any error or omission in the work, whether such an error or omission is the result of negligence or any other cause. Any and all such liability is disclaimed.

The moral rights of the authors to be identified as authors of this work have been asserted by them in accordance with the Copyright, Designs and Patents Act 1988.

British Library Cataloguing in Publication Data
A catalogue record for this product is available from the British Library

ISBN 978-1-78561-398-2 (hardback)
ISBN 978-1-78561-399-9 (PDF)

Typeset in India by MPS Ltd

Contents

Preface

A brain–machine interface (BMI) or brain–computer interface (BCI) provides a new pathway for communication and control through decoding information directly extracted from the neurophysiological signals. Such technology holds great promise as a basis for assisting people with severe disabilities. In addition, BMI can serve as a man–machine interaction channel for healthy users. Despite successful proof-of-concept demonstrations in the recent years, the technology underlying BMI is not yet mature enough for daily use out of the laboratory. To make BMI a reliable technology for daily use, several improvements are required in the areas of usability, signal acquisition, system integration, machine learning and signal processing. This book focuses on machine learning and signal processing part with the aim of improving the accuracy, robustness and usability of BMI.

In the majority of the current BMI systems, brain activity is measured through the electroencephalography (EEG), due to its low cost and high temporal resolution. However, EEG signals have a poor spatial resolution. A signal of interest is mixed with a set of irrelevant signals from concurrent brain activities while they may have similar activation frequencies and amplitude. Moreover, EEG signals are non-stationary, and they may be distorted by artifacts such as electrooculogram or electromyogram. These drawbacks can result in inaccurate and deteriorated BMI performances.

This present book covers numerous examples of advanced machine-learning and signal processing algorithms to robustly decode EEG signals, despite their low spatial resolution, their noisy and nonstationary nature. These algorithms are based on a number of advanced techniques including optimal spatial filtering, tangent-space mapping, neural networks and deep learning, transfer learning, parametric modeling, supervised connectivity analysis, supervised and unsupervised adaptation, and incorporating signal structures, among many others. Importantly, this book goes beyond the EEG-decoding challenge and discusses the importance of using signal processing and machine-learning methods to model and update the user's sates and skills over time. These user's models could help design a better EEG decoder (features and classifier) that not only leads to a good discrimination of BMI commands but also facilitates user learning.

It is good to note that in the present book a variety of different BMI types are used to evaluate the discussed signal processing and machine-learning algorithms, such as motor imagery, steady-state evoked potentials, rapid serial visual presentation (RSVP), P300 and other evoked related potentials. Furthermore, apart from traditional applications, such as controlling assistive devices and spellers, a number of advanced BMI applications including neurofeedback gaming and music perception are presented.

The book starts with the introduction to BCI/BMI in Chapter 1. Then, the rest of the book consists of five parts as follows.

The first part of the book is devoted to feature extraction methods, which is a core part of any BMI algorithms. In Chapter 2, Li et al. propose a discriminative learning of connectivity pattern of EEG data. In Chapter 3, Sartori et al. proposed a feature extraction of EEG using covariance matrices on the Riemannian manifold. Furthermore, in Chapter 4, Higashi and Tanaka utilize graph structures to extracting features. These three chapters are related to the recent advances in signal processing.

The second part deals with adaptation. The adaptation is very important to reduce dependency on each subject. A key of adaptation is transfer learning. Chapter 5 by Azab et al. reviews transfer learning algorithms and in Chapter 6, Kindermans et al. propose a method of unsupervised learning to circumvent the conventional supervised calibration process. In Chapter 7, Raza and Rathee describe a novel approach to covariance shift detection in terms of EEG-based BCI. In Chapter 8, Lotte et al. address a challenge of BCI concerning the modeling of the user and understanding and improving how and what the user is learning.

The third part is devoted to detection of event-related potentials (ERPs), which are very important in BMI/BCI as well as neurophysiology. In Chapter 9, Cecotti investigates the performance of feedforward deep neural networks to extract ERPs. In Chapter 10, Marghi et al. describe parametric modeling in high-speed ERP-based BCIs.

The fourth part has two chapters that address BCI/BMI using visual stimulation. In Chapter 11, Nakanishi et al. describe how to decode steady-state visually evoked potentials. In Chapter 12, Wang *et al.* review BCI systems based on RSVP.

The final part of this paper describes new and interesting applications of BCI/BMI. Stober and Sternin present music-based BCI using deep learning in Chapter 13. Vinod and Thomas explore state-of the-art BCI technology in neuro-feedback games in Chapter 14.

As the editors, we hope the chapters in the book will enrich the knowledge of our readers and stimulate them. We hope this book will help move toward a more efficient and effective BMI/BCI technology that can be used in practice, outside laboratories.

In the end, we would like to express our gratitude to all the chapter contributors. The editing process was very exciting and stimulating. We would like to thank Kaori Suefusa, a PhD student in the Tokyo University of Agriculture and Technology for her great assistance throughout editing this book. Moreover, we owe our deepest gratitude to a team of book editors in IET for a kind and helpful support. We could not have published this book without their support.

Toshihisa Tanaka and Mahnaz Arvaneh

Chapter 1

Brain–computer interfaces and electroencephalogram: basics and practical issues

Mahnaz Arvaneh[1] *and Toshihisa Tanaka*[2]

Abstract

This chapter is a comprehensive overview of noninvasive brain–machine interface (BMI). Though intended for nonspecialists, it contains some technical details and the background materials for the rest of the book. After introducing the core components of the BMI systems, this chapter describes various possibilities for brain activity measurements. It then emphasizes on electroencephalogram (EEG), which will be used as the source of the signals for BMI in the rest of the book. Next, possible standard preprocessing algorithms commonly used in EEG-based BMIs are illustrated along with the main categories of features extracted from EEG and used for classifications. Finally, some possible applications of BMI are described.

1.1 Introduction

Over the last four decades, brain–computer interface (BCI) has attracted a lot of attention, triggered by new scientific progress in understanding brain functions and by impressive applications. The definition of BCI as quoted from the first BCI international meeting in 1999 is: "A brain–computer interface is a communication system that does not depend on the brain's normal output pathways of peripheral nerves and muscles" [1]. Following this definition, a BCI system is supposed to work for completely paralyzed patients that are restricted to the use of brain activity signals and cannot control their muscles. The idea of a BCI system is to exploit brain patterns associated to mental activities (conscious or unconscious) in order to provide a means of communication for patients that do not have enough control over their motor system (muscles) to communicate in a normal way.

[1]Department of Automatic Control and Systems Engineering, University of Sheffield, United Kingdom
[2]Department of Electronic and Information Engineering, Tokyo University of Agriculture and Technology, Japan

Before the BCI technology can reach to its full potential, there are many critical issues that have to be addressed. The main challenges include improvement of signal acquisition techniques, development of feature extraction methods and translation algorithms, development of the hardware, improvement of operational protocols, and user training strategies [2–4]. In order to address these challenges, many BCI approaches have been proposed. Even though there is no standard design that represents a BCI, core components, techniques for measuring the brain activity, and neurophysiological signals used to drive them are similar, and they are briefly described in this chapter.

We would like to mention that BCI is nowadays also referred to as brain–machine interface (BMI), which is an interface between the human brain and a machine. Two terminologies, BMI and BCI, do not have clear difference. However, in general, BMI has a wider concept than BCI. In the case of invasive recording of the brain activity, BMI is more often used in the literature. Therefore, in the title of this book and in the rest of this chapter, BMI is adopted. In the following, this chapter is devoted to a comprehensive overview of noninvasive BMI. Technical details and the background materials to understand the rest of the book are described. More precisely, the rest of this chapter includes

- the introduction of the core components of the BMI systems;
- various possibilities for brain activity measurements;
- the basic concepts of EEG;
- standard preprocessing, feature extraction, and machine-learning algorithms commonly used in EEG-based BMIs;
- some possible applications of BMI.

1.2 Core components of a BMI system

The whole architecture of an online BMI system is summarized in Figure 1.1. As shown in Figure 1.1, the BMI input is the brain signals carrying informative neural features. BMI outputs can be letter or icon selection on a computer screen, a wheelchair control, neuroprosthesis, etc. [5–7]. Each BMI uses specific algorithms to translate its input into command signals to control the output device. The core components of a BMI system are as follows:

1. Data acquisition unit: This part is responsible for recording brain activities using various types of sensors. After amplification and digitization, the recorded brain signals serve as BMI inputs.
2. Preprocessing unit: This unit reduces noise and artifacts present in the brain signals in order to enhance the relevant information hidden in the input signals.
3. Feature extraction: The feature extractor transforms the preprocessed signals into feature values that correspond to the underlying neurological mechanism. These features are employed by BMI for controlling the output device.
4. Classification unit: This part is responsible for identifying the intention of the user from the extracted features.

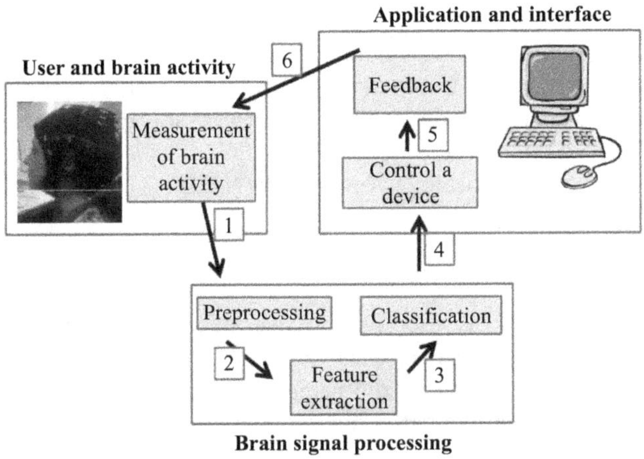

Figure 1.1 General architecture of an online brain–computer interface

5. Output device: The output device can be a computer, a wheelchair or a robotic arm, etc. The output of the classifier is used as a command to control the output device.
6. Feedback: Ideally, BMI should be a closed loop system such that the system shows the output (the identified mental state) to the user after processing the brain signals. A feedback can be in visual, auditory, or tactile form and helps the user control his brain activities and adapt accordingly to enhance the overall performance of BMI.

All the aforementioned units are highly important in the development of an efficient BMI and affect the BMI performance in terms of accuracy, speed, and information transfer rate. BMI must be designed such that it is comfortably carried out without any harm to the user's health. The choice of brain signal measurement technique is of great importance regarding the health, safety, and comfort of BMI users. The following section presents a few noninvasive brain signal recording techniques exploited in BMI applications.

1.3 Signal acquisition

Numerous techniques have been used to measure brain activity. These techniques fall into two groups: invasive and noninvasive [8]. The noninvasive techniques acquire data through the scalp. On the other hand, the invasive techniques record data directly from the skull and they require surgical interventions to implant the electrodes. Implanted electrodes give a better signal quality and spatial resolution than noninvasive techniques. But, the main drawback of the invasive BMI is that the implanted electrodes have limited lifetime which necessitates regular surgery operations for electrode replacement. They might be harmful to the health of the subject too. Therefore, the noninvasive techniques are preferred in the BMI systems than invasive measurement techniques [9]. A number of noninvasive techniques explored in BMI applications are briefly described here including electroencephalogram (EEG), positron emission tomography (PET), magnetoencephalography (MEG), functional magnetic resonance imaging (fMRI), and near-infrared spectroscopy (NIRS).

1.3.1 Electroencephalography

EEG [10] is the most widely used noninvasive technique for recording electrical brain activity. EEG records the electrical signals generated by action potentials of the neurons in the brain from the scalp using small metal electrodes. EEG is often called scalp EEG to distinguish from noninvasive electrical recordings, which is called intracranial EEG or electrocorticogram. The advantages of EEG are essentially the relatively low cost, high temporal resolution, and the noninvasiveness. Low spatial resolution due to volume conduction [11] and a high level of noise and variability in the EEG signal [12] are some of the disadvantages [13]. EEG is used in a variety of BMI systems.

1.3.2 Positron emission tomography

PET [14] indirectly measures metabolism on a cellular level by tracking injected radioactive isotopes. It is based on the principle that in areas of increased activity the metabolism is on a higher level, and more isotopes are supplied by the blood flow. This knowledge can be used to determine which areas are generating activity. Good spatial resolution is an advantage of PET. The really bad temporal resolution (about 2 min) is a distinct disadvantage. This is due to the fact that metabolism is a relatively slow process. Moreover, ionizing radiation makes this method harmful for the human body and thus unusable for daily applications like BMI.

1.3.3 Magnetoencephalography

MEG [15] directly measures the cortical magnetic fields produced by electrical currents. This technique is noninvasive and has good spatial and temporal resolutions. However, the equipment is extremely expensive, and due to the very weak magnetic fields, it requires a very impractical isolation/shielding room (cage of Faraday).

1.3.4 Functional magnetic resonance imaging

fMRI [16] provides information on brain metabolism by detecting changes in blood oxygen level. It uses strong fluctuating magnetic field to measure the location of dipoles. The advantages are the good spatial resolution and the noninvasiveness. But the temporal resolution is poor (about 1 s), and the required equipment is expensive.

1.3.5 Near-infrared spectroscopy

Near-infrared light penetrates to sufficient depths of the head to allow functional mapping of the cerebral cortex. Changes in tissue oxygenation cause the modulation of absorption and the scattering of photons, which can be measured [17]. NIRS can measure the attenuation changes due to cerebral hemodynamics that yield localized blood volume and oxygenation changes. This response has a poor temporal resolution (around 5–8 s). However, noninvasiveness, no interference, nonionizing, and not requiring any gel are some advantages of this method.

1.3.6 Commonly used method in BMI—why EEG?

The ideal brain measurement method needs to have high temporal and spatial resolution. Moreover, it should be cheap, portable, and easy to use. This method does not yet exist. The characteristics of the introduced signal acquisition techniques are summarized in Table 1.1. Among the listed techniques EEG is by far the most commonly used in BMI, due to its high temporal resolution and the ease of use. Furthermore, EEG is also a relatively low-cost method, when compared to methods like PET, MEG, or fMRI which require expensive equipment and skilled professionals to operate.

EEG signals are composed of different oscillations named rhythms [10]. These rhythms can have distinct properties in terms of spatial and spectral localization. There are six classical brain rhythms described in Table 1.2.

Table 1.1 Noninvasive techniques for measuring brain activity

Method	**Measured quantity**	**Spatial res.**	**Temporal res.**	**Portable?**
EEG	Electric pot.	−	++	Yes
MEG	Magnetic fields	+	++	No
fMRI	Hemodynamic act.	++	−	No
NIRS	Hemodynamic act.	−−	−	Yes

−−: very low to ++: very high, Res.: Resolution, Pot.: Potential

Table 1.2 EEG rhythms

Rhythm	**Frequency (Hz)**	**Generally found**
Delta	1–4	In adults during a deep sleep
Theta	4–7	During drowsiness, and in young children
Alpha	8–12	In healthy relaxed adults and with closed eyes, in occipital lobe
Mu	8–13	In motor and sensorimotor cortex, changes by doing motor tasks
Beta	13–30	In awaken and conscious people over the frontal and central regions, also affected by the performing movements in the motor areas
Gamma	30–40	When deep concentration is obtained

1.4 Measurement of EEG

To correctly measure the EEG, we should be aware of basic knowledge of the principles and the measurement. In order to get clean data, not only reliable devices but also good experience and know-how are needed. In the following, the principle of EEG and practical issues are reviewed to achieve good quality of analysis.

1.4.1 Principle of EEG

EEG is the voltage variation on the scalp due to electrical activity of neurons in the cerebral cortex. A type of neurons in the cerebral cortex is a pyramidal cell that has the single axon extending perpendicular to the cortical surface. The potential changes (postsynaptic potentials) occurring in pyramidal cells yield current flows through the cell. The electrical activity of the population of pyramidal cells appears as a relatively large potential. Since there are cerebrospinal fluid, skull, muscle, blood vessels, and skin from the cerebral cortex to the scalp, the observed voltage became greatly attenuated to the order of several tens of μV from μV. Moreover, the voltage variation of the EEG represents the electrical activity from a relatively wide area of brain.

As is well known, since the cortex has lots of folds (gyrification), EEG does not accurately represent the potential change immediately below the electrode. As described above, the spatial resolution of the EEG is very low, and it is almost impossible to know the localized activities EEG. However, since EEG reflects the electrical activity of neurons, temporal resolution is very high. It is possible to follow the changes in brain activity in the order of several milliseconds. This is the biggest advantage of using EEG.

1.4.2 How to measure EEG

A set of system for measuring EEG is called electroencephalography. The amplification part of the electroencephalography, which is a core of the measurement, is based on a differential amplifier. The output impedance with respect to the amplifier corresponds to the impedance caused by the contact between the scalp and the electrode. This impedance should be low as much as possible to accurately capture the waveform of the EEG. There are two types of electrodes: passive and active. The latter can reduce the noise contaminated through a cable.

In a standard electroencephalograph, to remove the power noise, a 50- or 60-Hz notch filter is applied (the frequency depends on the power plant). Further, in order to remove the drift component caused by the amplifier, a high-pass filter is usually applied. The electroencephalograph can be classified roughly into an analog and digital types (analog EEG and digital EEG). The analog EEG is the amplified signal and a hard-printed recording on paper by a plotter or a printer. This is now not used in the field of engineering applications; however, it is still widely used in clinics and hospitals. Mostly, the digital is used, where the amplified signal is digitized by an A/D converter and is stored in the computer as EEG digital data. Since the most of digital EEG equipment are packaged as a single system, there is no access for accessing an analog signal. If it is desired to obtain an amplified analog signal, a possible way is to use a general-purpose amplifier for biological signals. For example, one of the authors of this chapter uses the analog amplifier manufactured by Nihon Kohden connected to a genetic A/D converter to measure the EEG. For applications requiring real-time processing such as BMI, the use of analog amplifiers can ensure freedom of measurement.

1.4.3 Practical issues

Based on the principle understanding of the previous section, this section describes the common rules and procedure for measuring the EEG in practice.

1.4.3.1 Electrode positions

Usually, to obtain multichannel EEG signals, there is an international standard in positioning the electrodes. The most common electrode arrangement is the international 10–20 system [18], in which each electrode has an alphabetical label and a number as shown in Figure 1.2. The name of "10–20" comes from the ratio of distances between adjacent electrodes measured and marked in 10% and 20% to define the electrode positions.

The electrode positions are defined as follows. First, divide the median line that connects the nasion (the intersection of the frontal bone and two nasal bones of the human skull, i.e., the deepest part of the base of the nose) to the inion (a point at the external occipital protuberance of the skull) by the ratio of 10:20:20:20:20:10 to label Fpz, Fz, Cz, Pz, and Oz from the nasion to the inion. Second, divide the line passing through Cz from the left to right preauricular points by the same ratio to label T3, C3, Cz, C4, and T4 from left to right. Third, divide the line on the left hemisphere passing through T3 from Fpz to Oz by this ratio to label Fp1, F7, T3, T5, and O1. For the

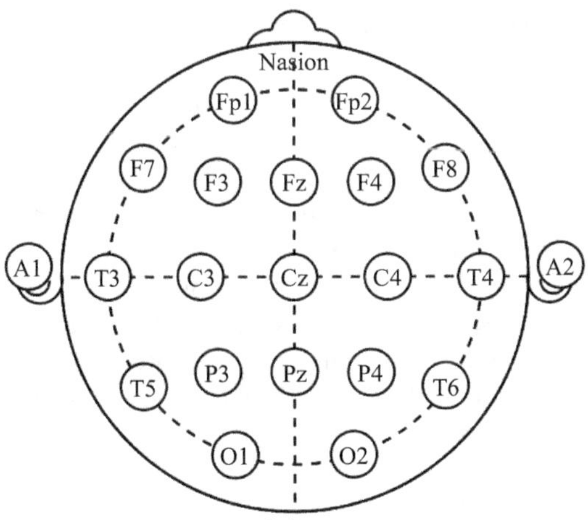

Figure 1.2 International 10–20 placement (from Wikipedia, public domain image)

right hemisphere, in the same manner, define the labels Fp2, F8, T4, T6, and O2. It should be noted that the left hemisphere has odd number labels and the right has even number labels. Furthermore, equally divide the line passing through C3 from Fp1 to O1 to label Fp1, F3, C3, P3, and O1. Similarly, with respect to the right hemisphere, label Fp2, F4, C4, P4, and O2. The left and right earlobes are labeled A1 and A2, respectively, which are used as the reference electrodes. The current EEG studies use more dense system such as 10-10 system used in the 64-electrode measurement and 10-5 system used in the 128-electrode measurement methods [19].

1.4.3.2 Reference electrode

As described above, since the EEG is measured through a differential amplifier it is necessary to choose the reference electrode (reference). The method for differentially amplifying the adjacent electrodes as a reference electrode is referred to as a bipolar derivation. This method can be used when it is necessary to weaken the common component over adjacent electrodes. In this method, letting A and B be the potentials at electrodes A and B, respectively, and then, EEG at A to B is given as

$$EEG_{A-B} = A - B. \tag{1.1}$$

On the other hand, a method for differential amplification with respect to the common reference electrode is referred to as a monopolar derivation. That is, letting R be the potential at the reference electrode, EEG at electrode A to the reference is given by

$$EEG_A = A - R. \tag{1.2}$$

During EEG measurement, a reference electrode is typically chosen as either the left or right earlobe. Currently, many of the digital EEG uses the monopolar derivation method because the bipolar derivation can be recalculated on a computer.

1.4.3.3 Eye artifacts and electrooculogram

EEG at frontal electrodes such as Fp1 and Fp2 close to the eye are easily contaminated by artifacts due to the movement of the eyeball and the eye blinking. The eye movement can cause electrooculogram (EOG) and the blinking can cause a change of the muscle potential and eye potential. These artifacts contaminate not only EEG at frontal electrodes but also at central electrodes. Incorporation of EOG can induce a waveform similar to event-related potentials described later and also induce pulse-like signals. Because EEG of patients with brain disorders may produce a pulse-like waveform, the signal changes in EEG must be determined whether or not the pulse is derived from a blink. Therefore, it is strongly recommended to measure the EOG whenever the EEG is simultaneously measured. By looking at EOG, it can be confirmed whether or not pulses in the EEG is due to the eye movement.

1.4.3.4 Attachment of electrodes

Since the contact impedance between the scalp and the electrode significantly affects the measurement result, the electrodes must be properly attached to the head. First,

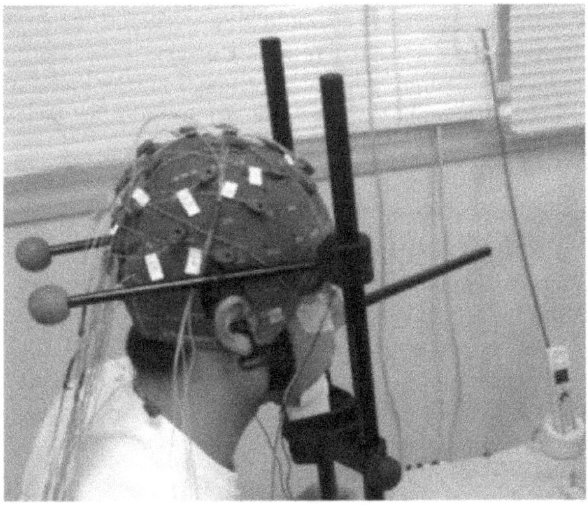

Figure 1.3 The head should be fixed at a chin rest to measure the EEG

a subject should be shampooed to remove dirt of the scalp, then remove the horny by using abrasive gel. Typical digital EEG system has a function to check the impedance, which should be lowered enough.

1.4.3.5 Fixing head and chin

Moving the head during the measurement can produce noise due to the electrode bonding surface and the shacked cables. Moreover, head movement produces the potential by muscle (electromyogram, EMG), which can largely contaminate the EEG. To avoid these artifacts, head and chin should be fixed to a chin rest (Figure 1.3) as much as possible.

1.5 Neurophysiological signals in EEG for driving BMIs

A neurophysiological signal suitable for BMI should be relatively easy to identify and also easy to control by the user. The neurophysiological signals used in BMI can be divided into two main categories [11,20]:

- **Evoked signals** are generated due to an external stimulus.
- **Spontaneous signals** are voluntarily generated by the user following an internal cognitive process, without any external stimulations.

1.5.1 Evoked potentials

The main advantage of evoked potentials over spontaneous signals is that evoked potentials do not require a specific training for the user, as they are automatically generated by the brain in response to a stimulus. As such, they can be used efficiently to drive a BMI from the first use [11,20]. Nevertheless, as these signals are evoked, they require external stimulations, which can be uncomfortable and tiring for the user. The steady-state evoked potentials (SSEPs) and P300 [11,20] are two well-known signals belonging to this category.

1.5.1.1 Steady-state evoked potentials

SSEPs are brain potentials that appear when the subject perceives a periodic stimulus such as a flickering picture or a sound. SSEP yields increases in the power of the EEG signals in the frequencies being equal to the stimulation frequency and its harmonics and/or subharmonics [21,22]. In an SSEP-based BCI application, there are several stimuli simultaneously flickering at different frequencies. Each stimulus is corresponding to a task. To activate a task, the user should draw his continuous attention to the corresponding stimulus [23].

1.5.1.2 P300

P300 consists of a positive potential appearing approximately 300 ms after an infrequent and relevant stimulus [20]. This potential is mainly located in the parietal area of the brain. In BMI, P300 is usually used for communication. A famous example is a speller device with a 6×6 matrix containing 26 letters of the alphabet and 10 digits [24,25]. The user is asked to focus on his intended letter/digit, while the rows and columns are randomly highlighted. Thus, the BMI system can identify the user's intention, since the highlighted rows and columns containing the intended letter/digit are responded by a large P300 potential. As other evoked potential signals, P300 has the advantage of not requiring any training for the subject in order to be used. On the other hand, P300-based BMI applications require the user to constantly focus on fast and repetitive visual stimuli, which can be eye-tiring and inconvenient.

1.5.2 Spontaneous signals

Among all the spontaneous signals, the most commonly used signals are undoubtedly sensorimotor rhythms. However, other neurophysiological signals such as slow cortical potentials (SCPs) are also used in BMI.

Before operating a BMI system based on the spontaneous signals, sufficient amount of training work is necessary. During the training phase, the brain signals are recorded, while the user is performing mental tasks [e.g., motor imagery (MI)]. Two strategies are used in BMI for training. In the old strategy, the signal processing algorithms and the classifier are fixed. Thus, the user should learn to adapt his brain signals to the system using the received feedback. This strategy needs a very long training period (i.e., several weeks). In contrast, the new strategy puts the load of learning on the computer as much as possible before human training is involved. Thus, the recorded brain signals are used to fine-tune the signal processing algorithms and train the employed machine-learning algorithms, classifiers, etc. This second strategy has a considerable shorter training time, since the BMI system is allowed to adapt itself to the subject-specific EEG characteristics of the user. However, there are still many data analysis problems as well as problems related to the experimental paradigms to be addressed.

1.5.2.1 Slow cortical potentials

SCPs are slow voltage shifts in the frequency range of 1–2 Hz, which can last from hundreds of milliseconds up to several seconds. Positive shifts in SCPs are associated with the cortical relaxation and slow negative shifts are associated with the cortical activation while evoked by either a movement execution or a mental task [26]. SCPs have been exploited to develop BMIs. For example, in [26], SCPs were used to move an object on a computer screen. However, a user generally needs a very long training period to learn the control of SCPs . This training is even longer than motor rhythms training [27]. However, it seems that SCPs would be a more stable signal if trained [27].

1.5.2.2 Event-related desynchronization (ERD) and event-related synchronization (ERS) in sensorimotor rhythms

Motor execution and motor imagination generally lead to changes in the sensory motor rhythms (i.e., Mu and Beta with 8–13 Hz and 13–30 Hz frequency ranges, respectively, over the sensorimotor cortex). When the brain is activated by either execution or imagination of motor movements, typically a decrease in Mu and Beta rhythms, called event related desynchronization (ERD), happens. Followed by ERD, an increase in the Mu and Beta rhythms, called event related synchronization (ERS), happens after the movement or imagination. ERD and ERS happen particularly contralateral to the movement. For instance, ERD happens over the right motor cortex by left-hand imagination, while right-hand imagination causes ERD over the left motor cortex.

The specific characteristics of the ERD/ERS patterns make them relatively easy to be automatically recognized [28,29]. Consequently, ERD/ERS patterns during MI are the most commonly used neurophysiological signals to drive EEG-based BMI. In the MI-based BMI, imagination of right and left hands, feet, and tongue are mainly used [30]. Advanced signal processing and machine-learning algorithms enable the MI-based BMI to work efficiently with only a few sessions of training [31].

1.6 Commonly used EEG processing methods in BMI

This section is dedicated to preprocessing, feature extraction, and classification of EEG signals. These three BMI components could be denoted as “EEG processing.” EEG processing is a key element in the design of a BMI as it aims at transforming the input brain signals into a command for a given application. Consequently, a wide majority of BMI studies aims at improving this component to make the whole system more efficient.

1.6.1 Preprocessing

The acquired EEG signals are highly contaminated by noise and artifacts. Thus, different preprocessing algorithms are applied on the EEG signals prior to the extraction of features to reduce the noise. Below some preprocessing steps commonly used in MI-based BMIs are introduced.

1.6.2 Re-referencing

When a reference electrode is set on one side of the earlobe (A1 or A2), spatially symmetric responses appear as asymmetric spatial distribution. A method for avoiding this is to record both A1 and A2 and set the average of A1 and A2 as the reference.

Another method is recalculating the voltage of each electrode at a potential average over all electrodes is commonly used. Changing the reference electrode is called re-referencing electrodes. In this case, all electrodes should be symmetrically located over left and right hemispheres.

Now, let N be the number of electrodes and EEG_n the measured EEG at the nth electrode ($n = 1, \ldots, N$). If the potential of each electrode is V_n and the potential at the reference electrode is R, then

$$EEG_n = V_n - R$$

Therefore, the EEG $REEG_n$ rereferenced by the average over all electrodes is given as

$$\begin{aligned} REEG_n &= V_n - \frac{1}{N}\sum_n V_n \\ &= (V_n - R) - \frac{1}{N}\sum_n (V_n - R) \\ &= EEG_n - \frac{1}{N}\sum_n EEG_n. \end{aligned}$$

The point is that without using unmeasured V_n and R, it is possible to reset the reference potential.

In addition, it is also possible to convert the monopolar derivation to the bipolar derivation method. Let A and B be the potentials at electrodes A and B, respectively. By definition, the bipolar derivation $EEG_{A-B} = A - B$ amounts to $EEG_A - EEG_B$, because the EEGs at A and B in the monopolar derivation are given as $EEG_A = A - R$ and $EEG_B = B - R$, respectively.

1.6.2.1 Channel selection

Multichannel EEG is generally used in BMIs whereby performing EEG channel selection (1) improves BMI performance by removing irrelevant channels and (2) enhances user convenience from the use of lesser channels [32,33]. However, how to perform optimal channel selection is not a trivial task, since selecting channels manually based on neurophysiologic knowledge does not necessarily yield optimal results compared to using all the EEG channels in the international 10–20 system [34].

1.6.2.2 Spectral filtering

Appropriate spectral filters, such as band-pass or low-pass filters, can enhance the accuracy and robustness of BMI by reducing the influence of activities that are lying outside of the frequency regions of interest [35,36]. For example, band-pass filters can reduce the negative effects of the power line noise, EMG, EOG, and other brain activities lying outside the desired frequency band. Furthermore, recent electroencephalograph may also be configured to acquire data at a sampling frequency exceeding 1,000 Hz. Indeed, since the required bandwidth is from 1 Hz up to 50 Hz or less, a high sampling rate such as 1,000 Hz is not necessary. Therefore, there is a case of thinning out the signal at downsampling prior to analysis. In this case, so as not to cause aliasing, it must be applied to low-pass filter having a cutoff frequency above two times the frequency of the preanalysis range. A filter widely used in preprocessing of EEG is a Butterworth filter, which has an infinite impulse response (IIR). Since an IIR filtering can bring phase distortion, zero phase filtering should be implemented. for example, using filtfilt function in MATLAB® and Python/SciPy.

1.6.2.3 Spatial filtering

Various simple spatial filters are used in order to isolate the relevant spatial information embedded in the signals. This is achieved by selecting or weighting the contributions from the different electrodes (and as such from the different spatial regions) [37].

Two simple and popular spatial filters are the common average reference (CAR) [37,40] and the surface Laplacian filters [37]. These two filters make it possible to reduce the background activity. The CAR filter subtracts the average value of all the electrodes from the electrode of interest, while the Laplacian filter subtracts the average of the surrounding channels from the channel of interest.

Numerous other spatial filtering algorithms which are more complex and more advanced have been proposed and used. The most popular spatial filtering algorithm, which is increasingly used for preprocessing in BMI and has proved to be very efficient, is the common spatial pattern (CSP) algorithm [38,40]. This algorithm is based on the decomposition of the EEG signals into spatial patterns. These patterns are selected in order to maximize the differences between the classes involved once the data have been projected onto these patterns. Determining these patterns is performed using a joint diagonalization of the covariance matrices of the EEG signals from each class [39,40]. These filters have proved to be very efficient, especially during BMI competitions [38,41,42].

1.6.2.4 Other preprocessing algorithms

Numerous other preprocessing algorithms have been proposed and used for BMI design. Among these algorithms, we can quote various regularized CSP algorithms [43], stationary subspace analysis [44] as well as numerous spectro-spatial filters [45–47]. In addition to filtering methods, other relatively simple methods are also used as preprocessing, such as moving average filtering, subsampling (in order to reduce the dimensionality of the problem), or baseline correction.

1.6.3 Feature extraction

In order to identify the performed MI tasks, informative features have to be extracted from the recorded EEG signals. To accurately classify the performed mental tasks, the feature extraction process is required to form the discriminative set of features.

Features used in BMI can be divided into three main groups: temporal features [48,49], frequency domain features [28,50], and hybrid features that exploit both temporal and frequential information [51,52]. These three groups of features are briefly described in this subsection.

It is noted that the features exploiting spatial information can be also added here as the fourth group. However, these features are mainly based on temporal or frequential features extracted after spatial filtering. The most common features in this group is the features of CSP, which are indeed band power features extracted after CSP spatial filtering [40].

1.6.3.1 Temporal features

Temporal methods use the temporal variations of the signals as features. These methods are particularly adapted to describe neurophysiological signals with a specific time signature, such as P300 or ERD/ERS. Among the temporal feature extraction methods, the amplitude of raw EEG signals [49,53], autoregressive parameters [30,48], and Hjorth parameters [30] are mainly used.

1.6.3.2 Frequency domain features

As the ERD/ERS patterns occur in subject-specific reactive frequency bands, the power of the signal filtered in that range is used as the band power feature for classifying MI patterns [28]. The band power features are also used with synchronization features to achieve a better discrimination during multiclass MI [50,54].

1.6.3.3 Time-frequency features

Time-frequency feature extraction algorithms are based on various time-frequency representations such as the short-time Fourier transform or wavelets to extract information that are both frequential and temporal. The main advantage of these time-frequency representations is that they can catch relatively sudden temporal variations of the signals, while still keeping the frequency information.

1.6.4 Classification

There are many classification algorithms employed for BMIs [55]. Among these classification algorithms, linear classifiers are probably the most popular algorithms for BMI applications. Linear classifiers use linear functions to distinguish classes. Linear discriminant analysis is one of the classification algorithms widely used in BMI experiments.

1.7 Feedback

Users need feedback from the BMI system to improve their performance. Feedback is the response of the system after an action from the user. This can be visual, auditory, tactile, etc.

Feedback can speed up the training process and improve overall performance. It can also motivate and hold the subject's attention. However, feedback may have some possible drawbacks. A feedback can affect the brain rhythms itself caused by visual changes in the environment. It can distract the user from concentrating on producing the brain patterns. False classification can cause frustration, and thus in this case, feedback negatively affects the BMI performance.

1.8 BMI applications

Once the class of the signals have been identified, the system can associate a specific command to this identified mental state and send this command to a given application. Available BMI applications can be divided into two main categories. The first and most important category is the medical domain [20,56]. Indeed, the main objective of BMI is to serve disabled people with providing a new communication and control channel. The second category is the nonmedical domain [3]. Thus, even if BMIs are mainly designed for disabled people, they can also be interesting for healthy people [57], for instance by proposing video games based on BMI [58,59]. Some of the applications possible with current BMIs are described below.

Personal communication: This group of applications comprises the spelling of text, composing and sending emails and letters. If a patient is severely paralyzed, the only way of communication without a BMI (if any) is with the help of a caretaker. Introducing BMI-based communication devices enables the patient to send confidential messages that provides a great amount of independence and self-determination.

Environmental control: Opening the front door to visitors, turning lights on or off, controlling shades, regulating the room temperature, and changing the backrest position are examples for this group of applications [60]. Their use increases the independence of the patient and decreases the workload of caretakers.

Rehabilitation training: Recently, MI-based BMI has been proposed as a rehabilitation tool to facilitate motor recovery in stroke [61,62]. Several studies have demonstrated that MI has a positive effect on motor rehabilitation after stroke [63–65] through activation of the affected sensorimotor networks [66,67]. Since the performance of MI is internal to the subject, and thus not directly observable, BMI can facilitate the MI-based stroke rehabilitation by providing direct and immediate feedback on the MI performance.

Control of paralyzed limbs: Using a BMI system can lead to the control of a limb orthosis or even allow for direct control of body parts via functional electrical stimulation (e.g., for bladder and swallow control or for paralysis due to different types of palsy) [68].

Mobility: Mobility is of inherent interest for a paralyzed patient. A straightforward application of BMI is the control of electrical wheelchairs [5,69]. If classifiable signals can be reliably translated into control sequences for the wheelchair, the paralyzed patient regains a certain degree of mobility.

Treatment for attention deficiency: BMI can quantify one's attention level as measured by EEG waves, thereby allowing users to employ their attention to play some neurofeedback games [70]. This new treatment provides a safe and interactive tool to guide and regulate the brain from the disorder. This treatment can be effectively used for attention-deficit hyperactivity disorder without side effects that medication may cause.

Gaming and virtual reality: In addition to medical and rehabilitation applications, there is an increasing number of BMI applications for multimedia, such as for simple 2D video games [71] to more advanced 3D video games [46]. There are BMI systems used for navigating virtual worlds [72] and BMI systems used for selecting and/or manipulating virtual objects [73,74].

1.9 Summary

In this chapter, the existing methods for developing BMIs as well as the existing BMI applications were reviewed. At first, the basic building units of a BMI including brain signal measuring unit, preprocessing and feature extraction units, classification algorithms, and experimental protocols were presented. The noninvasive techniques for brain signal acquisition were also discussed, and EEG as the most commonly used techniques for brain signal acquisition was explained in detail. Thereafter, various neurophysiological signals used to drive EEG-based BMIs were presented, from which the focus was on the ERD/ERS patterns accompanied with the mental imagination of motor movements. Thus, this chapter mainly provided a brief review of the BMI technology and existing algorithms to develop a BMI framework.

A large number of BMI research studies aimed at improving various issues in the development of BMI. MI has been extensively explored in developing BMIs operated by healthy as well as paralyzed patients. Despite this large number of BMI studies, the appropriate signal processing, classification, and translational algorithms are yet to be developed so that BMI could be available in market. The main signal processing issues in MI-based BMIs are the noise and nonstationarity inherent in EEG signals. Thus, in the next chapters, novel signal processing and machine-learning algorithms are proposed to deal with noisy and nonstationary EEG signals in MI-based BMIs.

References

[1] Wolpaw JR, Birbaumer N, Heetderks WJ, *et al.* Brain–computer interface technology: a review of the first international meeting. IEEE Trans Rehabil Eng. 2000;8(2):164–173.

[2] Vaadia E, Birbaumer N. Grand challenges of brain computer interfaces in the years to come. Front Neurosci. 2010;3(2):151–154.

[3] Blankertz B, Tangermann M, Vidaurre C, *et al.* The Berlin brain–computer interface: non-Medical uses of BCI technology. Front Neurosci. 2010;4:1–17.

[4] van ERP J, Lotte F, Tangermann M. Brain computer interface for non-medical applications: how to move forward. IEEE Comput. 2012;45(4):26–34.

[5] Vanacker G, del R Millán J, Lew E, *et al.* Context-based filtering for assisted brain-actuated wheelchair driving. Intell Neurosci. 2007 January;2007:1–12. Available from: http://dx.doi.org/10.1155/2007/25130.

[6] Pfurtscheller G, Neuper C, Guger C, *et al.* Current trends in Graz brain–computer interface (BCI) research. IEEE Trans Rehabil Eng. 2000;8(2): 216–219.

[7] Blankertz B, Krauledat M, Dornhege G, *et al.* In: Stephanidis C, editor. A note on brain actuated spelling with the Berlin brain–computer interface. vol. 4555 of Lecture Notes in Computer Science. Springer Berlin/Heidelberg; 2007. p. 759–768.

[8] Wolpaw JR, Loeb GE, Allison BZ, *et al.* BCI meeting 2005 – workshop on signals and recording methods. IEEE Trans Neural Syst Rehabil Eng. 2006;14(2): 138–141.

[9] Haynes J, Rees G. Decoding mental states from brain activity in humans. Nat Rev Neurosci. 2006;7:523–534.

[10] Niedermeyer E, Lopes da Silva FH. Electroencephalography: basic principles, clinical applications, and related fields, chapter The normal EEG of the waking adult. E. Niedermeyer and F. Lopes da Silva eds., Philadelphia, London: Lippincott Williams & Wilkins; 2005.

[11] Curran EA, Stokes MJ. Learning to control brain activity: a review of the production and control of EEG components for driving brain–computer interface (BCI) systems. Brain Cogn. 2003;51(3):326–336.

[12] Krauledat M. Analysis of nonstationarities in EEG signals for improving brain–computer interface performance. Technische Universitat Berlin, Berlin, Fakultät IV- Elektrotechnik und Informatik; 2008.

[13] Krauledat M, Dornhege G, Blankertz B, *et al.* Robustifying EEG data analysis by removing outliers. Chaos Complexity Lett. 2007;2(3):259–274.

[14] Phelps ME. Emission computed tomography. Semin Nucl Med. 2007; 7:337–365.

[15] Rampp S, Stefan H. On the opposition of EEG and MEG. Clin Neurophys. 2007;118(8):1658–1659.

[16] Hinterberger T, Weiskopf N, Veit R, *et al.* An EEG-driven brain–computer interface combined with functional magnetic resonance imaging (fMRI). IEEE Trans Biomed Eng. 2004;51(6):971–974.

[17] Coyle S, Ward T, Markham C, McDarby G. On the suitability of near-infrared (NIR) systems for next-generation brain–computer interfaces. Physiol Meas. 2004;25(4):815.

[18] Klem GH, Lüders HO, Jasper H, *et al.* The ten–twenty electrode system of the International Federation. Electroencephalogr Clin Neurophysiol. 1999;52(3):3–6.

[19] Jurcak V, Tsuzuki D, Dan I. 10/20, 10/10, and 10/5 systems revisited: their validity as relative head-surface-based positioning systems. NeuroImage. 2007; 34(4):1600–1611.

[20] Wolpaw JR, Birbaumer N, McFarland DJ, *et al.* Brain–computer interfaces for communication and control. Clin Neurophysiol. 2002;113(6):767–791.

[21] Nijholt A, Tan D. Playing with your brain: brain–computer interfaces and games. In: Int. Conf. Advances in Computer Entertainment Technology. ACE'07. New York, NY, USA: ACM; 2007. p. 305–306. Available from: http://doi.acm.org/10.1145/1255047.1255140.

[22] Müller-Putz GR, Scherer R, Neuper C, *et al.* Steady-state somatosensory evoked potentials: suitable brain signals for brain–computer interfaces? IEEE Trans Neural Syst Rehabil Eng. 2006;14(1):30–37.

[23] Middendorf M, McMillan G, Calhoun G, *et al.* Brain–computer interfaces based on the steady-state visual-evoked response. IEEE Trans Rehabil Eng. 2000 Jun;8(2):211–214.

[24] Farwell LA, Donchin E. Talking off the top of your head: toward a mental prosthesis utilizing event-related brain potentials. Electroencephalogr Clin Neurophysiol. 1988;70(6):510–523. Available from: http://www.sciencedirect.com/science/article/B6SYX-4834R11-BR/2/c5f34c77e7efd0fadeee6fddc114640a.

[25] Donchin E, Spencer KM, Wijesinghe R. The mental prosthesis: assessing the speed of a P300-based brain–computer interface. IEEE Trans Rehabil Eng. 2000;8(2):174–179.

[26] Birbaumer N. In: Slow cortical potentials: their origin, meaning, and clinical use. G. J. M. van Boxtel and K. B. E. Böcker eds. Tilburg: Tilburg University Press; 1997.

[27] Birbaumer N. Breaking the silence: brain–computer interfaces (BCI) for communication and motor control. Psychophysiology. 2006;43(6):517–532.

[28] Pfurtscheller G, Neuper C, Flotzinger D, *et al.* EEG-based discrimination between imagination of right and left hand movement. Electroencephalogr Clin Neurophysiol. 1997;103:642–651.

[29] Pfurtscheller G, Brunner C, Schlögl A, *et al.* Mu rhythm (de)synchronization and EEG single-trial classification of different motor imagery tasks. Neuro Image. 2006;31(1):153–159. Available from: http://www.sciencedirect.com/science/article/B6WNP-4J4HK9W-3/2/e2ea7ba6446cd2f634604f1cc42b6080.

[30] Pfurtscheller G, Neuper C. Motor imagery and direct brain–computer communication. Proc IEEE. 2001;89(7):1123–1134.

[31] Blankertz B, Dornhege G, Krauledat M, *et al.* The Berlin brain–computer interface: EEG-based communication without subject training. IEEE Trans Neural Syst Rehabil Eng. 2006;14(2):147–152.

[32] Arvaneh M, Guan C, Ang KK, *et al.* Optimizing EEG channel selection by regularized spatial filtering and multi band signal decomposition. In: IASTED Int. Conf. Biomedical Engineering; 2010. p. 86–90.

[33] Lal TN, Schroder M, Hinterberger T, *et al.* Support vector channel selection in BCI. IEEE Trans Biomed Eng. 2004;51(6):1003–1010.

[34] Blankertz B, Losch F, Krauledat M, *et al.* The Berlin brain–computer interface: accurate performance from first-session in BCI-naive subjects. IEEE Trans Biomed Eng. 2008;55(10):2452–2462.

[35] Ang KK, Chin ZY, Zhang H, *et al.* Mutual information-based selection of optimal spatial-temporal patterns for single-trial EEG-based BCIs. Pattern Recognit. 2011;45(6):2137–2144.

[36] Lemm S, Blankertz B, Curio G, *et al.* Spatio-spectral filters for improving the classification of single trial EEG. IEEE Trans Biomed Eng. 2005;52(9):1541–1548.

[37] McFarland DJ, McCane LM, David SV, *et al.* Spatial filter selection for EEG-based communication. Electroencephalogr Clin Neurophysiol. 1997;103(3):386–394.

[38] Blankertz B, Tomioka R, Lemm S, *et al.* Optimizing spatial filters for robust EEG single-trial analysis. IEEE Signal Process Mag. 2008;25(1):41–56.

[39] Dornhege G, Blankertz B, Curio G, *et al.* Boosting bit rates in noninvasive EEG single-trial classifications by feature combination and multiclass paradigms. IEEE Trans Biomed Eng. 2004;51(6):993–1002.

[40] Ramoser H, Muller-Gerking J, Pfurtscheller G. Optimal spatial filtering of single trial EEG during imagined hand movement. IEEE Trans Rehabil Eng. 2000;8(4):441–446.

[41] Blankertz B, Müller KR, Curio G, *et al.* The BCI competition 2003: progress and perspectives in detection and discrimination of EEG single trials. IEEE Trans Biomed Eng. 2004;51(6):1044–1051.

[42] Blankertz B, Müller KR, Krusienski DJ, *et al.* The BCI competition III: validating alternative approaches to actual BCI problems. IEEE Trans Neural Syst Rehabil Eng. 2006;14(2):153–159.

[43] Lotte F, Guan C. Regularizing common spatial patterns to improve BCI designs: unified theory and new algorithms. IEEE Trans Biomed Eng. 2011;58(2):355–362.

[44] von Bünau P, Meinecke FC, Kiraly F, *et al.* Finding stationary subspaces in multivariate time series. Phys Rev Lett. 2009;103(21):1–4.

[45] Dornhege G, Blankertz B, Krauledat M, *et al.* Combined optimization of spatial and temporal filters for improving brain–computer interfacing. IEEE Trans Biomed Eng. 2006;53(11):2274–2281.

[46] Lalor EC, Kelly SP, Finucane C, *et al.* Steady-state VEP-based brain–computer interface control in an immersive 3D gaming environment. EURASIP J Appl Signal Process. 2005 Jan;2005:3156–64. Available from: http://dx.doi.org/10.1155/ASP.2005.3156.

[47] Tomioka R, Dornhege G, Nolte G, *et al.* Spectrally Weighted Common Spatial Pattern Algorithm for Single Trial EEG Classification. Mathematical engineering technical reports. Department of Mathematical Informatics, Graduate School of Information Science and Technology, the University of Tokyo; 2006. Available from: http://books.google.ie/books?id=02S3XwAACAAJ.

[48] Schloegl A, Lugger K, Pfurtscheller G. Using adaptive autoregressive parameters for a brain–computer-interface experiment. In: 19th Annual Int. Conf. IEEE Engineering in Medicine and Biology Society. vol. 4; 1997. p. 1533–1535.

[49] Kaper M, Meinicke P, Grossekathoefer U, *et al.* BCI competition 2003-data set IIb: support vector machines for the P300 speller paradigm. IEEE Trans Biomed Eng. 2004;51(6):1073–1076.

[50] Gysels E, Celka P. Phase synchronization for the recognition of mental tasks in a brain–computer interface. IEEE Trans Neural Syst Rehabil Eng. 2004;12(4):406–415.

[51] Bostanov V. BCI competition 2003-data sets Ib and IIb: feature extraction from event-related brain potentials with the continuous wavelet transform and the *t*-value scalogram. IEEE Trans Biomed Eng. 2004;51(6):1057–1061.
[52] Fatourechi M, Bashashati A, Ward RK, *et al.* A hybrid genetic algorithm approach for improving the performance of the LF-ASD brain computer interface. In: IEEE International Conference on Acoustics, Speech, and Signal Processing. vol. 5; 2004. p. v/345–v/348.
[53] Hoffmann U, Garcia G, Vesin JM, *et al.* A boosting approach to P300 detection with application to brain–computer interfaces. In: 2nd International IEEE EMBS Conference on Neural Engineering; 2005. p. 97–100.
[54] Brunner C, Scherer R, Graimann B, *et al.* Online control of a brain–computer interface using phase synchronization. IEEE Trans Biomed Eng. 2006;53(12):2501–2506.
[55] Lotte F, Congedo M, Lécuyer A, Lamarche F, Arnaldi B. A review of classification algorithms for EEG-based brain–computer interfaces. J Neural Eng. 2007;4(2):R1.
[56] Kubler A, Mushahwar VK, Hochberg LR, *et al.* BCI meeting 2005-workshop on clinical issues and applications. IEEE Trans Neural Syst Rehabil Eng. 2006;14(2):131–134.
[57] Allison B, Graimann B, Gräser A. Why use a BCI if you are healthy? In: ACE Workshop – Brain–Computer Interfaces and Games; 2007. p. 7–11.
[58] Müller KR, Blankertz B. Toward noninvasive brain–computer interfaces. IEEE Signal Processing Magazine. 2006;23(5):128–126.
[59] Kleber B, Birbaumer N. Direct brain communication: neuroelectric and metabolic approaches at Tübingen. Cogn Process. 2005;6(1):65–74.
[60] Xiaorong G, Dingfeng X, Ming C, *et al.* A BCI-based environmental controller for the motion-disabled. IEEE Trans Neural Syst Rehabil Eng. 2003;11(2):137–140.
[61] Ang KK, Guan C, Chua KSG, *et al.* A large clinical study on the ability of stroke patients to use EEG-based motor imagery brain–computer interface. Clin EEG Neurosci. 2011;42(4):253–258.
[62] Pfurtscheller G, Müller-Putz GR, Scherer R, *et al.* Rehabilitation with brain–computer interface systems. Computer. 2008;41(10):58–65.
[63] Page SJ, Levine P, Leonard A. Mental practice in chronic stroke: results of a randomized, placebo-controlled trial. Stroke. 2007;38:1293–1297.
[64] Butler AJ, Page SJ. Mental practice with motor imagery: evidence for motor recovery and cortical reorganization after stroke. Arch Phys Med Rehabil. 2006;87:2–11.
[65] Johnson-Frey SH. Stimulation through simulation? Motor imagery and functional reorganization in hemiplegic stroke patients. Brain Cogn. 2004;55: 328–331.
[66] Sharma N, Pomeroy VM, Baron JC. Motor imagery: a backdoor to the motor system after stroke? Stroke. 2006;37:1941–1952.
[67] Vries S, Mulder T. Motor imagery and stroke rehabilitation: a critical discussion. J Rehabil Med. 2007;39(3):5–13.

[68] Taylor D, Tillery S, Schwartz A. Direct cortical control of 3D neuroprosthetic devices. Science. 2002;296(5574):1829–1832.

[69] Rebsamen B, Burdet E, Guan C, *et al.* A brain-controlled wheelchair based on P300 and path guidance. In: the 1st IEEE Int. Conf. Biomedical Robotics and Biomechatronics; 2006. p. 1101–1106.

[70] Lim CG, Lee TS, Guan C, *et al.* Effectiveness of a brain–computer interface based programme for the treatment of ADHD: a pilot study. Psychopharmacol Bull. 2010;43:73–82.

[71] Krepki R, Blankertz B, Curio G, *et al.* The Berlin brain–computer interface (BBCI): towards a new communication channel for online control in gaming applications. J Multimedia Tools Appl. 2007;33(1):73–90.

[72] Friedman D, Leeb R, Dikovsky L, *et al.* Controlling a virtual body by thought in a highly-immersive virtual environment – a case study in using a brain–computer interface in a virtual-reality cave-like system. In: The 2nd Int. Conf. Computer Graphics Theory and Applications; 2007. p. 83–90.

[73] Bayliss JD. Use of the evoked potential P3 component for control in a virtual apartment. IEEE Trans Neural Syst Rehabil Eng. 2003;11(2):113–116.

[74] Leeb R, Lee F, Keinrath C, *et al.* Brain–computer communication: motivation, aim, and impact of Exploring a virtual apartment. IEEE Trans Neural Syst Rehabil Eng. 2007;15(4):473–482.

Chapter 2
Discriminative learning of connectivity pattern of motor imagery EEG

Xinyang Li[1], *Cuntai Guan*[2], *and Huijuan Yang*[3]

Abstract

Different mental states result in different synchronizations or desynchronizations between multiple brain regions, and subsequently, electroencephalogram (EEG) connectivity analysis gains increasing attention in brain computer interfaces (BCIs). Conventional connectivity analysis is usually conducted at the scalp-level and in an unsupervised manner. However, due to the volume conduction effect, EEG data suffer from low signal-to-noise ratio and poor spatial resolution. Thus, it is hard to effectively identify the task-related connectivity pattern at the scalp-level using unsupervised method. There exist extensive discriminative spatial filtering methods for different BCI paradigms. However, in conventional spatial filter optimization methods, signal correlations or connectivities are not taken into consideration in the objective functions. To address the issue, in this work, we propose a discriminative connectivity pattern-learning method. In the proposed framework, EEG correlations are used as the features, with which Fisher's ratio objective function is adopted to optimize spatial filters. The proposed method is evaluated with a binary motor imagery EEG dataset. Experimental results show that more connectivity information are maintained with the proposed method, and classification accuracies yielded by the proposed method are comparable to conventional discriminative spatial filtering method.

2.1 Introduction

Brain computer interface (BCI) has been an effective technology that translates the brain signal to the command signal for external devices so that the communication

[1]School of Computer Science and Electronic Engineering, University of Essex, United Kingdom
[2]School of Computer Science and Engineering, College of Engineering, Nanyang Technological University, Singapore
[3]Neural and Biomedical Technology Unit, Institute for Infocomm Research (I2R), Agency for Science, Technology and Research (A*STAR), Singapore

between the brain and external world can be established. This is especially helpful for those who have motor disabilities, which thus promotes its use in stroke rehabilitation by driving the assistive devices via the brain signals [1–3]. Despite the rapid progress made in motor-imagery-BCI (MI-BCI), most MI-BCI systems still rely on spatial, temporal and spectral features of single channel to classify different brain patterns. For example, quantification of the event-related desynchronization or synchronization (ERD/ERS) at particular frequency bands such as mu and central beta is used to discriminate the MI movement pattern. There also exists extensive discriminative learning methods for electroencephalogram (EEG) feature extraction in BCI. Spatial filters such as independent component analysis and common spatial pattern (CSP) generally unmix the signals to recover the original sources. CSP analysis is one of the most successful discriminative frameworks for spatial filter design [4–6]. With the labeled data, a projection model could be trained to maximize powers of the projected EEG signals under one class while minimizing it under another class. There exist a lot of unsupervised spatial filtering methods similar to CSP [7–13]. However, the targeted features in those methods are mostly powers of the signal. In other words, most spatial filter-optimization frameworks are designed to extract discriminative power features.

Typical difficulties in using BCI technology lie in different responsive frequency bands for MI across subjects and sessions, different locations, frequencies and timings for ERD/ERS [14]. The most popular electrodes used to capture EEG patterns for MI of limb movements (MI-LM) are C3, Cz and C4, which are based on the observations that MI-LM normally activates the contralateral side of somatosensory cortex. However, this is not always the case; different brain regions are generally collaborated and communicated together for a cognitive task, e.g., the supplementary motor area, prefrontal area, premotor cortex and primary motor cortex are activated for MI of hand movements. While the activated brain regions for MI of foot/walking (MI-FW) and MI of swallow (MI-SW) are different, e.g., Cz for MI-FW [15], and superior temporal gyrus, the premotor and supplementary motor cortex, and the subcentral area for MI-SW and motor execution of swallow [16,17]. Although ERD/ERS are significant components of EEG during MI and other cognitive tasks, connectivity pattern is also of great research value and has drawn a lot of attentions.

Brain connectivity consists of neuroanatomical (structural), functional connectivity (FC) and effective connectivity (EC) [14]. Typical methods to measure FC include the linear methods, e.g., cross correlation of pairs of EEG channels in spatial domain and the magnitude squared coherence or coherence (COH) of phases of two signals in frequency domain. COH is sensitive to both power and phases, whereas correlation is more sensitive to phase and polarity [14]. Nonlinear methods include the phase synchronization (PS) and phase locking value (PLV). The binarized PLVs of all the electrodes for the overlapped time windows were used as the features, which were subsequently selected based on fisher ratio before being fed to a two-level support vector machine (SVM) classifier [18]. The classification performance showed better results for some subjects demonstrating the useful information contained in phases [18]. It was argued that phase dynamics subserved all mental processes including motor planning and MI. It is noted that PS and PLV are sensitive to volume conduction; hence, the weighted phase lag index which was based on the imaginary component of the cross-spectrum was proposed to reduce the noise [19].

EC represents the direction and strength of information flow and reflects the causal interaction between activated brain areas [14,20,21]. EC can be measured by model-based method [e.g., dynamic causal modeling (DCM)] and data-driven method [e.g., granger causality (GC)]. The former is based on the theoretical models on how the brain areas interact with each other, whereas the latter does not assume any underlying model or prior knowledge on the spatial and temporal relationships [14,20,21]. GC and DCM models measure the influence of neural activity exerted from one to another. Such statistical dependence between different brain regions is important for understanding the functional network engaged in the cognitive and perceptive processing [21]. DCM has been widely used to model the nonlinear couplings, whereas linear GC has been used to analyze the EC for EEG, MEG and functional magnetic resonance imaging (fMRI) [21]. Due to the nonlinearity of the physiological basis of the brain signal, e.g., the cerebral blood flow, cerebral blood volume in fMRI signal, nonlinear extension of DCM and GC are thus investigated to capture the nonlinear interaction between brain regions. GC does not need to be reciprocal, which allows the evaluation of the flow direction between elements.

COH, phase locking, GC and coefficient of autoregressive model have been used to measure and analyze the EEG connectivities [14,20–25]. However, compared to discriminative framework for power feature extraction, most of the existing connectivity analysis methods are unsupervised. Possible discriminative connectivity features are usually selected based on prior neurophysiological knowledge or machine-learning method. EEG data are mixture of signals of interest and background noises, with low signal-to-noise ratio and significant artifact contamination [26,27]. Conducting the connectivity and synchronization measurements in the scalp space cannot address the volume conduction effects. Due the low signal-to-noise ratio and poor spatial resolution of EEG, it is difficult to identify the task-related connectivity patterns [14].

To address these issues, we propose a discriminative learning of connectivity pattern of EEG data. In the proposed method, the correlations of the EEG data after spatial filtering are used as features, with which Fishers' ratio objective function is adopted for spatial filtering optimization. Compared to objective functions in the conventional spatial filter optimization, connectivity information could be taken into consideration. Based on the above discussion, we highlight the contributions of this work as follows:

1. A discriminative learning of EEG connectivity pattern based on Fisher's ratio is proposed.
2. EEG data after spatial filtering based on the proposed method are compared with that based on conventional spatial filtering method.
3. Differences in EEG connectivity pattern between MI and passive movement are discussed.

2.2 Discriminative learning of connectivity pattern of motor imagery EEG

2.2.1 Spatial filter design for variance feature extraction

For ERD/ERS analysis, the second-order feature is usually used, i.e., the power of the EEG signal over certain time samples. Thus, in conventional spatial filtering methods aiming at enhancing ERD/ERS effects, the spatial filters are optimized to maximize the differences of the powers of the spatially filtered signals between different mental conditions.

Take CSP for instance, let $X \in \mathbb{R}^{n_c \times n_t}$ be the EEG signal recorded from n_c channels with n_t time points per channel. The covariance matrix of EEG signal R with normalization could be written as

$$R = \frac{XX^\top}{\mathbf{tr}(XX^\top)} \tag{2.1}$$

where $\mathbf{tr}(\cdot)$ indicates the trace of a matrix. The objective function of CSP could be expressed in the form of generalized eigenvalue decomposition, i.e.,

$$\mathbf{E}[R|\omega_1]\mathbf{w} = \lambda \mathbf{E}[R|\omega_2]\mathbf{w} \tag{2.2}$$

where $\mathbf{E}[R|\omega_c]$ is the expectation of R for class ω_c, $c = 1, 2$, and $\mathbf{w}$ is the CSP spatial filter. Multiple solutions could be obtained by solving (2.2), and usually the spatial filters $\mathbf{w}$ that correspond to the r largest and r smallest eigenvalues λ are selected to form the projection matrix $W \in \mathbb{R}^{n_c \times d}$ with $d = 2r$. The spatial projection matrix W jointly diagonalize $\mathbf{E}[R|\omega_1]$ and $\mathbf{E}[R|\omega_2]$:

$$\bar{\Lambda}_1 \equiv W^\top \mathbf{E}[R|\omega_1]W \tag{2.3}$$

$$\bar{\Lambda}_2 \equiv W^\top \mathbf{E}[R|\omega_2]W \tag{2.4}$$

where $\bar{\Lambda}_1$ and $\bar{\Lambda}_2$ are diagonal matrices with

$$\bar{\Lambda}_1 + \bar{\Lambda}_2 = I \tag{2.5}$$

The diagonal elements of $\bar{\Lambda}_1$ and $\bar{\Lambda}_2$ are the variances of the EEG signals after the spatial filtering. Thus, with (2.5), the spatial projection in CSP could minimize the power of the EEG signals under one class while maximizing it under the other class. Thus, the CSP features consist of the variances of the EEG signals after projection, as the following:

$$\mathbf{f}_{csp} \equiv \mathbf{diag}(W^\top RW) \tag{2.6}$$

As shown in (2.2)–(2.6), W is obtained by jointly diagonalizing $\mathbf{E}[R|\omega_1]$ and $\mathbf{E}[R|\omega_2]$ so that variance features $\mathbf{f}_{csp}$ are discriminative. However, as $\bar{\Lambda}_1$ and $\bar{\Lambda}_2$ are diagonal matrices, the EEG signals after projection become uncorrelated and no longer contain information of the connectivity patterns. Usually, the connectivity analysis is conducted in the scalp space without spatial filtering, which is prone to noise contamination and the volume conduction effect of the EEG recording. Moreover, as most of the connectivity feature extraction is performed in an unsupervised manner, it is difficult to effectively extract connectivity information related to task of interest. To address the issue, in the next section, we will introduce the discriminative learning of connectivity pattern.

2.2.2 Discriminative learning of connectivity pattern

To take the interaction between different channels into consideration, in this work, instead of the feature formed by diagonal elements, feature that consists of elements from the upper triangular part of the covariance matrix after projection is proposed as the following:

$$\begin{aligned} \mathbf{f} &\equiv \mathbf{upper}(\Lambda) \\ &= \left[\lambda_{11}, \sqrt{2}\lambda_{12}, \lambda_{22}, \sqrt{2}\lambda_{13}, \sqrt{2}\lambda_{23}, \lambda_{33}, \ldots, \lambda_{dd}\right]^T \end{aligned} \tag{2.7}$$

The off-diagonal elements λ_{ij} $(i \neq j,\ i,j < d)$ are multiplied by $\sqrt{2}$ so that $||f||_2 = ||\Lambda||_F$, where $||\cdot||_F$ denotes the Frobenius norm, and the details could be found in [28]. With (2.7), the discriminative connectivity pattern could be reflected by the feature elements λ_{ij} $(i \neq j, i,j < d)$ in feature **f**. The Fisher's ratio objective function based on **f** in (2.7) is adopted to optimize spatial filters. In this way, interclass feature distance could be maximized while the within-class feature distances being minimized, and joint diagonalization of EEG covariance matrices could be avoided. In the rest parts for this section, the optimization of the spatial filters with the Fisher's ratio objective function will be presented.

Let $I_d \in \mathbb{R}^{d \times n_c}$

$$I_d = \begin{pmatrix} 1 & 0 & \ldots & \ldots & \ldots & 0 \\ 0 & 1 & \ldots & \ldots & \ldots & 0 \\ \vdots & \vdots & 1 & \vdots & \vdots & 0 \\ 0 & 0 & \ldots & 1 & \ldots & 0 \end{pmatrix} \tag{2.8}$$

so that

$$W^T = I_d U^T \tag{2.9}$$

where $U \in \mathbb{R}^{n_c \times n_c}$ is the rotation matrix with $U^T U = I$. The feature **f** in (2.7) could be written as

$$\mathbf{f} = \mathbf{upper}\left(I_d U^T R U I_d^T\right) \tag{2.10}$$

Given R_k and R_l as covariance matrices calculated for trials k and l, the feature distance $d_{ij}(k,l)$ for the off-diagonal feature element $\sqrt{2}\lambda_{ij}$ $(i \neq j, i,j < d)$ could be written as

$$d_{ij}(k,l) = \sqrt{2}(\mathbf{1}_{n_c}^i)^T U^T (R_k - R_l) U (\mathbf{1}_{n_c}^j) \tag{2.11}$$

where $\mathbf{1}_{n_c}^m \in \mathbb{R}^{n_c}$ be the vector with the mth element as 1 and the rest as 0:

$$\mathbf{1}_{n_c}^m \equiv [0, 0, \ldots, 1, \ldots, 0]^T \tag{2.12}$$

Similarly, for the diagonal feature element $\sqrt{2}\lambda_{ii}$ $(i < d)$, the feature distance $d_{ii}(k, l)$ is

$$d_{ii}(k,l) = (\mathbf{1}^i_{n_c})^T U^T (R_k - R_l) U (\mathbf{1}^i_{n_c}) \tag{2.13}$$

With (2.11) and (2.13), the feature distance between trial k and l could be calculated as

$$\begin{aligned} \mathbf{d}^2_{\mathbf{f}}(k,l) &\equiv ||\mathbf{f}_k - \mathbf{f}_l||_2 \\ &= \sum_i^d \sum_j^d d^2_{ij}(k,l) \end{aligned} \tag{2.14}$$

The average feature $\bar{\mathbf{f}}_{\omega_c}$ for class ω_c could be obtained as

$$\bar{\mathbf{f}}_{\omega_c} = \frac{1}{|Q_c|} \sum_{l \in Q_c} \mathbf{f}_l \tag{2.15}$$

where Q_c is the set contained all the trial indices for class ω_c. Then, the feature distance between trial k and the average feature for class ω_c, $\mathbf{d}_{\mathbf{f}}(k)$ could be easily obtained as

$$\mathbf{d}^2_{\mathbf{f}}(k) \equiv ||\mathbf{f}_k - \bar{\mathbf{f}}_{\omega_c}||_2 \tag{2.16}$$

Thus, the within-class feature dissimilarity $S_{\mathbf{w}}$ in the Fisher's ratio could be obtained as

$$S_{\mathbf{w}} = \frac{1}{2n_{\mathbf{f}}} \sum_{c=\{1,2\}} \frac{1}{||Q_c||} \sum_{k \in Q_c} \mathbf{d}^2_{\mathbf{f}}(k) \tag{2.17}$$

The interclass feature dissimilarity $S_{\mathbf{b}}$ could be obtained similarly, as the following

$$S_{\mathbf{b}} = \frac{1}{n_{\mathbf{f}}} ||\bar{\mathbf{f}}_{\omega_1} - \bar{\mathbf{f}}_{\omega_2}||_2 \tag{2.18}$$

With $S_{\mathbf{w}}$ and $S_{\mathbf{b}}$, U is optimized to minimize the Fisher's ratio discriminant objective function

$$\begin{aligned} \hat{U} &= \arg\min_U \mathcal{J}(U) = \frac{S_{\mathbf{w}}}{S_{\mathbf{b}}} \\ &s.t.\ U^T U = I \end{aligned} \tag{2.19}$$

Upon the optimization U, the spatial projection matrix W and the feature $\mathbf{f}$ could be calculated and classified accordingly.

2.3 Experimental study

2.3.1 Experimental setup and data processing

In the experimental study, 27 channels of EEG signals were recorded using Nuamps EEG acquisition hardware, where the electrodes were unipolar Ag/AgCl channels, the sampling rate was 250 Hz and the resolution was 22 bits for the voltage range of ± 130 mV. In the acquisition hardware, a prefiltering bandpass filter is set as 0.05–40 Hz.

Totally 16 subjects attended the experiment, which consisted of two parts: calibration part and test part. The calibration part included one MI session and one passive movement session, and both sessions contained 2 runs. In the MI session, there were two classes of trials, i.e., MI trials and idle condition trials. Similarly, in the passive movement session, the two classes of trials were passive movement trials and idle condition trials. Each run lasted for approximately 16 minutes and comprised 40 trials of MI or passive movement, and 40 trials of idle state. During the MI trials, the subjects were asked to perform kinesthetic MI of the chosen hand, while during the passive movement trials, the subjects performed passive movement of the chosen hand conducted by a haptic knob robot. During the entire EEG recording process, the subjects were asked to avoid physical movement and eye blinking. In the test session, there was one MI session consisting of 2–3 runs. Thus, there are 80 trials per class in the training session and 80–120 trials per class in the test session, yielding totally 160 training trials and 160–240 test trials for each subject. Details of the experimental setup can be found in [29].

2.3.2 Correlation results

2.3.2.1 Comparison between covariance matrices

To compare the proposed spatial filtering method with the conventional discriminative spatial filtering method, i.e., CSP, the difference of covariance matrices after spatial filtering between the two classes are calculated as

$$\Lambda_{diff} \equiv \bar{\Lambda}_1 - \bar{\Lambda}_2 \tag{2.20}$$

Λ_{diff} obtained with CSP and the proposed method are illustrated in Figures 2.1–2.3, where subfigure (a) and subfigure (b) correspond to Λ_{diff} with CSP and the proposed methods, respectively. As CSP seeks the joint diagonalization of $\mathbf{E}[R|\omega_1]$ and $\mathbf{E}[R|\omega_2]$, $\bar{\Lambda}_1$, $\bar{\Lambda}_2$ and Λ_{diff} are all diagonal matrices. By contrast, in the proposed method, the covariance matrices after filtering are not diagonal anymore. Subfigures (b) in Figures 2.1–2.3 correspond to $d = 4$, 6 and 8, respectively. It could be seen that with the proposed method, the elements in the d-by-d submatrix in the upper-left corner show the enhanced discriminative power between the two classes, which is different from CSP where the first and last diagonal elements in Λ_{diff} are the most discriminative features. Moreover, with the proposed method, the off-diagonal elements in the d-by-d submatrix in the upper-left corner represent the discriminative connectivity features, which could not be found in the Λ_{diff} obtained with CSP.

To further compare the EEG signal connectivities after spatial filtering, we have calculated the EEG signal coherences after spatial filtering in the proposed method and CSP. Coherence is the quantification of the frequency-based relationship between

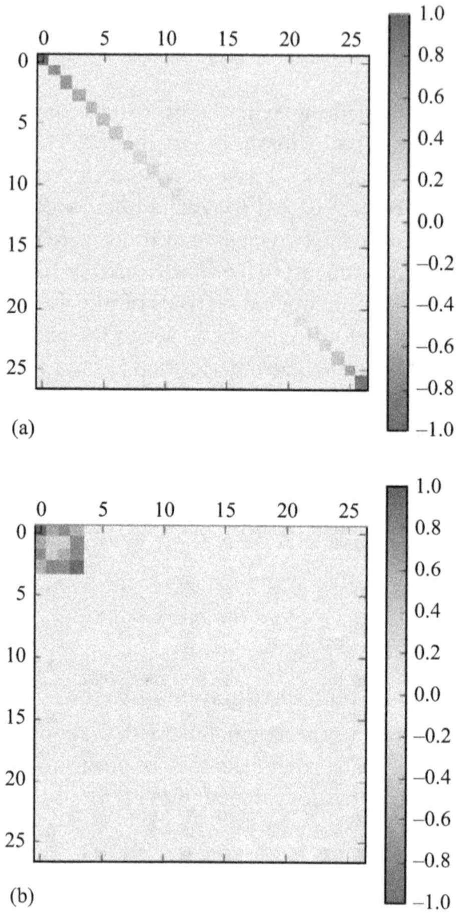

Figure 2.1 Λ_{diff} *obtained with (a) CSP and (b) the proposed method with* $d = 4$

two signals and has been widely used as a measurement of EEG connectivities [24]. The coherence COH_{ij} between EEG signals i and j at frequency f is given by

$$COH_{ij}(f) = \frac{|P_{ij}(f)|^2}{P_{ii}(f)P_{jj}(f)} \tag{2.21}$$

where $P_{ii}(f)(P_{jj}(f))$ is the power spectral density (PSD) for channel $i(j)$, and $P_{ij}(f)$ is the cross PSD between channels i and j at frequency f.

For both CSP and the proposed method, the average coherence between the d signals after spatial filtering is calculated as

$$\overline{COH} \equiv \frac{1}{|Q|} \sum_{k \in Q} \sum_{j,j \neq i}^{d} \sum_{i}^{d} \sum_{f \in \{f_l, f_h\}} C_{ij,k}(f) \tag{2.22}$$

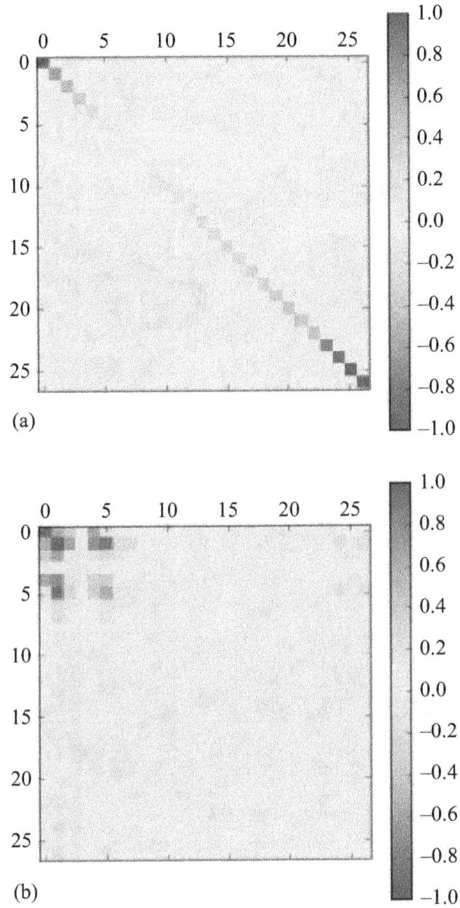

Figure 2.2 Λ_{diff} *obtained with (a) CSP and (b) the proposed method with* $d = 6$

where Q is the set containing all trial indices and $\{f_l, f_h\}$ the frequency range to be investigated. Let $\overline{COH}_{CSP}$ and $\overline{COH}_p$ be $\overline{COH}$ calculated with CSP and the proposed method. $\overline{COH}_{CSP}$ and $\overline{COH}_p$ are obtained for each subject, and subsequently, paired t-test is applied for the comparison of these two values. The mean $\overline{COH}_{CSP}$ and $\overline{COH}_p$ over all subjects and the paired t-test results could be found in Table 2.1.

As shown in Table 2.1, for α-band, β-band and 8–32 Hz band, the average coherences $\overline{COH}_p$ are higher than $\overline{COH}_{CSP}$, and the significance of the differences are validated by paired t-test. In [30], it is found that there is a high degree of comparability between EEG signal correlations and coherences. Therefore, by keeping the signal correlations in time domain, the proposed method could maintain stronger coherences between the signals after spatial filtering, compared to CSP jointly diagonalizing EEG signals.

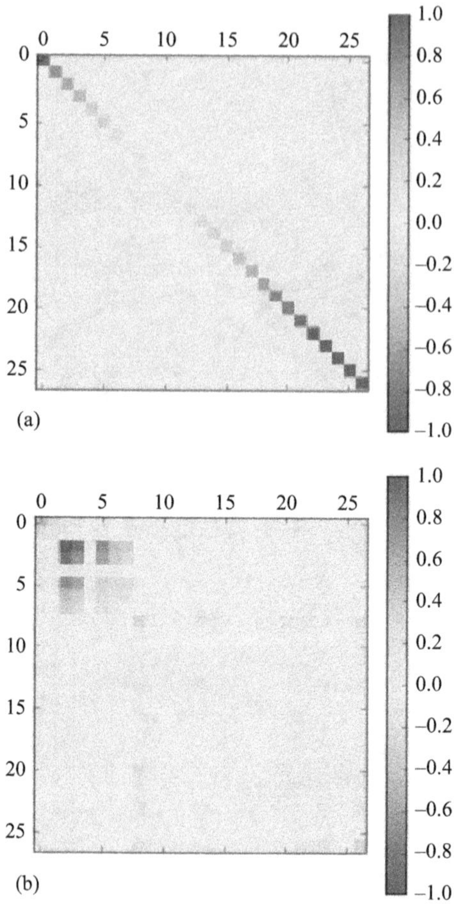

Figure 2.3 Λ_{diff} *obtained with (a) CSP (d = 4, subject 1) and (b) the proposed method (d = 8, subject 1) with d = 8*

Table 2.1 Comparing connectivity between motor imagery and passive movement (%)

Band	$\overline{COH}_p$	$\overline{COH}_{CSP}$	***p*-Value**
8–32 (Hz)	0.025	0.021	<0.001
α-Band (8–12 Hz)	0.026	0.022	<0.001
β-Band (12–30 Hz)	0.025	0.022	<0.001

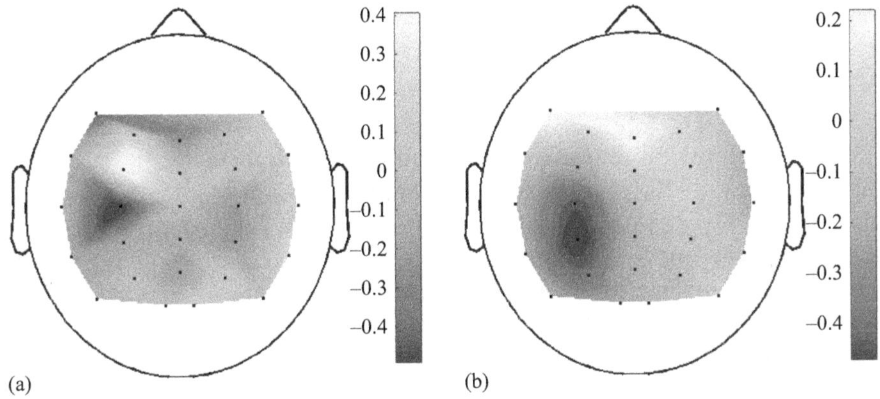

Figure 2.4 Source pair ((a) and (b)) obtained in the proposed method (subject 3)

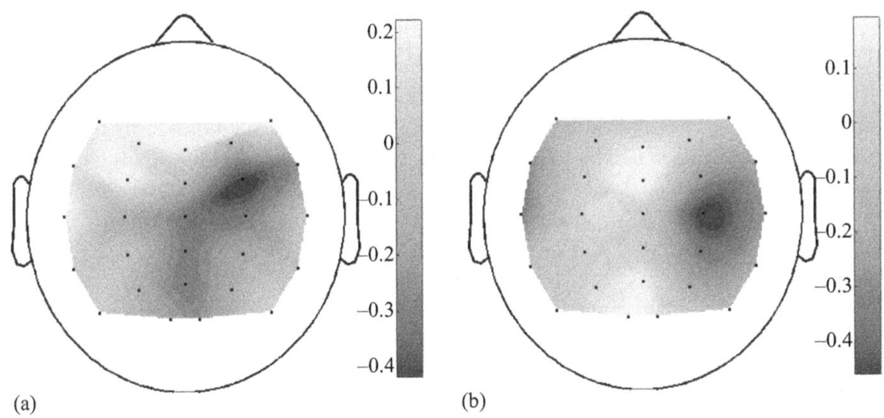

Figure 2.5 Source pair ((a) and (b)) obtained in the proposed method (subject 7)

2.3.2.2 Spatial patterns of source pair

Examples of the spatial patterns of the source pair obtained by the proposed method are illustrated in Figures 2.4 and 2.5, which correspond to subjects 3 and 7, respectively. For subject 3 in Figure 2.4, the source pair are both located in the left hemisphere, with one of the source close to frontal area and the other in the parietal area. Subject 3 performed right-hand MI, and it is found that the connectivity between the two sources is decreased for the MI class compared with the idle class. Moreover, it is found that these two sources are ERD sources, whose powers are decreased for the MI class compared with the idle class. For subject 7 who performed left-hand MI, there are two ERD sources located in the right hemisphere, as illustrated in Figure 2.5. In this case, the connectivity between the two sources is also decreased for the MI class compared with the idle class. These two examples show that besides ERD effects, source desynchronization existing during MI could also be captured by the proposed method.

2.3.2.3 Comparing motor imagery and passive movement

For further analysis, the connectivity patterns of MI and passive movement are compared. With the proposed method, the number of source pairs whose connectivity during MI/passive movement class that is higher than its connectivity during the idle class is calculated and denoted as n_{in}. Similarly, the number of source pairs whose connectivity during MI/passive movement class is lower than its connectivity during the idle class is calculated and denoted as n_{de}. In this analysis, $d = 6$ which yields 15 connectivity pairs. The results are summarized in Table 2.2, where ***p***-values of the paired ***t***-tests between n_{in} and n_{de} of all subjects are also presented.

It is found that for the MI case, n_{in} is equal to n_{de} on the average, while n_{in} is significantly larger than n_{de} for the passive movement case. We also calculate the number of ERD and ERS sources for both MI passive movement cases. We found that on average, for MI among six sources, the number of ERD and ERS sources are 2.9 and 3.1, respectively. For passive movements, the average number of ERS sources is 3.6 which is higher than that of ERD sources, i.e., 2.4, although the difference is not significant. From Table 2.2, it could be seen that more enhanced source synchronization can be found in the passive movement case compared with the

Table 2.2 Comparing connectivity between motor imagery and passive movement (%)

	Motor imagery		**Passive movement**	
SubID	n_{in}	n_{de}	n_{in}	n_{de}
1	4	11	9	6
2	12	3	3	12
3	8	7	11	4
4	6	9	8	7
5	6	9	9	6
6	0	15	15	0
7	7	8	9	6
8	7	8	11	4
9	6	9	10	5
10	11	4	8	7
11	8	7	9	6
12	10	5	8	7
13	8	7	12	5
14	7	8	5	10
15	11	4	10	5
16	9	6	13	2
Mean	7.5	7.5	9.3	5.6
p-Value	–	0.10	–	**0.02** (p-Value < 0.05)

MI case. This could be possibly related to the fact that the accuracies of classifying passive movement from idle condition is higher than that of classifying MI from idle condition [29]. In our future endeavor, more studies would be conducted to explore the pattern differences between MI and passive movement.

2.3.3 *Classification results*

To validate that the connectivities extracted by the proposed methods are discriminative, feature classification is performed using $\mathbf{f}$ in (2.10) with linear SVM classifiers. For the purpose of comparison, we also conduct classification with the SVM classifier using CSP features $\mathbf{f}_{csp}$ in (2.6) and the feature that consists of elements from the upper triangular part of the covariance matrix without any spatial projection, i.e.,

$$\mathbf{f}_{raw} = \mathbf{upper}(R) \tag{2.23}$$

In the classification analysis, the numbers of spatial filters for both CSP and the proposed methods are six. The session-to-session classification results are summarized in Table 2.3.

As shown in Table 2.3, the proposed method achieves results that are comparable to CSP with the same average accuracies and higher median accuracies. More importantly, the classification accuracies obtained by the proposed methods are higher than that obtained with $\mathbf{f}_{raw}$ for most of the cases. The feature $\mathbf{f}_{raw}$ could be regarded the raw

Table 2.3 Classification accuracy

	$\mathbf{f}_{csp}$		$\mathbf{f}_{raw}$		$\mathbf{f}$	
Subject ID	**Training**	**Test**	**Training**	**Test**	**Training**	**Test**
0	0.86	0.62	0.62	0.51	0.82	0.66
1	0.83	0.56	0.58	0.52	0.61	0.53
2	0.82	0.58	0.62	0.55	0.78	0.63
3	0.93	0.51	0.84	0.75	0.93	0.50
4	0.93	0.51	0.58	0.58	0.92	0.50
5	0.95	0.72	0.63	0.64	0.93	0.68
6	0.87	0.75	0.62	0.66	0.77	0.72
7	0.97	0.94	0.92	0.88	0.97	0.95
8	0.92	0.73	0.66	0.60	0.83	0.72
9	0.78	0.57	0.66	0.55	0.82	0.57
10	0.73	0.50	0.54	0.50	0.67	0.51
11	0.95	0.79	0.86	0.73	0.89	0.81
12	0.71	0.50	0.53	0.55	0.67	0.50
13	0.94	0.73	0.89	0.69	0.88	0.74
14	0.87	0.64	0.59	0.55	0.86	0.63
15	0.89	0.73	0.69	0.60	0.88	0.72
Mean	0.87	0.65	0.68	0.62	0.83	0.65
Median	0.88	0.63	0.63	0.59	0.85	0.65

connectivity features at the scalp-level. Therefore, it could be seen that the proposed method is able to enhance the discriminative power of the connectivity features.

2.4 Relations with existing methods

In [31,32], source power correlation analysis (SPoC) and canonical SPoC (cSPoC) were proposed, where the authors used oscillatory correlations between EEG signals to form the objective functions for spatial filter design. Different from SPoC and cSPoC, the connectivity features used in the proposed method is the correlations between the EEG signals. When cSPoC is applied to the one EEG dataset, it could not be used to find out sources with maximized the oscillatory correlations but only with minimized correlations, because the trivial solutions would be obtained when maximizing the source correlations [32]. Therefore, to address the problem of trivial solutions, instead of maximizing or minimizing the connectivities of the EEG signals after projection, Fisher's ratio objective function is adopted in the proposed method.

In [33], a general model of multiway optimization for EEG is proposed as subspace-constrained uncorrelated multilinear discriminant analysis (SMLDA), which combines spatial and temporal filter optimization. SMLDA could be used to conduct discriminative FC analysis of EEG by imposing certain subspace constraint in the optimization function. SMLDA aims at first-order EEG feature analysis which is different from the proposed method aiming at ERD/ERS effects-related connectivity analysis. However, it is possible to extend the objective function in SMLDA for second-order feature analysis.

2.5 Conclusion

In this work, a discriminative learning method of EEG connectivity is proposed. On the one hand, for MI EEG analysis, the targeted features in the conventional spatial filter optimization methods are mostly powers of the signals. In particular, for methods jointly diagonalizing the EEG covariance matrices, the connectivity patterns of EEG signals are no longer kept after spatial filtering. On the other hand, most of the existing connectivity analysis methods, such as COH, GC and coefficients of autoregressive model, are at the scalp-level and in an unsupervised manner. As the spatial resolution of EEG is poor, such connectivity pattern analysis may not reflect the metal task of interest. To address the issue, in this work, correlations of EEG signals after spatial filtering are taken into consideration for spatial filter design. To be specific, a Fishers' ratio optimization objective function is used to optimize spatial filters with features that include EEG correlations. The proposed method is evaluated with a real-work binary MI EEG dataset. As shown by experiment results, more connectivity information could be maintained in the proposed method compared to convention spatial filtering method CSP. Moreover, classification results show that the correlation features extracted by the proposed method are discriminative.

References

[1] Wolpaw JR, Birbaumer N, McFarland DJ, *et al.* Brain-computer interfaces for communication and control. Clinical Neurophysiology. 2002;113(6):767–91.

[2] Leeb R, Friedman D, Müller-Putz GR, *et al.* Self-paced (asynchronous) BCI control of a wheelchair in virtual environments: a case study with a tetraplegic. Computational Intelligence and Neuroscience. 2007;2007:79642.

[3] Müller-Putz GR, Pfurtscheller G. Control of an electrical prosthesis with an SSVEP-based BCI. IEEE Transactions on Biomedical Engineering. 2008;55(1):361–4.

[4] Ramoser H, Müller-Gerking J, Pfurtscheller G. Optimal spatial filtering of single trial EEG during imagined hand movement. IEEE Transactions on Rehabilitation Engineering. 2000;8(4):441–6.

[5] Blankertz B, Tomioka R, Lemm S, *et al.* Optimizing spatial filters for robust EEG single-trial analysis. IEEE Signal Processing Magazine. 2008; 25(1):41–56.

[6] Wang D, Miao D, Blohm G. Multi-class motor imagery EEG decoding for brain-computer interfaces. Frontiers in Neuroscience. 2012;6:151.

[7] Li X, Guan C, Zhang H, *et al.* A unified Fisher's ratio learning method for spatial filter optimization. IEEE Transactions on Neural Networks and Learning Systems. 2016;PP(99):1–11.

[8] Li X, Guan C, Zhang H, *et al.* Adaptation of motor imagery EEG classification model based on tensor decomposition. Journal of Neural Engineering. 2014;11:056020.

[9] Li X, Zhang H, Guan C, *et al.* Discriminative learning of propagation and spatial pattern for motor imagery EEG analysis. Neural Computation. 2013;25(10):2709–33.

[10] Li X, Guan C, Zhang H, *et al.* Discriminative ocular artifact correction for feature learning in EEG analysis. IEEE Transactions on Biomedical Engineering. August 2017;64(8):1906–13.

[11] Zhang H, Chin ZY, Ang KK, *et al.* Optimum spatio-spectral filtering network for brain–computer interface. IEEE Transactions on Neural Networks. 2011 Jan;22(1):52–63.

[12] Zhang H, Yang H, Guan C. Bayesian learning for spatial filtering in an EEG-based brain–computer interface. IEEE Transactions on Neural Networks and Learning Systems. 2013;24(7):1049–60.

[13] Ang KK, Chin ZY, Zhang H, Guan C. Mutual information-based selection of optimal spatial-temporal patterns for single-trial EEG-based BCIs. *Pattern Recognition*. 2012;45:2137–44.

[14] Hamedi M, Salleh SH, Noor AM. Electroencephalographic motor imagery brain connectivity analysis for BCI: a review. Neural Computation. 2016;28(6): 999–1041.

[15] Yang H, Guan C, Wang C, *et al.* Detection of motor imagery of brisk walking from electroencephalogram. Journal of Neuroscience Methods. 2015; 244:33–44.

[16] Kober SE, Wood G. Changes in hemodynamic signals accompanying motor imagery and motor execution of swallowing: a near-infrared spectroscopy study. NeuroImage. 2014;93(Pt 1):1–10.

[17] Yang H, Guan C, Chua KSG, *et al.* Detection of motor imagery of swallow EEG signals based on the dual-tree complex wavelet transform and adaptive model selection. Journal of Neural Engineering. 2014;11(3):035016.

[18] Song L, Gordon E, Gysels E. Phase Synchrony Rate for the Recognition of Motor Imagery in Brain-Computer Interface. In: Weiss Y, Schölkopf PB, Platt JC, editors. Advances in Neural Information Processing Systems 18. MIT Press; 2006. p. 1265–72.

[19] Vinck M, Oostenveld R, van Wingerden M, *et al.* An improved index of phase-synchronization for electrophysiological data in the presence of volume-conduction, noise and sample-size bias. NeuroImage. 2011;55(4):1548–65.

[20] Sakkalis V. Review of advanced techniques for the estimation of brain connectivity measured with EEG/MEG. Computers in Biology and Medicine. 2011;41(12):1110–17.

[21] Marinazzo D, Liao W, Chen H, *et al.* Nonlinear connectivity by Granger causality. NeuroImage. 2011;58(2):330–8.

[22] Kus R, Kaminski M, Blinowska KJ. Determination of EEG activity propagation: pair-wise versus multichannel estimate. IEEE Transactions on Biomedical Engineering. 2004;51(9):1501–10.

[23] Wei Q, Wang Y, Gao X, *et al.* Amplitude and phase coupling measures for feature extraction in an EEG-based brain – computer interface. Journal of Neural Engineering. 2007;4(2007):120–9.

[24] Krusienski DJ, McFarland DJ, Wolpaw JR. Value of amplitude, phase, and coherence features for a sensorimotor rhythm-based brain–computer interface. Brain Research Bulletin. 2012;87(1):130–4.

[25] Billinger M, Brunner C, Muller-Putz GR. Single-trial connectivity estimation for classification of motor imagery data. Journal of Neural Engineering. 2013;2013(046006):8pp.

[26] Luck SJ. An introduction to the event-related potential technique. Cambridge, MA: MIT Press; 2014.

[27] Rugg MD, Coles MG. Electrophysiology of mind: event-related brain potentials and cognition. Oxford, UK: Oxford University Press; 1995.

[28] Barachant A, Bonnet S, Congedo M, *et al.* Classification of covariance matrices using a Riemannian-based kernel for {BCI} applications. Neurocomputing. 2013;112:172–8. Available from: http://www.sciencedirect.com/science/article/pii/S0925231213001574.

[29] Ang KK, Guan C, Wang C, *et al.* Calibrating EEG-based motor imagery brain-computer interface from passive movement; Conference of the IEEE Engineering in Medicine and Biology Society. 2011:4199–202.

[30] Guevara MA, Corsi-Cabrera M. EEG coherence or EEG correlation? International Journal of Psychophysiology. 1996;23(3):145–53. Available from: http://www.sciencedirect.com/science/article/pii/S0167876096000384.

[31] Dähne S, Meinecke FC, Haufe S, *et al.* SPoC: a novel framework for relating the amplitude of neuronal oscillations to behaviorally relevant parameters. NeuroImage. 2014;86:111–22.

[32] Dähne S, Nikulin VV, Ramírez D, *et al.* Finding brain oscillations with power dependencies in neuroimaging data. NeuroImage. 2014;96:334–48.

[33] Higashi H, Rutkowski TM, Tanaka T, *et al.* Multilinear discriminant analysis with subspace constraints for single-trial classification of event-related potentials. IEEE Journal of Selected Topics in Signal Processing. 2016;10(7): 1295–305.

Chapter 3

An experimental study to compare CSP and TSM techniques to extract features during motor imagery tasks

Matteo Sartori[1], *Simone Fiori*[2], *and Toshihisa Tanaka*[3]

Abstract

Common spatial pattern (CSP) is a well-established technique to extract features from electroencephalographic recordings for classification purpose in motor imagery brain–computer interface (BCI). The CSP algorithm is a mathematical procedure used for separating a multivariate signal into additive components which have maximum differences in variance between two windows; in other terms, CSP increases the signal variance for one condition while minimizing the variance for the other condition. Features computed by means of CSP are fed to a data classifier in order to discriminate between two mental tasks. A novel technique to achieve feature extraction is tangent-space mapping (TSM) that insists on spatial covariance matrices computed from the recorded electroencephalogram signals (EEG). TSM is based on Riemannian geometry, which allows one to estimate statistical features of data distributions over non-Euclidean spaces.

The aim of this chapter is twofold: first, to provide a new data-visualization tool to visually inspect data distributions on the Riemannian space of spatial covariance matrices and its tangent bundle; second, to present an experimental comparison of CSP and TSM feature extraction, in conjunction with two classification methods, namely, support-vector machine and linear discriminant analysis. In particular, the experimental comparison performed on a number of data sets will show the superiority of TSM-based feature extraction over CSP.

[1]Department of Information Engineering and Graduate School of Information and Automation Engineering, Università Politecnica delle Marche, Italy

[2]Department of Information Engineering and Graduate School of Information and Automation Engineering, Università Politecnica delle Marche, Italy

[3]Department of Electrical and Electronic Engineering, Tokyo University of Agriculture and Technology, Japan

3.1 Introduction

Brain–computer interface (BCI) is a fast-paced neuroengineering technology. BCI is sometimes referred to as mind–machine interface, direct neural interface or brain–machine interface. BCIs are systems that use brain signals (electric, magnetic, metabolic) to control external devices such as computers, switches, wheelchairs or neuroprosthetic tools. Originally, the motivation for developing BCIs has been to provide severely disabled individuals with a basic communication system. A motor imagery-based brain–computer interface (MI-BCI) is a technology which provides a communication between the brain and external world by decoding the brain activities into commands in order to control computer or peripheral devices. This type of direct-brain interfaces would increase an individual's independence, leading to a dramatically improved quality of life and reduce social costs. BCI research is dependent upon a variety of fields including psychology, psycho-physiology, cognitive neuroscience, computer science, electrical and automation engineering, rehabilitation engineering, cognitive computation and neurology. Recent application of BCI with appropriate feedback offers neurorehabilitation which assists stroke patients to restore impaired motor functions. Patients receive visual, auditory or kinesthetic feedback in order to promote the brain response to the MI task after stroke.

Brain–computer interface research builds on 20 years of efforts in the brain sciences and on recent developments in adaptive computing and cognitive computation. In the 1970s, it was discovered that subtle changes occur in the brain signals when humans plan movements. Such changes are called movement-related desynchronizations (MRDs) because, when movements are planned, the activity of neurons in the motor cortex becomes desynchronized [1]. An internally paced or externally paced event results not only in the generation of an event-related potential but also in a change in the brain signals in the form of an event-related desynchronization or event-related synchronization, which are responses of different neuronal structures in the brain [2]. The MRD signals are rarely stronger than a few tens of μV and are often buried beneath other signals. It is, therefore, necessary to employ advanced signal processing and pattern recognition methods, such as machine learning techniques, to detect the MRD signals.

Desirable features of a BCI system are [3]: (1) the BCI device should rely on signals recorded directly from the brain of the user; (2) there should be at least one, among the capturable brain signals, which the user can intentionally modulate; (3) the BCI device should be able to operate real-time processing; (4) the user should be able to obtain a feedback from the BCI device. In fact, BCI relies on electrical measures of brain activity, through sensors placed over the head. Electroencephalogram, for instance, refers to the electrical activity recorded through electrodes positioned on the scalp. Brain signals are evaluated at multiple channels and undergo dimension reduction to remove extraneous information, thereby reducing computational costs. During the last decades, EEG-based BCI became one of the most popular noninvasive techniques. A major challenge in the development of brain–computer interfaces is to identify cortical regions and related functions that an individual can reliably and consciously manipulate. The motor system is currently one of the primary focuses of signal acquisition, where signals are obtained during imagined motor responses. In MI, the subject imagines performing an action, the resulting recorded EEG data are classified online, and the result is graphically presented to the subject (e.g., as a horizontal bar on the screen that moves right if right-hand MI is detected or moves left if left-hand MI is detected).

The processing of the acquired signals in BCI systems is usually divided into four stages. It starts with an initial preprocessing stage in which the EEG signals are filtered, and some of the possible artifacts superimposed on the signal of interest (flickering, eye movements, electrocardiogram, muscle movements) are eliminated. This is followed by a second stage in which certain specific characteristics of the EEG signal are extracted. Subsequently, feature-selection methods are used to choose the most significant features, i.e., those which codify the user's intention. Classification and translation algorithms identify and transform EEG features into command-signals to perform user intentions. Brain–computer interfaces recognize and classify different brain signaling patterns into a group based on its traits. In the classification stage, the BCI system attempts to learn the user intention by the characteristics that distinguish different brain activities. Among the different kinds of algorithms that are currently in use, the most popular ones are regression and classification. Regression algorithms directly use EEG signal features as independent variables in inferring the user's intention. In classification algorithms, the EEG signal features are employed to outline the perimeters between two or more targets in the target space.

A well-established approach to extract features from observations of MI is the *common spatial pattern* (CSP) method [4–6]. This method performs a linear combination of the EEG signals by means of coefficients determined so that the variance of the signal extracted by the spatial filter differs between two different tasks (such as, for example, left-hand and right-hand movement). An emerging alternative approach that makes use of the covariance structure of the recorded EEG signals has been proposed recently [7]. Since covariance matrices are symmetric and positive-definite, they belong to a curved Riemannian manifold [8–12]. Tangent-space mapping (TSM) extracts features from EEG signals for classification purpose by using Riemannian geometry. Since the observed EEG signals normally include outlier trials strongly affected by the artifacts such as eye blink and muscle movement, robust estimators of covariance matrices have been proposed [13–17]. In this chapter, we discuss the use of robust estimators of covariance matrices based on the notion of *geometric median* and of *trimmed averaging*. The geometric median is one of the multivariate extensions of the median (in the scalar case) which is widely used as a robust estimator of centrality instead of the mean. Likewise, the trimmed averaging removes those samples that lay too far from the average of all matrices and compute an average of the remaining samples. In the present chapter, we focus on TSM-based feature extraction for MI-BCI and compare the classification accuracy of TSM and CSP on two-class problems [7,18,19]. The experimental evaluation on real EEG data illustrates a comparison between CSP and TSM, the performances of the geometric median and the performance of "trimmed averages."

3.2 Theoretical concepts and methods

This section regards theoretical concepts and explains the features of the BCI stages with the support of Riemannian geometry applied to sample covariance matrices (SCM). Within this section, we present the stages of features extraction and features classification based on the CSP and the TSM methods together with linear discriminant analysis (LDA) and support vector machine (SVM) learning models. The comparison between CSP and TSM methods plays a central role in this contribution and the success of those methods depends on the choice of the correct reference points by which the spatial filters or the tangent spaces were calculated. Therefore, Section 3.2.1 discusses several averaging techniques of SCMs to estimate reference points, like means, medians and trimmed averages, while Section 3.2.2 explains the role played by averages in data classification. The present section ends by presenting the theory of multidimensional scaling (MDS) algorithm (Section 3.2.3), which allows to display the MI-trials, summarized into SCMs, like a cloud of points in a 2D chart.

3.2.1 Averaging techniques of SCMs

Some averaging techniques of SCMs that we adopted are explained in [20,21]. These methods use different concepts to achieve averaging over the space of symmetric, positive-definite (SPD) matrices. Below we make a quick recall of averaging methods and their formulas for a set of N SPD matrices $P_1, \ldots, P_N$.

- *Arithmetic mean*: This is the most straightforward way of taking the average of a set of matrices and does not make any reference to the SPD nature of the spatial covariance matrices being averaged. The arithmetic mean is defined as

$$P_A = \frac{1}{N} \sum_{i=1}^{N} P_i. \tag{3.1}$$

  Nevertheless, the arithmetic mean is consistent in that P_A is still an SPD matrix. It has the advantage of possessing a closed-form expression.
- *Riemannian Geometric mean*: This type of mean is fully compatible with the geometry of the SPD manifold and depends on the metrics that this manifold is endowed with. The Riemannian geometric mean does not possess a closed-form expression and may be evaluated only by means of an iterative algorithm [22] expressed by

$$P_R^{(k+1)} = \mathrm{Exp}_{P_R^{(k)}} \left(\frac{1}{N} \sum_{i=1}^{N} \mathrm{Log}_{P_R^{(k)}}(P_i) \right), \quad k = 0, 1, \ldots, \tag{3.2}$$

  where the initial guess $P_R^{(0)}$ is set to the identity matrix I. In the iteration formula, $\mathrm{Exp}(\cdot)$ and $\mathrm{Log}(\cdot)$ denote the exponential map and the logarithmic map, respectively (for technical details and a derivation, see [22]).

- *Log-Euclidean mean*: This type of mean is known from the literature and may be also obtained from the Riemannian geometric mean stopped at the first iteration [22]. It has, therefore, a closed-form expression given by

$$P_L = \exp\left(\frac{1}{N}\sum_{i=1}^{N}\log P_i\right), \tag{3.3}$$

where $\exp(\cdot)$ and $\log(\cdot)$ denote matrix exponential and principal matrix logarithm, respectively.

- *Harmonic mean*: This is a transliteration of the harmonic mean of positive scalars and is defined by

$$P_H = \left(\frac{1}{N}\sum_{i=1}^{N}P_i^{-1}\right)^{-1}. \tag{3.4}$$

The inverse of the Harmonic mean may be regarded as the arithmetic mean of the inverses of the data. The Harmonic mean is consistent with the SPD-nature of the data P_i, in fact, it is well-known that the inverse of a symmetric positive-definite matrix is still symmetric and positive-definite.

- *Resolvent mean*: The resolvent mean combines the features of the Harmonic mean with the concept of *resolvent* associated to a matrix into the closed-form expression

$$P_R(\mu) = \left(\sum_{i=1}^{N}(P_i + \mu^{-1}I)^{-1}\right)^{-1} - \mu^{-1}I, \tag{3.5}$$

where $\mu > 0$ is a parameter that controls the structure of the resolvents being averaged.

- *Euclidean geometric median*: A median is a robust version of a mean. The Euclidean geometric median does not admit any closed-form expression and may be approximately inferred by means of the following iterative algorithm:

$$P_{A*}^{(k+1)} = \left(\sum_{i=1}^{N}\frac{P_i}{d_E(P_{A*}^{(k)}, P_i)}\right)\left(\sum_{i=1}^{N}\frac{1}{d_E(P_{A*}^{(k)}, P_i)}\right)^{-1}, \quad k = 0, 1, \ldots, \tag{3.6}$$

where the initial guess $P_{A*}^{(0)}$ is set to the identity matrix I and $d_E(\cdot,\cdot)$ denotes a Euclidean distance between a pair of matrices.

- *Riemannian geometric median*: This median also does not admit any closed-form expression and may be approximately inferred by means of the following two-stage iterative algorithm:

$$V^{(k)} = \left(\sum_{i=1}^{N}\frac{\mathrm{Log}_{P_G^{(k)}}(P_i)}{d_R(P_G^{(k)}, P_i)}\right)\left(\sum_{i=1}^{N}\frac{1}{d_R(P_G^{(k)}, P_i)}\right)^{-1}, \quad k = 0, 1, \ldots, \tag{3.7}$$

$$P_G^{(k+1)} = \mathrm{Exp}_{P_G^{(k)}}(V^{(k)}), \quad k = 0, 1, \ldots, \tag{3.8}$$

where the initial guess $P_G^{(0)}$ is set to the identity matrix I. In the above formulas, the function $d_R(\cdot, \cdot)$ denotes a Riemannian distance in the manifold of SPD matrices.

- *Log-Euclidean geometric median*: Similarly to the two previous medians, this median also does not possess any closed-form expression and may be approximated by means of the following two-stage iterative algorithm:

$$V^{(k)} = \left(\sum_{i=1}^{N} \frac{R_e^{-1}(P_{L*}^{(k)}, P_i)}{d_R(P_{L*}^{(k)}, P_i)} \right) \left(\sum_{i=1}^{N} \frac{1}{d_R(P_{L*}^{(k)}, P_i)} \right)^{-1}, \quad k = 0, 1, \ldots, \tag{3.9}$$

$$P_{L*}^{(k+1)} = R_e(P_{L*}^{(k)}, V^{(k)}), \quad k = 0, 1, \ldots, \tag{3.10}$$

where R_e and R_e^{-1} denote a retraction map and its inverse lifting map, respectively, and the initial guess $P_{L*}^{(0)}$ is set to the identity matrix I. We recall that a retraction map is a generalization of the exponential map $\mathrm{Exp}(\cdot)$ [22].

The aim of introducing trimmed methods is to remove outliers from a dataset in conjunction with the averaging techniques explained before. The leading concept in data trimming is that we can compute the distance between SPD matrices and their averages; hence, we can detect which samples are outliers according to the distance of the samples from their averages. Removing the outliers we will have a restricted group of SPD matrices to calculate the average again. The trimmed methods which have been used are

- *Trimmed Riemannian mean*
- *Trimmed log-Euclidean mean*
- *Trimmed Riemannian median*
- *Trimmed log-Euclidean median*

They are defined as

$$P_{\hat{\mathscr{G}}} = \mathscr{G}\left[\mathrm{Trim}_d^{\mathscr{G}}(P_1, \ldots, P_N)\right], \tag{3.11}$$

$$P_{\hat{\mathscr{L}}} = \mathscr{L}\left[\mathrm{Trim}_d^{\mathscr{L}}(P_1, \ldots, P_N)\right], \tag{3.12}$$

where $\mathscr{G}[\cdot]$ and $\mathscr{L}[\cdot]$ are a Riemannian geometric mean or median and a Log-Euclidean mean or median operator, and $\mathrm{Trim}_d^{\mathscr{G}}(\cdot)$ and $\mathrm{Trim}_d^{\mathscr{L}}(\cdot)$ are the trimming operators discarding $d\%$ matrices that exhibit the largest distance to the mean or median in terms of Riemannian and Log-Euclidean metrics, respectively.

3.2.2 SCM averages in CSP and TSM methods

TSM is a powerful method of feature extraction and strongly depends on the selection of reference matrix considered as a covariance matrix. The SCM averages are important for the selection of the reference matrix in TSM-based MI-BCI. A correct choice of reference point to calculate the relative tangent plane leads to an easier classification approach and consequently a better result in terms of classification accuracy. A reference point should be selected carefully to obtain in the tangent plane a set of samples that can be linearly separable [23]. The averaging techniques explained before determine the success of the TSM method because the average of SPDs on the manifold is the base point for the tangent space on which the other SPDs will be projected. In the CSP method, the averages of the observed signals are the reference point to calculate the variance of the signals extracted (for more details, see, e.g., [24–26]).

3.2.3 Multidimensional scaling (MDS) algorithm

In MI-BCI, the spatial covariance matrices are of very high dimensionality. For example, in the experiments illustrated in the following sections, the dimensions of the SPD matrices range from 29×29 to 118×118. Unlike in 2D or 3D data analysis, in MI-BCI it is not immediate, therefore, to visualize the datasets as point-clouds. In order to achieve a possible data visualization, we made use of a special technique, referred to as manifold multidimensional scaling (MMDS) [27]. For the implementation of this algorithm, it is sufficient to follow the theoretical concepts and the algorithmic steps explained in [27]. In general terms, the algorithm takes a set of SCM matrices, computes their pairwise distances and finally associates each SCM to a single point.

An MMDS algorithm allows to reduce the size of the high-dimensional matrices up to visualize them like single points in a two- or three-dimensional chart [27]. We can use this technique to visualize the SCMs, therefore the MI-trials, on the Riemannian manifold or on tangent planes like single points. In this way, we can find and understand some relationships between the trials that are part of the same MI-class. This kind of visualization could underline the separation between classes as well as their overlapping and, therefore, the connections with a better or worse classification accuracy. Figure 3.1 shows a typical result of execution of MMDS

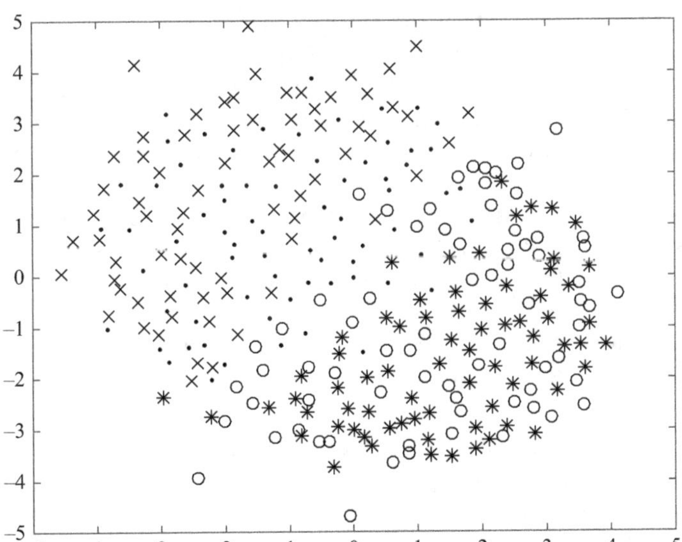

Figure 3.1 An exemplary result of MMDS algorithm applied to a dataset of 288 SCMs divided into four classes

algorithm applied to a set of SCMs. The aim is to visualize each SCM as a point in a 2D chart, in a way that preserves the distances between SCMs. In Figure 3.1, each mark represents a single SCM/MI-trial; therefore, we can have an idea about all SCMs distribution on the Riemannian manifold. If the trials are labeled, we can also differentiate the points with different symbols to recognize the classes and study their distribution.

3.3 Experimental results

This section is divided into two parts: Section 3.3.1 presents an experimental comparison of CSP and TSM for two-classes problems using both LDA and SVM and applying a 10-fold crossvalidation procedure to compute the classification accuracy; Section 3.3.2 shows results of application of the MMDS data-visualization tool in order to visualize how the high-dimensional spatial covariance matrices, mapped to a given tangent space, are distributed spatially on these tangent space. In the experiments, we have used publicly available datasets from the BCI Competition III and the BCI Competition IV (freely available at http://www.bbci.de/competition/) and the dataset JK-HH 1 containing EEG signals recorded during left hand and foot motor imageries [17]. It is important to notice that some four-classes datasets were restricted to two-classes datasets in this study, since the CSP method can only cope with two-classes problems. The bold values in the tables represent the highest values of each column.

3.3.1 Classification accuracy

3.3.1.1 Results obtained on the Dataset III_IIIa

In this experiment, we applied the TSM and the CSP methods on a partial set of classes, namely, "feet" and "tongue" pertaining to three subjects labeled *k3b*, *k6b* and *l1b*. Table 3.1 shows the classification accuracy of the TSM SVM/CSP method. We observed a difference of about 3% between the LDA and SVM methods in favor of the SVM one, mostly for the subject *k3b* and relatively to the trimmed-average matrices as reference points for the tangent spaces. In this experiment, the accuracy of the classification based on the trimmed averages is greater than those relative to the traditional averages. Table 3.2 shows the results of CSP SVM/LDA method applied on the same classes of Table 3.1. It can be readily appreciated again how the CSP method performs worse than the TSM method in terms of classification accuracy. In fact, we observed a difference of about 8% between them for the subjects *k3b* and *l1b*.

3.3.1.2 Results obtained on the Dataset III_IIIb

Table 3.3 shows the results obtained by TSM SVM/LDA method on the classes "left hand" and "right hand" for the subjects labeled *S4b* and *X11b*. There appear to be no relevant differences between the SVM and the LDA method. It is important to underline that the classification is useless for the subject *S4b*. Table 3.4 shows the results of CSP SVM/LDA method. The reader may note that the performance of the CSP method is worse than the performance of the TS method for the subject *X11b*

Table 3.1 Classification accuracy (%) of tangent space mapping SVM/LDA on a partial set of classes. Dataset III_IIIa, classes "feet" and "tongue."

	Subject					
	k3b		k6b		l1b	
	SVM	LDA	SVM	LDA	SVM	LDA
Arithmetic mean	90.0	91.1	91.6	91.6	73.3	73.3
Riemannian geometric mean	88.8	90.0	91.6	95.0	68.3	**78.3**
Log-Euclidean mean	93.3	88.8	95.0	93.3	65.0	66.6
Harmonic mean	91.1	91.1	93.3	91.6	73.3	70.0
Resolvent mean	94.4	91.1	93.3	95.0	75.0	71.6
Euclidean geometric median	88.8	91.1	93.3	93.3	68.3	68.3
Riemannian geometric median	88.8	82.2	91.6	95.0	68.3	68.3
Log-Euclidean geometric median	88.8	88.8	95.0	93.3	60.0	65.0
Trimmed Riemannian mean	**96.6**	90.0	95.0	95.0	**76.6**	73.3
Trimmed Log-Euclidean mean	95.5	**92.2**	**96.6**	95.0	75.0	73.3
Trimmed Riemannian median	95.5	90.0	95.0	95.0	**76.6**	**78.3**
Trimmed Log-Euclidean median	94.4	**92.2**	95.0	**96.6**	73.3	71.6
Identity matrix	90.0	88.8	91.6	91.6	68.3	70.0

Table 3.2 Classification accuracy (%) of common spatial pattern SVM/LDA on a partial set of classes. Dataset III_IIIa, classes "feet" and "tongue."

	Subject					
	k3b		k6b		l1b	
	SVM	LDA	SVM	LDA	SVM	LDA
Arithmetic mean	**92.2**	**90.0**	90.0	78.3	63.3	63.3
Riemannian geometric mean	74.4	73.3	93.3	80.0	63.3	**65.0**
Log-Euclidean mean	78.8	78.8	95.0	86.6	63.3	53.3
Harmonic mean	85.5	84.4	93.3	85.0	60.0	53.3
Resolvent mean	85.5	85.5	**96.6**	90.0	**68.3**	**65.0**
Euclidean geometric median	86.6	85.5	95.0	91.6	63.3	45.0
Riemannian geometric median	74.4	73.3	93.3	80.0	65.0	**65.0**
Log-Euclidean geometric median	80.0	80.0	73.3	75.0	65.0	55.0
Trimmed Riemannian mean	83.3	84.4	93.3	88.3	65.0	**65.0**
Trimmed Log-Euclidean mean	81.1	78.8	95.0	**93.3**	66.6	63.3
Trimmed Riemannian median	83.3	83.3	93.3	90.0	65.0	**65.0**
Trimmed Log-Euclidean median	81.1	81.1	78.3	83.3	**68.3**	**65.0**
Identity matrix	50.0	51.1	58.3	58.3	45.0	41.6

Table 3.3 Classification accuracy (%) of tangent space mapping SVM/LDA. Dataset III_IIIb, classes "left hand" and "right hand."

	Subject			
	S4b		X11b	
	SVM	LDA	SVM	LDA
Arithmetic mean	50.9	49.8	65.0	54.8
Riemannian geometric mean	50.9	49.8	65.2	65.0
Log-Euclidean mean	50.7	49.8	65.0	65.0
Harmonic mean	50.5	49.6	**65.3**	65.0
Resolvent mean	50.5	49.6	**65.3**	65.0
Euclidean geometric median	51.1	49.6	65.2	65.0
Riemannian geometric median	50.9	50.0	**65.3**	65.0
Log-Euclidean geometric median	50.7	49.8	65.0	65.0
Trimmed Riemannian mean	50.9	50.0	**65.3**	**65.2**
Trimmed Log-Euclidean mean	50.9	50.0	65.2	**65.2**
Trimmed Riemannian median	50.9	50.0	**65.3**	**65.2**
Trimmed Log-Euclidean median	50.9	50.0	**65.3**	**65.2**
Identity matrix	**52.4**	**53.1**	64.6	65.0

Table 3.4 Classification accuracy (%) of common spatial pattern SVM/LDA. Dataset III_IIIb, classes "left hand" and "right hand."

	Subject			
	S4b		X11b	
	SVM	LDA	SVM	LDA
Arithmetic mean	54.8	54.0	65.0	64.2
Riemannian geometric mean	54.8	53.1	64.8	63.7
Log-Euclidean mean	55.3	52.9	64.8	63.8
Harmonic mean	53.7	52.0	64.8	63.7
Resolvent mean	54.0	53.1	64.8	63.7
Euclidean geometric median	53.7	51.8	64.6	63.5
Riemannian geometric median	55.5	51.8	64.8	63.7
Log-Euclidean geometric median	54.6	51.6	64.6	63.7
Trimmed Riemannian mean	55.0	55.1	**65.5**	64.6
Trimmed Log-Euclidean mean	55.5	55.5	**65.5**	**65.0**
Trimmed Riemannian median	55.9	55.0	65.1	64.2
Trimmed Log-Euclidean median	**57.5**	**56.1**	65.3	63.8
Identity matrix	52.9	53.8	64.6	64.2

Table 3.5 Classification accuracy (%) of tangent space mapping SVM/LDA. Dataset III_IVa, classes "left hand" and "right hand."

	Subject			
	aa		al	
	SVM	LDA	SVM	LDA
Arithmetic mean	77.1	77.8	97.5	**97.1**
Riemannian geometric mean	77.5	79.2	97.1	**97.1**
Log-Euclidean mean	77.1	79.6	97.1	**97.1**
Harmonic mean	78.9	80.3	97.1	**97.1**
Resolvent mean	81.0	81.0	97.5	**97.1**
Euclidean geometric median	80.4	78.5	97.1	**97.1**
Riemannian geometric median	77.8	64.6	97.1	55.0
Log-Euclidean geometric median	78.2	80.0	97.1	**97.1**
Trimmed Riemannian mean	82.8	82.8	**97.8**	**97.1**
Trimmed Log-Euclidean mean	81.4	82.5	97.5	**97.1**
Trimmed Riemannian median	**83.2**	82.8	**97.8**	96.0
Trimmed Log-Euclidean median	82.1	**83.2**	97.5	**97.1**
Identity matrix	74.2	75.0	96.7	**97.1**

of about 1%; conversely, the CSP method performs better than the TSM method for the subject *S4b* of about 6% (e.g., the result of the trimmed Log-Euclidean median is 50.9% for the TSM-SVM versus 57.5% for the CSP-SVM): a possible reason for this behavior is that the trials on tangent space are confused and difficult to separate out.

3.3.1.3 Results obtained on the Dataset III_IVa

Table 3.5 shows the results of TSM SVM/LDA method on the classes "right hand" and "feet" for two subjects labeled *aa* and *al*. The Riemannian geometric median as reference point for the TSM-LDA model results in a lower classification accuracy of about 13% for the subject *aa* (from 77.8% to 64.6%) and of about 42% for the subject *al* (from 97.1% to 45%). Conversely, the best classification result was obtained for the subject *al* (97%). Table 3.6 shows the results of CSP SVM/LDA method, which appears to perform worse than the TSM method for all subjects of about 2%. As an interesting comparative result, it is readily appreciated how choosing the identity matrix as reference point for the CSP model results in a lower classification accuracy of about 30–40%.

3.3.1.4 Results obtained on the Dataset IV_IIa

In this experiment, we applied the TSM and the CSP methods to a partial set of classes, namely, "foot" and "tongue," for five subjects labeled *A01T*, *A03T*, *A07T*, *A08T* and *A09T*. Table 3.7 shows the classification accuracy of the TSM SVM/CSP

Table 3.6 Classification accuracy (%) of common spatial pattern SVM/LDA. Dataset III_IVa, classes "left hand" and "right hand."

	Subject			
	aa		**al**	
	SVM	**LDA**	**SVM**	**LDA**
Arithmetic mean	75.7	74.2	94.6	94.6
Riemannian geometric mean	80.0	80.0	94.2	93.5
Log-Euclidean mean	81.4	77.8	94.6	95.3
Harmonic mean	77.5	75.7	95.3	95.0
Resolvent mean	78.2	77.1	95.3	95.0
Euclidean geometric median	**81.8**	**81.4**	94.2	93.5
Riemannian geometric median	80.0	80.0	94.2	93.5
Log-Euclidean geometric median	80.3	78.2	94.2	94.6
Trimmed Riemannian mean	81.4	80.0	95.3	95.3
Trimmed Log-Euclidean mean	81.4	78.9	**96.0**	**95.7**
Trimmed Riemannian median	81.4	80.0	95.3	95.3
Trimmed Log-Euclidean median	80.3	80.7	95.3	95.3
Identity matrix	52.8	47.8	61.4	60.7

Table 3.7 Classification accuracy (%) of tangent space mapping SVM/LDA on a partial set of classes. Dataset IV_II, classes "foot" and "tongue."

	Subject									
	A01T		**A03T**		**A07T**		**A08T**		**A09T**	
	SVM	**LDA**	**SVM**	**LDA**	**SVM**	**LDA**	**SVM**	**LDA**	**SVM**	**LDA**
Arithmetic mean	82.6	84.0	84.0	85.4	89.6	89.5	93.7	92.3	76.4	76.3
Riemannian geometric mean	80.5	82.6	87.5	87.5	90.2	88.1	91.0	93.7	74.3	73.6
Log-Euclidean mean	83.3	86.8	86.1	86.8	89.6	89.5	93.0	93.7	74.3	75.0
Harmonic mean	79.8	83.3	83.3	84.0	88.9	89.5	93.0	93.0	**81.9**	81.2
Resolvent mean	79.8	84.0	87.5	**88.8**	90.3	89.5	93.7	93.7	**81.9**	**82.6**
Euclidean geometric median	79.1	79.8	84.0	86.1	89.6	90.2	93.7	93.0	79.8	79.1
Riemannian geometric median	80.5	73.6	**88.2**	52.0	90.3	88.8	91.0	57.6	74.3	54.8
Log-Euclidean geometric median	**85.4**	84.7	86.1	85.4	89.6	89.5	91.6	93.0	77.0	74.3
Trimmed Riemannian mean	84.0	83.3	**88.2**	87.5	91.0	90.9	94.4	**95.8**	79.9	73.6
Trimmed Log-Euclidean mean	**85.4**	**87.5**	87.5	88.1	**92.3**	**92.3**	93.7	93.7	80.5	79.8
Trimmed Riemannian median	84.0	84.0	**88.2**	71.5	91.7	90.2	**95.1**	95.1	79.9	70.1
Trimmed Log-Euclidean median	**85.4**	86.1	**88.2**	88.1	91.7	91.6	93.7	95.1	79.9	77.7
Identity matrix	81.2	81.2	84.0	83.3	87.5	86.1	90.9	93.0	79.8	78.4

Table 3.8 Classification accuracy (%) of common spatial pattern SVM/LDA on a partial set of classes. Dataset IV_II, classes "foot" and "tongue."

	Subject									
	A01T		**A03T**		**A07T**		**A08T**		**A09T**	
	SVM	**LDA**	**SVM**	**LDA**	**SVM**	**LDA**	**SVM**	**LDA**	**SVM**	**LDA**
Arithmetic mean	73.6	75.0	84.0	84.0	81.9	77.7	88.1	88.1	72.9	66.6
Riemannian geometric mean	76.3	77.7	83.3	82.6	81.9	81.2	**94.4**	93.7	71.5	72.9
Log-Euclidean mean	80.5	79.8	83.3	82.6	**83.3**	77.7	93.0	93.0	72.2	72.2
Harmonic mean	77.0	77.0	79.8	81.2	79.8	80.5	93.7	**94.4**	72.2	74.3
Resolvent mean	80.5	79.8	81.9	81.9	79.8	81.2	**94.4**	**94.4**	72.9	74.3
Euclidean geometric median	75.6	76.3	84.0	84.0	80.5	80.5	86.1	87.5	68.0	65.2
Riemannian geometric median	77.7	77.7	83.3	83.3	81.2	79.8	**94.4**	93.0	71.5	70.8
Log-Euclidean geometric median	73.6	75.0	81.9	81.9	81.2	79.8	87.5	88.8	71.5	71.5
Trimmed Riemannian mean	**81.2**	79.9	84.0	84.0	81.9	**82.6**	**94.4**	**94.4**	**74.3**	73.6
Trimmed Log-Euclidean mean	**81.2**	**81.9**	84.0	**84.7**	**83.3**	**82.6**	**94.4**	**94.4**	**74.3**	**75.0**
Trimmed Riemannian median	80.5	81.2	**84.7**	**84.7**	81.2	81.2	**94.4**	93.7	73.6	72.2
Trimmed Log-Euclidean median	76.3	77.7	83.3	84.0	**83.3**	81.9	89.5	90.0	72.9	72.9
Identity matrix	54.1	57.6	51.3	54.1	50.6	48.6	60.4	57.6	63.8	59.0

method. Table 3.8 shows the obtained results of the CSP SVM/LDA method applied on the same two classes of Table 3.7. The CSP method compares unfavorably with the TS method. According to the obtained results, in this case, the LDA learner behaves comparably to the SVM learner in terms of classification accuracy, both for the TSM method and for the CSP method. The results of the TSM method are better than the respective results obtained by means of the CSP method for all subjects. An example is the dataset pertaining to the subject *A01T*, whose classification accuracy increases from an average of 78.8% of CSP-SVM method to an average of 83% of TSM-SVM method. It is worth underlining how the classification accuracy decreases of about 30% in the CSP method by choosing the identity matrix as reference point.

3.3.1.5 Results obtained on the Dataset JKHH-1

In this classification experiment, we applied the TSM and the CSP methods on a partial set of classes, namely, "left hand" and "foot," for five subjects labeled *sa*, *sb*, *sc*, *sd* and *se*. Table 3.9 shows the classification accuracy of TSM SVM/LDA method. It is readily appreciated how, for the subjects *sa* and *se*, the classification accuracy is higher than

Table 3.9 Classification accuracy (%) of tangent space mapping SVM/LDA on a partial set of classes. Dataset JKHH-1, classes "left hand" and "foot."

	Subject									
	sa		sb		sc		sd		se	
	SVM	**LDA**	**SVM**	**LDA**	**SVM**	**LDA**	**SVM**	**LDA**	**SVM**	**LDA**
Arithmetic mean	79.5	**82.0**	66.5	63.5	53.5	53.0	61.5	62.0	88.5	89.5
Riemannian geometric mean	78.5	80.0	66.5	63.5	52.5	54.0	59.5	59.5	89.5	90.0
Log-Euclidean mean	78.0	77.5	60.0	60.5	53.5	51.5	57.0	59.0	89.5	89.0
Harmonic mean	81.5	79.5	63.5	66.0	55.0	53.5	63.0	60.0	89.5	88.5
Resolvent mean	**83.0**	**82.0**	68.0	**67.5**	55.0	54.5	64.0	64.0	**92.0**	**92.0**
Euclidean geometric median	78.5	81.0	65.5	65.0	49.0	51.0	57.5	59.5	89.0	90.5
Riemannian geometric median	79.0	79.5	65.5	64.0	52.0	51.0	59.5	58.5	90.0	90.0
Log-Euclidean geometric median	78.5	77.5	60.5	56.5	53.5	51.0	58.0	60.5	89.5	89.0
Trimmed Riemannian mean	82.5	81.0	**68.5**	**67.5**	57.0	57.0	64.0	64.5	91.0	90.0
Trimmed Log-Euclidean mean	80.5	79.0	67.0	66.5	59.0	**59.5**	**64.5**	**65.5**	**92.0**	91.0
Trimmed Riemannian median	82.0	81.5	67.5	65.5	55.5	57.0	62.0	65.0	91.0	90.0
Trimmed Log-Euclidean median	82.0	79.5	68.0	66.0	**59.5**	57.0	**64.5**	64.5	91.5	90.5
Identity matrix	73.5	74.5	63.5	63.5	47.5	48.5	62.5	61.5	90.5	88.5

for the other subjects (respectively of 82% and 91%). Table 3.10 shows the results obtained by the CSP SVM/LDA method applied to the same classes of Table 3.9. A direct comparison with the results reported in Table 3.9 reveals that the CSP method perform unfavorably compared to the TS method. From Tables 3.9 and 3.10, it can be observed that the performance of the SVM classification method and the LDA classification method are very similar (absolute differences in the range ±1%).

3.3.2 SCMs distributions on tangent spaces

Projected SCMs distributions relative to the Dataset III_IIIa

Figure 3.2 shows the distribution of the SCMs projected to the tangent space relative to the best average in terms of classification accuracy, pertaining to the subjects *k3b* and *l1b* and relative to the results of TSM-SVM summarized in Table 3.1. The "feet" trials are visualized by open circles (o), while the "tongue" trials are visualized by stars (*). It is interesting to see how a high classification accuracy obtained by TSM-SVM corresponds an apparent separation of the classes on tangent spaces (subject *k3b*).

Table 3.10 Classification accuracy (%) of common spatial pattern SVM/LDA on a partial set of classes. Dataset JKHH-1, classes "left hand" and "foot."

	Subject									
	sa		sb		sc		sd		se	
	SVM	**LDA**	**SVM**	**LDA**	**SVM**	**LDA**	**SVM**	**LDA**	**SVM**	**LDA**
Arithmetic mean	78.0	75.0	56.5	57.5	51.5	52.0	43.5	44.0	88.0	88.0
Riemannian geometric mean	**79.0**	77.0	60.5	62.0	52.0	52.0	59.5	58.0	86.5	87.5
Log-Euclidean mean	76.5	75.5	58.0	54.5	50.5	52.5	59.0	58.0	88.0	87.0
Harmonic mean	75.5	74.0	62.5	60.0	**54.0**	50.5	54.5	55.0	87.0	87.0
Resolvent mean	78.5	77.0	64.0	64.5	53.5	52.0	60.5	60.5	89.0	88.5
Euclidean median	76.5	76.0	61.5	61.0	51.0	49.5	53.0	50.0	87.5	87.0
Riemannian geometric median	78.5	74.5	56.5	57.5	**54.0**	53.5	60.0	59.5	86.0	86.5
Log-Euclidean geometric median	73.5	73.5	61.0	55.5	**54.0**	52.5	59.0	59.0	88.0	88.0
Trimmed Riemannian mean	**79.0**	**78.5**	**65.0**	**65.5**	53.5	**55.5**	**63.5**	**61.5**	89.5	89.0
Trimmed Log-Euclidean mean	77.5	76.5	**65.0**	63.0	**54.0**	53.0	61.0	60.5	89.0	**89.5**
Trimmed Riemannian median	78.5	**78.5**	63.5	62.5	**54.0**	54.5	61.5	60.5	**90.0**	89.0
Trimmed Log-Euclidean median	73.5	73.5	64.0	64.0	**54.0**	54.0	60.0	60.5	88.5	89.0
Identity matrix	44.5	46.0	51.5	52.0	40.5	40.0	61.5	58.5	58.5	60.0

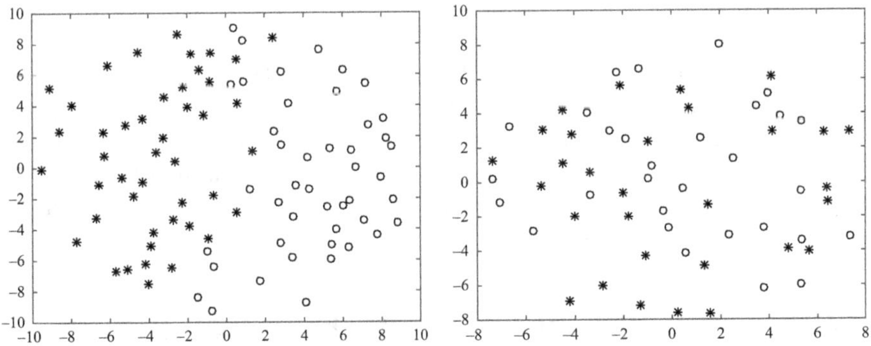

Figure 3.2 Left panel: subject k3b—trimmed Riemannian geometric mean. Right panel: subject l1b—trimmed Riemannian geometric mean

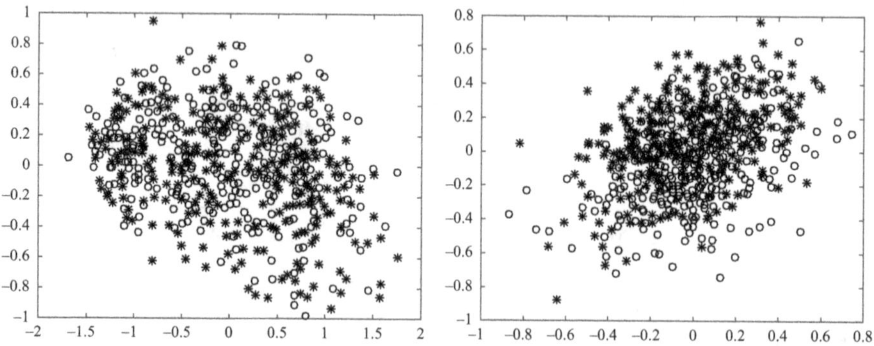

Figure 3.3 *Left panel: subject S4b—identity matrix. Right panel: X11b—trimmed Log-Euclidean geometric median*

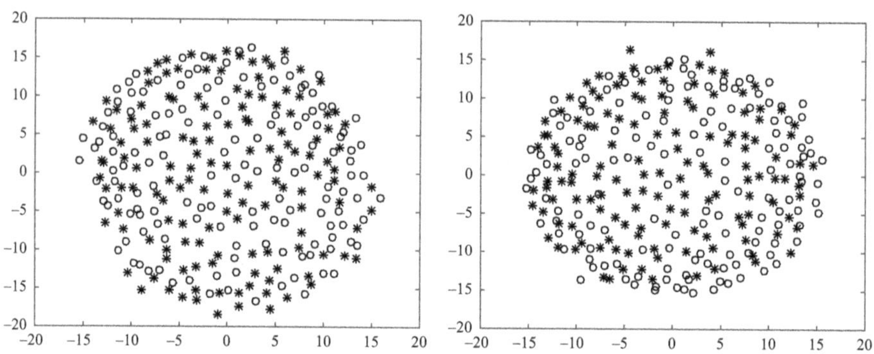

Figure 3.4 *Left panel: subject aa—trimmed Riemannian geometric median. Right panel: subject al—trimmed Riemannian Geometric median*

3.3.2.1 Projected SCMs distributions relative to the Dataset III_IIIb

Figure 3.3 shows the distributions of SCMs projected to the tangent space relative to the best reference points in terms of classification accuracy, pertaining to the subjects *S4b* and *X11b* and relative to the results of TSM-SVM summarized in Table 3.3. The "left-hand" trials are visualized by open circles (o), while the "right-hand" trials are visualized by stars (*). The MI-trials are confused; in fact, the classification accuracy is not very good.

3.3.2.2 SCMs distributions relative to the Dataset III_IVa

Figure 3.4 shows the distribution of SCMs projected to the tangent space relative to the best average SCMs in terms of classification accuracy, pertaining to the subjects *aa* and *al* and relative to the results of TSM-SVM summarized in Table 3.5. The "right-hand" trials are visualized by open circles (o), while the "foot" trials are visualized by stars (*). The visual separation of the samples is not good, despite the good

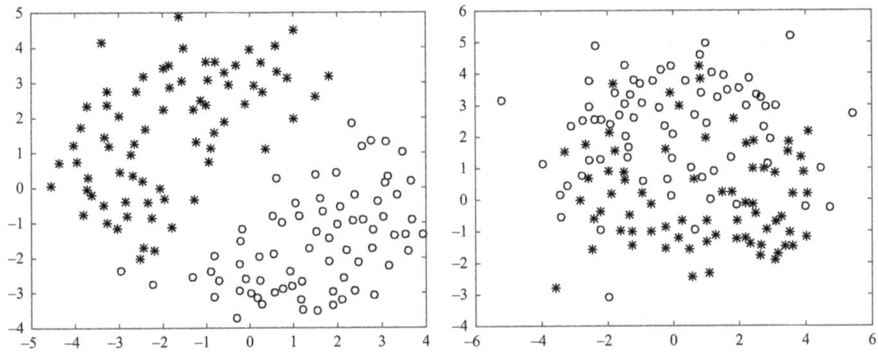

Figure 3.5 *Left panel: A01T—trimmed Riemannian geometric median. Right panel: A07T—trimmed Log-Euclidean geometric median*

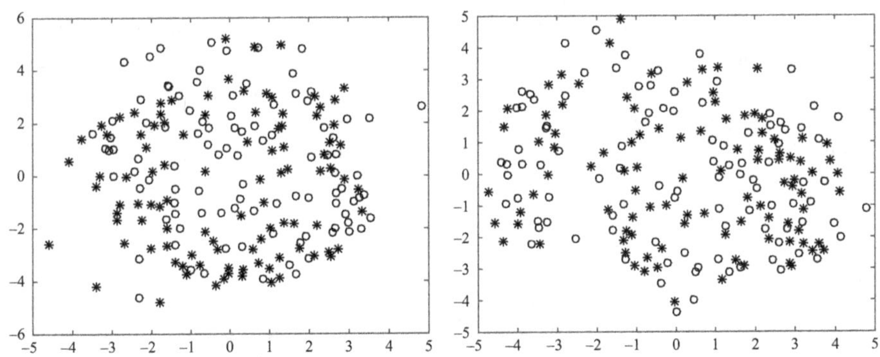

Figure 3.6 *Left panel: Subject sa—trimmed Log-Euclidean geometric median. Subject sb—trimmed Log-Euclidean geometric median*

classification accuracy, because it is the result of projection from the high-dimensional Riemannian manifold of SCMs to the two-dimensional visual space, so the points get confused even if they are spatially separated.

3.3.2.3 SCMs distributions relative to the Dataset IV_IIa

Figure 3.5 shows the distributions of SCMs projected to the tangent space relative to the best average techniques, pertaining to the subjects *A01T* and *A07T*, according to Table 3.7. The "foot" trials are visualized by open circles (o), while the "tongue" trials are visualized by star marks (*). We can notice that the circles are clearly separated from the star marks.

3.3.2.4 SCMs distributions relative to the Dataset JKHH-1

Figure 3.6 shows the distributions of SCMs mapped to the tangent space relative to the best average techniques in terms of classification accuracy, pertaining to the subjects *sa* and *sb* and relative to the results of TSM-SVM showed in Table 3.9. The "left-hand" trials are visualized by open circles (o), while the "foot" trials are visualized by star marks (*). It is interesting to see how some SCMs, the outliers, lay away from most of the other SCMs (subject *sb*).

3.4 Conclusions

All the tables exposed the numerous examples and results with various datasets for a total of 17 subjects studied have proved the superiority of TSM-based feature extraction over CSP in terms of classification accuracy even if we need larger datasets for generalizing the results. Furthermore, the choice of the reference point by which we calculate the tangent plane to project the SCMs is crucial to obtain a better result. This is proved by the comparison of the results obtained with a nonoptimal point, like the identity matrix, or with the trimmed methods that allow to remove the outliers. Another important part of this chapter is covered by the concepts and the results of MDS algorithm; it allows to visualize an MI-trial as a point in a chart that represents the Riemannian manifold; therefore, each SCM can be converted into a single point, and all points can be represented in the same chart. It gives us the opportunity to discern the different classes at a glance and hazard an MI-task prediction.

References

[1] Thomas KP, Guan C, Lau CT, *et al.* A new discriminative common spatial pattern method for motor imagery brain–computer interfaces. IEEE Transactions on Biomedical Engineering. 2009;56:2730–2733.

[2] Pfurtscheller G, Lopes FH. Event-related EEG/MEG synchronization and desynchronization: basic principles. Clinical Neurophysiology. 1999;110:1842–1857.

[3] Pfurtscheller G, Allison BZ, Brunner C, *et al.* The hybrid BCI. Frontiers in Neuroscience. 2010;4:1–11.

[4] Arvaneh M, Guan C, Ang K, *et al.* Facilitating motor imagery-based brain–computer interface for stroke patients using passive movement. Neural Computing and Applications. 2017;28(11):3259–3272.

[5] Muller-Gerking J, Pfurtscheller G, Flyvbjerg H. Designing optimal spatial filters for single-trial EEG classification in a movement task. Clinical Neurophysiology. 1999;110(5):787–798.

[6] Ramoser H, Müller-Gerking J, Pfurtscheller G. Optimal spatial filtering of single trial EEG during imagined hand movement. IEEE Transactions on Rehabilitation Engineering. 2000;8(4):441–446.

[7] Barachant A, Bonnet S, Congedo M, *et al.* Multiclass brain–computer interface classification by Riemannian geometry. IEEE Transactions on Biomedical Engineering. 2012;59(4):920–928.

[8] Fiori S. Learning the Fréchet mean over the manifold of symmetric positive-definite matrices. Cognitive Computation. 2009;1:279–291.

[9] Moakher M. A differential geometric approach to the geometric mean of symmetric positive-definite matrices. SIAM Journal on Matrix Analysis and Applications. 2005;26(3):735–747.

[10] Yger F. A review of kernels on covariance matrices for BCI applications. In: 2013 IEEE International Workshop on Machine Learning for Signal Processing (MLSP); 2013. p. 1–6.

[11] Yger F, Sugiyama M. Supervised Log-Euclidean Metric Learning for Symmetric Positive Definite Matrices; 2015. Available from: http://arxiv.org/pdf/1502.03505v1.

[12] Yger F, Berar M, Lotte F. Riemannian approaches in brain-computer interfaces: a review. IEEE Transactions on Neural Systems and Rehabilitation Engineering. 2017;25(10):1753–1762.

[13] Yong X, Ward RK, Birch GE. Sparse spatial filter optimization for EEG channel reduction in brain–computer interface. In: Proceedings of 2008 IEEE International Conference on Acoustics, Speech and Signal Processing (ICASSP); 2008. p. 417–420.

[14] Lotte F, Guan C. Regularizing common spatial patterns to improve BCI designs: Unified theory and new algorithms. IEEE Transactions on Biomedical Engineering. 2011;58(2):355–362.

[15] Samek W, Muller KR. Information geometry meets BCI spatial filtering using divergences. In: 2014 International Winter Workshop on Brain–Computer Interface (BCI); 2014. p. 1–4.

[16] Tomida N, Yamagishi M, Yamada I, *et al.* A reduced rank approach for covariance matrix estimation in EEG signal classification. In: 36th Annual International Conference of the IEEE Engineering in Medicine and Biology Society; 2014. p. 668–671.

[17] Tomida N, Tanaka T, Ono S, *et al.* Active data selection for motor imagery EEG classification. IEEE Transactions on Biomedical Engineering. 2015;62(2):458–467.

[18] Barachant A, Bonnet S, Congedo M, *et al.* Classification of covariance matrices using a Riemannian-based kernel for BCI applications. Neurocomputing. 2013;112(18):172–178.

[19] Llera A, Gómez V, Kappen HJ. Adaptive multiclass classification for brain computer interfaces. Neural Computation. 2014;26(6):1108–1127.

[20] Uehara T, Tanaka T, Fiori S. Robust averaging of covariance matrices for motor-imagery brain–computer interfacing by Riemannian geometry. In: Advances in Cognitive Neurodynamics (V) (Proceedings of the Fifth International Conference on Cognitive Neurodynamics – 2015); 2016. p. 347–353.

[21] Uehara T, Sartori M, Tanaka T, *et al.* Robust averaging of covariances for EEG recordings classification in motor imagery brain computer interfaces. Neural Computation. 2017;29(6):1631–1666.

[22] Fiori S, Tanaka T. An algorithm to compute averages on matrix Lie groups. IEEE Transactions on Signal Processing. 2009;57(12):4734–4743.

[23] Sen SK. Classification on Manifolds. Department of Statistics and Operations Research, University of North Carolina at Chapel Hill; 2008.

[24] Yang H, Guan C, Chua K, *et al.* Detection of motor imagery of swallow EEG signals based on the dual-tree complex wavelet transform and adaptive model selection. Journal of Neural Engineering. 2014;11(3):035016.

[25] Zhang H, Yang H, Guan C. Bayesian learning for spatial filtering in an EEG-based brain—computer interface. IEEE Transactions on Neural Networks and Learning Systems. 2013;24(7):1049–1060.

[26] Yang H, Guan C, Wang C, *et al.* Detection of motor imagery of brisk walking from electroencephalogram. Journal of Neuroscience Methods. 2015; 244:33–44.

[27] Fiori S. Visualization of Riemannian-manifold-valued elements by multidimensional scaling. Neurocomputing. 2011;74(6):983–992.

Chapter 4
Robust EEG signal processing with signal structures

Hiroshi Higashi[1] *and Toshihisa Tanaka*[2]

Abstract

Brain decoding has contributed to the development of cognitive neuroscience and the production of brain–machine interfaces/brain–computer interfaces (BCI/BMI). For brain decoding, electroencephalography (EEG), which allows the observation of the electrophysiological activities of neurons, is widely used to observe brain activity. In particular, an EEG read from electrodes installed on a scalp, which has some advantages when it comes to cost, size, and ease of measurement, is a promising recording method for producing noninvasive BMIs against magnetoencephalograms, functional magnetic resonance imaging, and so on. However, an EEG signal has low spatial resolution and is highly affected by noise. Moreover, the recoding of an EEG is time-consuming and tires BMI users. Therefore, signal processing techniques that can robustly extract brain activity patterns from EEG signals with a low signal-to-noise ratio and a small sample size are necessary. One approach to achieve such techniques is to incorporate additional information retrieved separately from an EEG in signal processing. This chapter describes some techniques that can accomplish this. The signal structures include physical structures, such as the location of the electrodes, and functional structures, such as synchronizing brain regions. The discussion on the importance of the signal structures in a source analysis of EEG signals that can improve the BMI performance will be discussed in this chapter. In contrast to a source analysis, which finds the brain patterns in a source domain, regularization incorporating the signal structures in a sensor domain will be discussed. Moreover, we will show the signal processing in a graph spectral domain that is a vector space derived from the signal structure.

4.1 Introduction

Brain decoding, an important and challenging technology for observing brain activities and extracting or estimating features or patterns from brain observations, is an

[1]Department of Computer Science and Engineering, Toyohashi University of Technology, Japan
[2]Department of Electrical and Electronic Engineering, Tokyo University of Agriculture and Technology, Japan

important technique in the field of brain–machine interfaces/brain–computer interfaces (BMI/BCI) [1]. One of the media for observing brain activities is through the electrical activities of neurons. These electrical measurements can be roughly categorized into invasive and noninvasive methods. Invasive measurements involve the installment of electrodes beneath the skull. Signals observed by invasive measurements have high signal-to-noise ratio, high time resolution, and high spatial resolution compared with noninvasive measurements. Noninvasive measurements install electrodes externally, usually on the scalp. Because these measurements do not require the surgical installation of electrodes, noninvasive measurements have been widely used to observe brain activities.

One type of the noninvasive measurement, scalp electroencephalography (EEG), offers advantages in cost, size, and ease of measurement. Because an electrode captures the electrical activities induced by thousands of neurons surrounding the electrode, the spatial and spectral resolutions are lower than those of invasive observations.

In the EEG recording of brain activities associated with given tasks, an experiment participant is asked to perform the tasks and not do anything else. For example, an experiment participant is often asked to suppress body movement and blinking because EEG signals are easily contaminated by the electrical activity in muscle tissues. We call these muscle noise components *artifacts*. Voluntary components that are spontaneously generated and that do not relate to tasks are mixed in the EEG signals. To remove nontask-related signals such as artifacts and voluntary components from the observed signals, averaging the signals of several trials is an effective technique. The averaging process can reduce noise components because these components are supposed to be generated by normal distributions. For averaging over trials, experimental participants must perform the tasks repeatedly. However, because the repetition of the tasks and preparation for the recording of EEG signals is time-consuming, it is difficult to collect the required number of samples.

A dense, multichannel spatial system for recording the EEG signals is an essential technology for accurate brain decoding. However, this type of system can sometimes make the decoding difficult because the number of the electrodes or channels for capturing brain activities increases. Large signal and small sample sizes mean that feature extraction and pattern recognition of the EEG signals can be overfitted and unstable. Overfitting is a condition that an estimator excessively fits for certain samples and does not work for the other samples. Therefore, signal processing techniques that can robustly extract brain activity patterns from EEG samples against a low signal-to-noise ratio, a large signal size, and a small sample size are required.

The incorporation of additional information for signal processing, collected separately from EEG signals, is the most promising avenue for achieving robust processing methods [2,3]. This chapter will introduce the methods for utilizing signal structures as additional information. Signal structures represent situations of measurements (e.g., geometric features of electrode arrangement) and structures of measurement objects (e.g., functional connectivity of the brain). By viewing a brain as an object with physical and functional structures, we can extract more information from the EEG signals.

Section 4.2 describes the role of signal structures in an EEG source analysis. We can estimate signals within the brain from scalp EEG signals by using a source analysis. Source analysis projects signals on a space are defined by the positions of the source currents (a source domain). Regularization in sensor domains (the original domain where EEG signals are observed as defined by the positions of electrodes) is introduced in Section 4.3. We will additionally discuss a graph spectral domain (defined by a graph derived from a signal structure) in Section 4.4, where we also discuss recent research on feature extraction and pattern recognition.

4.2 Source analysis

As mentioned, EEG signals have good time resolution but poor spatial resolution. One cause of the poor spatial resolution is the physical size of EEG electrodes. The other reason rests in the generative process of EEG signals. An EEG signal records neurons' electrical activities through cerebrospinal fluid, dura mater, the skull, the scalp, and so on. During this process, activities in the neurons surrounding an electrode are mixed before reaching the electrode. This mixing effect is called the volume conductance effect [1]. Therefore, an EEG originally is limited in its spatial resolution, even though smaller electrodes and a more dense arrangement are used [4]. To overcome this poor spatial resolution, a source analysis was developed. A source analysis estimates currents observed on the brain surface or within the brain from the scalp EEG signals. A source analysis can be a problem when converting the signals on the sensor domain to ones on the source domains.

The problem with a source analysis is illustrated in Figure 4.1. Let $\boldsymbol{y} \in \mathbb{R}^r$ be the currents observed in r sources within a brain. Then, EEG signals $\boldsymbol{x} \in \mathbb{R}^M$ observed in M electrodes on the scalp can be modeled by

$$\boldsymbol{x} = \boldsymbol{A}\boldsymbol{y} + \boldsymbol{n}, \tag{4.1}$$

where $\boldsymbol{A}$ is a lead field matrix, and $\boldsymbol{n}$ is noise. The currents $\boldsymbol{y}$ to be estimated is the current density in each compartment of a mesh that is defined on the cortex of the head model that is derived from structural magnetic resonance imaging (MRI) [5,6]. The lead field matrix $\boldsymbol{A}$ represents how to conduct the current densities to the locations of the electrodes on the scalp. There is cerebral spinal fluid, skull, and scalp between the cortex and electrodes, and those channels of the current have different conductances. Therefore, to estimate an appropriate lead field matrix $\boldsymbol{A}$, a head model that represents a geometric arrangement of the grid compartments and electrodes and the volume conductance effect model [7,8] is needed. Although we can use an existing head model such as the Colin27 MRI average brain [5], a head model derived from an individual structural MRI would be needed for an accurate estimation [6].

In the inverse problem of (4.1), because the number of sources is supposed to be significantly greater than the number of electrodes, the solution is not unique (an ill-posed problem). Therefore, some constraints are needed to solve the problem.

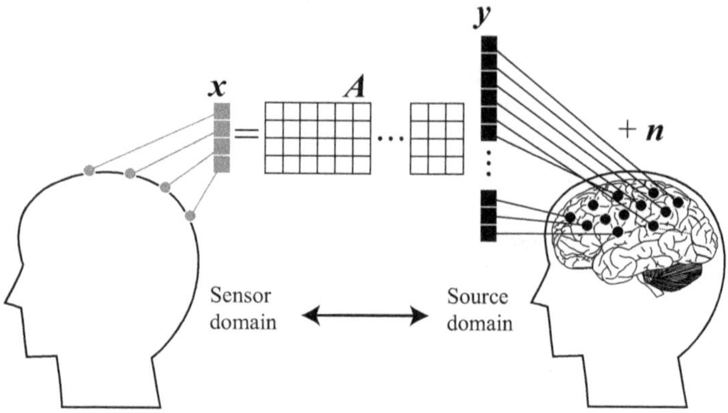

Figure 4.1 The model for the source analysis. The scalp EEG signals (the left side) are a projection of the current sources on the brain surface (the right side). The source analysis finds the current sources **y** *from the scalp EEG signals* **x**

For example, the minimum norm estimation (MNE) [9] finds the source currents $\boldsymbol{y}$ by solving the optimization problem:

$$\min_{\boldsymbol{y}} \ \|\boldsymbol{A}\boldsymbol{y} - \boldsymbol{x}\|_2^2 + \lambda\|\boldsymbol{y}\|_2^2, \tag{4.2}$$

where λ is a regularization parameter that adjusts the effect of the norm regularization. The solution of (4.2) is given by [10]

$$\boldsymbol{y} = (\boldsymbol{A}\boldsymbol{A}^T + \lambda\boldsymbol{I})^{-1}\boldsymbol{A}^T\boldsymbol{x}. \tag{4.3}$$

In addition to MNE, there are methods using Tikhonov regularization [10], which is a generalization of MNE, and spatial smoothness represented by low resolution electromagnetic tomography (LORETA) [11].

Congedo *et al.* [12] reported that classification accuracy in BMI can be improved. For example, by applying a source analysis (a 59-channel EEG signal was decomposed into 1,497 sources) and then using feature extraction and pattern recognition in motor-imagery-based BMIs, the classification accuracy turned out to be about 15% higher than that without source analysis [13]. Why does source analysis improve accuracy? One possible reason is feature selection and noise reduction by converting signals into a source domain [14]. Another possibility is the effect of incorporating electrode geometric structures by projection [15]. The lead field matrix $\boldsymbol{A}$ is designed with relative distances between the sources and electrodes. The source domain, therefore, is designed partially on the basis of the relative distances. Because the relative distances are incorporated into the signals in the source domain during the process of converting the domains, BMI can be more accurate. Although this example uses physical structures of electrodes, functional structures are also useful. Some studies [14,16] showed an improvement in BMI accuracy rates by using spatial features derived from other measurement equipment, such as functional MRIs. These accuracy improvements can be explained by the incorporation of functional structures.

4.3 Regularization

One problem with finding the projection from the sensor domain to the source domain is that if the number of the sources exceeds the number of sensors, the solution would be ill-posed. Moreover, if the head model and volume conductance models are not accurate, the estimated current densities would also not be accurate. Additionally, measuring individual structural MRIs and searching for the appropriate parameters would be costly. Instead of a source analysis, methods that can incorporate additional information in the sensor domain, as opposed to the source domain, are required. One of these methods is regularization.

To solve the inverse problem of a source analysis (4.2), the regularization term is added to the cost function. The regularization term works effectively to estimate or extract brain activity patterns because the term is constrained to the solution space. In general, a regularization is formulated by adding regularization terms $\Omega(\boldsymbol{x})$ into the cost function to be maximized or minimized $f(\boldsymbol{x})$:

$$\min_{x} f(\boldsymbol{x}) + \lambda\Omega(\boldsymbol{x}). \tag{4.4}$$

Regularizations using the l_2-norm and l_1-norm are useful for solving ill-posed problems and overfitting [17]. Figure 4.2 illustrates how these norm regularizations constrain a parameter space. The l_2-norm regularization works such that the solution does not have too much power [10]. Similarly, a solution solved with the l_1-norm regularization promotes the solution's sparsity [18,19].

In addition to the regularizations using the norms, we can constrain a solution space by using signal structures that are formulated as regularization terms [20]. For example, electrode arrangements, one of the geometric structures of EEG signals, can be incorporated into designed spatial filters for multichannel EEG signals [21]. The spatial filters are a method of generating a subset of quasichannel signals by using linear combinations of the multichannel signals and filter coefficients for each

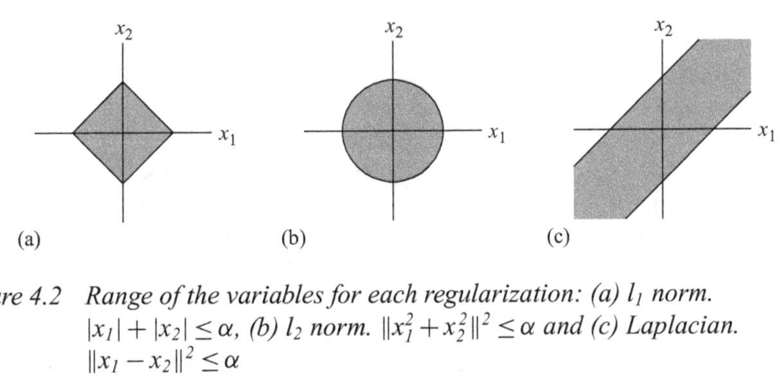

Figure 4.2 Range of the variables for each regularization: (a) l_1 norm. $|x_1| + |x_2| \leq \alpha$, (b) l_2 norm. $\|x_1^2 + x_2^2\|^2 \leq \alpha$ and (c) Laplacian. $\|x_1 - x_2\|^2 \leq \alpha$

channel [22]. The projection to the source domain formulated as (4.3) is one of the spatial filters for the multichannel EEG signal. Averaging the signals over the channels and using a Laplacian filter, which gives the averaged signal over the nearby channels [23], are also considered spatial filters. Unlike simple averaging in which filter coefficients are fixed and empirically decided, filtering techniques that optimize the filter coefficients for the observed signals have been developed [22]. By incorporating an electrode arrangement into the optimization problem for the filter coefficients as a regularization term, we can spatially smooth the filter coefficients.

First, we must define an adjacency matrix that gives the spatial structure of the electrodes, as follows: Let M be the number of electrodes. Then, the size of the adjacency matrix will be $M \times M$. The element at the ith line and jth column of the adjacency matrix represents the physical relative distance between the ith and jth electrodes. Because the larger coefficients in the adjacency matrix denote strong connections between electrodes, the coefficients are designed in such a way that the coefficient between the nearer two electrodes has the larger value than that between the farther two electrodes. For example, the physical distance between two electrodes can be converted by a threshold Gaussian kernel weighting function [21,24]:

$$[\boldsymbol{W}]_{ij} = \begin{cases} \exp\left(-\frac{d_{ij}^2}{2q^2}\right), & d_{ij} \le \kappa \\ 0, & \text{otherwise} \end{cases}, \quad \forall i,j = 1,\ldots,M \tag{4.5}$$

where $d_{i,j}$ is the distance between the ith and the jth electrodes, and κ is a threshold for the distance.

Next, the Laplacian matrix $\boldsymbol{L}$ is derived from the adjacency matrix as [20]:

$$\boldsymbol{L} = \boldsymbol{D} - \boldsymbol{W}, \tag{4.6}$$

where $\boldsymbol{D}$ is a diagonal matrix whose diagonal elements are given by $[\boldsymbol{F}]_{i,i} = \sum_{k=1}^{M} [\boldsymbol{W}]_{i,k}$. We can also define a normalized one [24]:

$$\hat{\boldsymbol{L}} = \boldsymbol{D}^{-1/2}\boldsymbol{L}\boldsymbol{D}^{-1/2}. \tag{4.7}$$

Then, the regularization for spatial smoothing can be formulated by

$$\Omega(\boldsymbol{x}) = \boldsymbol{x}^T\boldsymbol{L}\boldsymbol{x}. \tag{4.8}$$

This regularization term becomes large if the difference in the filter coefficients between nearby electrodes is large. Therefore, the regularization term works in such a way that the filter coefficients for the nearby electrodes take similar values. This effect eventually provides spatial smoothness for the spatial filters [21,25]. Consider a case of two closely located channels. With regularization using the Laplacian matrix (4.8), the parameter space for the coefficients of two electrodes is constrained to $\|x_1 - x_2\|^2 \le \alpha$, namely, the parameter space is limited to the range illustrated in Figure 4.2(c). On the other hand, the difference between two electrodes located far apart does not significantly affect the cost shown in (4.8).

Introducing spatial smoothness for the design of spatial filters corresponds to incorporating into optimization a hypothesis: Electrodes that are near each other observe similar signals. This hypothesis is possible because EEG electrodes observe the sum of electrical activity of a number of neurons around the electrodes [26]. Therefore, nearby electrodes measure the activities from the same neurons, and the observed signals are similar. The spatial smoothing contributes to the design of a spatial filter that robustly extracts brain patterns, even in small sample size settings [25]. We will again discuss smoothing in the spatial domain in Section 4.4.2.

Regularization using brain functional structures has been proposed for the design of spatial filters [3,21,27]. For example, stationary components, which are commonly observed for different participants, sessions, trials, and unstationary components, which change between them, can be represented as regularization [28]. Similarly, noise components are also used to constrain the parameter space [29]. This regularization is an incorporation of a functional signal structure that is estimated from the data observed from other participants, sessions, and trials [30,31].

4.4 Filtering in graph spectral domain

In Section 4.3, we used signal structures in a sensor domain for regularization. However, regularization could lead to difficulty in tuning the regularization parameter λ and finding the optimal solution to the problems [27]. Using a projection on a graph spectral domain is one alternative. A graph spectral domain is derived from a signal structure that we want to incorporate into the problems. By projecting signals on a graph spectral domain, we can easily incorporate the signal structure, applying to conventional techniques such as spatial filter designs. The idea of projecting a sensor domain to the other domains is the same as the source analysis that was discussed in Section 4.2. However, unlike the source domain, which is defined by observed signals and the relation between the electrodes and sources, the graph spectral domain is defined only by the relation between the electrodes.

In this section, we show the definition of the graph Fourier transform (GFT), which is a projection from a sensor domain to a graph spectral domain. Then, smoothing and dimensionality reduction by GFT are described in Section 4.4.2. We will illustrate the application of smoothing and dimensionality reduction in a principal component analysis (PCA) with artificial signals that model an EEG signal. Moreover, we introduce processing for multidimensional data, which has multiple graphs for different domains such as time and space.

4.4.1 Graph Fourier transform

GFT has been proposed for processing signals on graphs [24]. A graph is defined by a finite set of vertices and a set of edges denoted by $\mathcal{G} = \{\mathcal{V}, \mathcal{E}, \boldsymbol{W}\}$, where $\mathcal{V}$ is a finite set of vertices, $\mathcal{E}$ is a set of edges, and $\boldsymbol{W}$ is a weighted adjacency matrix. For EEG signals, we consider a graph in which the vertices are the electrodes and the edges connect the electrodes. Let M be the number of the vertices (the number of the electrodes); then, the weights of the edge can be represented by an $M \times M$ adjacency matrix. Same as with the regularization, the element at the ith line and jth column of the adjacency matrix can be considered the strength of the connection between the

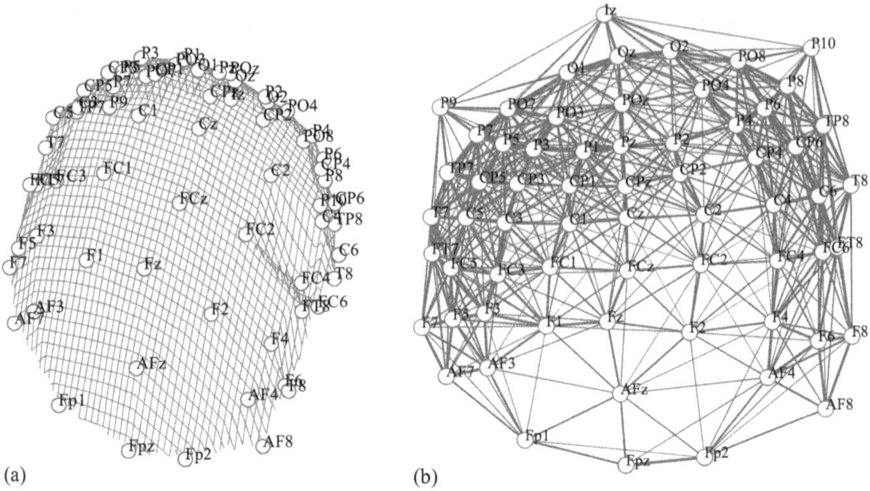

(a) (b)

Figure 4.3 The electrode arrangement shown in a three-dimensional coordinate and the graph based on the physical distance between the electrodes. The labels for the electrodes use the International 10–10 Method [32]. (a) Electrode arrangement and (b) graph

ith and jth electrodes. The strength of the connection can be derived on the basis of a relation between the electrodes, such as the physical distance and the similarity of the observed signals. In the graph shown in Figure 4.3(b), the line segments connecting the two electrodes represents the edges, and the width of the line segments represents the weights for the edges. Because we defined the weights for the edges based on the physical distances, the weights (line widths) for the two electrodes that are physically close are large.

GFT is a transform to a space whose bases consist of the eigenvectors of the Laplacian matrix [20] of an adjacency matrix. The Laplacian matrix is also called the graph Laplacian. Let $\{\lambda_n, \boldsymbol{u}_n\}_{n=0}^{M-1}$ be the M eigenvalues and the corresponding eigenvectors. Then, the indexes of the eigenvalues and eigenvectors are decided in such a way that $0 = \lambda_0 < \lambda_1 \leq \lambda_2 \leq \cdots \leq \lambda_{M-1}$. GFT transforms the signals observed on the vertices $\boldsymbol{x} \in \mathbb{R}^M$ by

$$\boldsymbol{y} = \boldsymbol{U}^T \boldsymbol{x}, \tag{4.9}$$

where $\boldsymbol{U} = [\boldsymbol{u}_0, \boldsymbol{u}_1, \ldots, \boldsymbol{u}_{M-1}]$. The column vectors of $\boldsymbol{U}$ are the basis vectors for the graph spectral domain.

In the basis of the graph spectral domain, the basis vectors corresponding to the small eigenvalues oscillate slowly. On the other hand, the basis vector corresponding to the large eigenvalues oscillate more rapidly. If the ith and jth vertices connect strongly, $[\boldsymbol{u}_l]_i$ and $[\boldsymbol{u}_l]_j$, which are the elements of the basis vector corresponding to small eigenvalues, result in a similar value. On the other hand, the difference

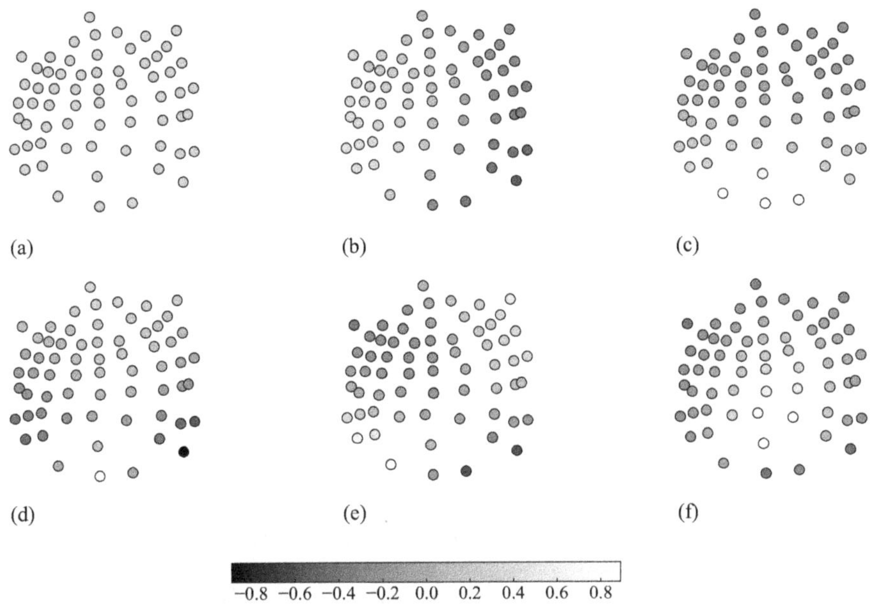

Figure 4.4 Eigenvectors of the graph Laplacian derived from the graph shown in Figure 4.3(b). The luminance of the vertices corresponds to the intensity of each element of the eigenvectors shown in the color bar. (a) $\boldsymbol{u}_0$, (b) $\boldsymbol{u}_1$, (c) $\boldsymbol{u}_2$, (d) $\boldsymbol{u}_3$, (e) $\boldsymbol{u}_4$, (f) $\boldsymbol{u}_5$

between $[\boldsymbol{u}_m]_i$ and $[\boldsymbol{u}_m]_j$ for the basis vector corresponding to a large eigenvalue is large. The relation between the eigenvectors and eigenvalues corresponds to that between the basis and frequencies in classical Fourier analysis [24]. As an example, Figure 4.4 shows the basis of the graph shown in Figure 4.3(b). The eigenvector $\boldsymbol{u}_0$ corresponding to the smallest eigenvalue does not have any zero-crossing points in its elements (Figure 4.4(a)). If the vertices are close, the elements of the eigenvectors corresponding to small eigenvalues take similar values, and the zero-crossing points in the spatial domain are less. As the eigenvalues become larger, the eigenvectors have more zero-crossing points and oscillate rapidly.

For a signal observed in every node, the spectrum in the graph spectral domain can be defined. For example, Figure 4.5 shows graph spectra of a component that is smooth over the graph shown in Figure 4.3 and of a noise component that is not smooth over the graph. The smooth component is illustrated in Figure 4.6(a). In the smooth component, the coefficients of GFT for the basis corresponding to the small eigenvalues are large. On the other hand, the noise component has large coefficients even in the largest eigenvalues. By this way, a graph Fourier spectrum is an indicator showing whether or not signals observed in nodes that are connected strongly in a graph are similar.

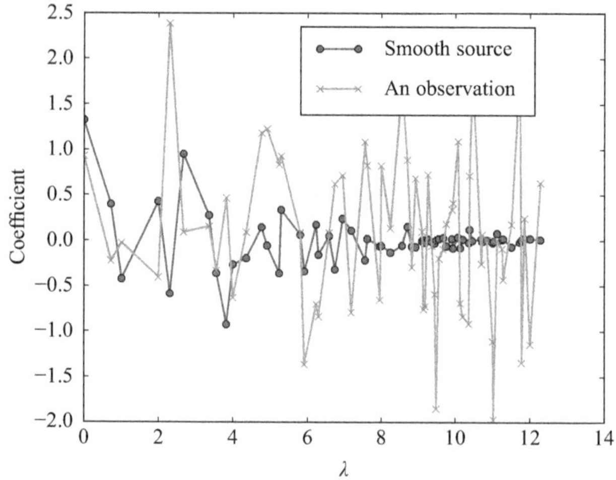

Figure 4.5 Representatives of the graph Fourier spectrum given by the GFT of Figure 4.3. Those spectra are given by two artificially generated signals: One is smooth over the graph and the other is a random component

4.4.2 Smoothing and dimensionality reduction by GFT

Here, we introduce a method for spatially smoothing a solution to an optimization problem, which will be done using a simulation with artificial signals. We consider a problem of a source signal that is spatially smooth, as shown in Figure 4.6(a), and white noise is mixed and observed in the electrodes arranged in Figure 4.3. We apply PCA for the signals to estimate the source signal.

Let $\{\boldsymbol{x}_i \in \mathbb{R}^M\}_{i=1}^N$ be N signals observed from M electrodes. To extract a spatial component that is common in all the samples, PCA extracts it by

$$\hat{\boldsymbol{p}} = \arg\min_{\boldsymbol{p}} \sum_{i=1}^{N} \|\boldsymbol{p}^T \boldsymbol{x}_i\|^2 = \arg\min_{\boldsymbol{p}} \boldsymbol{p}^T \boldsymbol{S} \boldsymbol{p}, \tag{4.10}$$

where $\boldsymbol{S}$ is a correlation matrix defined as

$$\boldsymbol{S} = \frac{1}{N-1} \sum_{i=1}^{N} \boldsymbol{x}_i \boldsymbol{x}_i^T. \tag{4.11}$$

The ordinary PCA estimates the principal component shown in Figure 4.6(c). The component shown in Figure 4.6(c) is estimated with 100 observed that have randomly generated white noise. The estimated results by GFT are shown in Figures 4.6(d)–4.6(f). The estimated components are given by top r points in GFT ($\tilde{\boldsymbol{y}} = [[\boldsymbol{y}]_0, [\boldsymbol{y}]_1, \ldots, [\boldsymbol{y}]_{r-1}]^T$) of the observed signals. With r, the principal component that is very similar to the source signal can be obtained. The reason why a solution becomes smooth in a spatial domain by using the eigenvectors corresponding to the

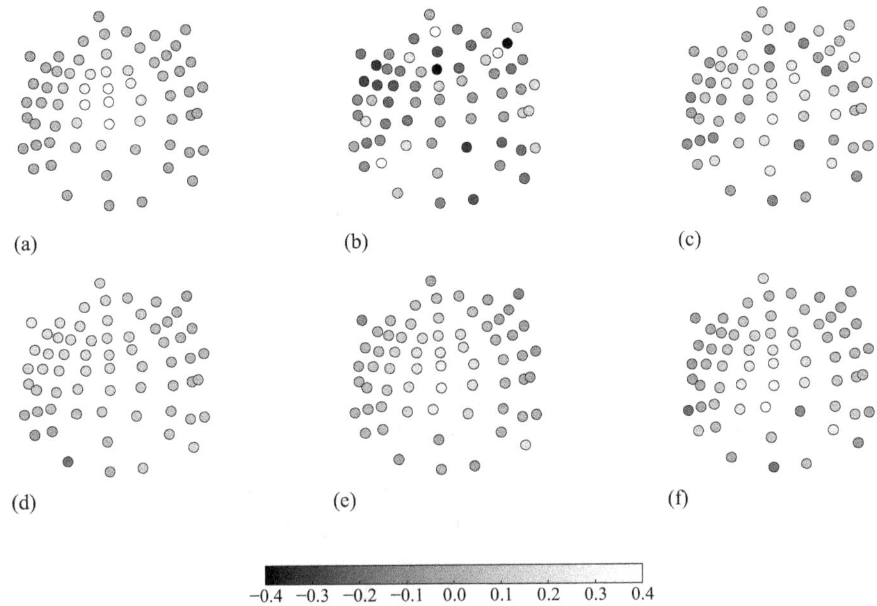

Figure 4.6 The source signal, observed signals, and estimated principal components in the simulation with artificial signals. The color of the vertices represents the value of the vertex element of a signal. The luminance of the vertices corresponds to the value of each vertex element of the signals shown in the color bar. (a) Original, (b) an observed signal, (c) PCA, (d) smoothing ($r = 5$), (e) smoothing ($r = 10$), (f) smoothing ($r = 20$)

smallest r eigenvalues is that the solutions are found in the subspace spanned by the bases that are smoothed over the graph. This scheme is similar to a band-limited Fourier series; a signal that has a higher frequency than those of the Fourier basis cannot be represented. So the smoothing by GFT can be interpreted as lowpass filtering in a graph spectral domain [33].

Lowpass filtering in a graph spectral domain works for dimensionality reduction. Here, the dimension of the signal decreases from M to P. Unlike the transform from a sensor domain to source domain, which increases the dimension of the signal according to the number of the sources, the filtering in a graph spectral domain can reduce the dimension by incorporating a signal structure. Moreover, unlike dimensionality reduction methods that use statistical methods such as PCA, which needs a number of samples, the method using signal structures does not depend on statistics and provides a more robust dimensionality reduction. This is an advantage of signal-structure-based dimensionality reduction methods in EEG signal processing, where it is difficult to collect a lot of samples.

We have applied the spatial smoothing method for some feature extraction techniques for BMIs. In [34], spatial smoothing is incorporated into common spatial pattern (CSP) methods. CSP method for feature extraction of brain activity associated with motor imagery [35]. The CSP methods are well-known designs spatial filters that maximize the ratio in variances between signals during two motor imagery tasks (e.g., left hand and right hand motor imagery tasks). The spatial filters effectively work to extract different brain spatial patterns according to two motor imagery tasks. The CSP method designs the spatial filters by using learning data. The extraction accuracy, however, degrades if the size of the learning samples is small. In [34], the authors show that spatial smoothing can solve the problem and achieve 9% improvement in classification accuracy in BMI classification problems.

A canonical correlation analysis (CCA), which is a method to design spatial filters for extracting steady-state evoked potentials (SSVEP) [36], benefits from spatial smoothing. SSVEP is an EEG component that is induced by looking at a flickering stimulus. The SSVEP has a peak at the same frequency as the frequency of the flickering stimulus. This feature can be utilized to build a BMI that detects the stimulus a user is looking at one of multiple stimuli flickering at different frequencies [37]. Furthermore, because the amplitude of the SSVEPs varies with the subject's attention, they are used to observe human cognition [38]. To extract these SSVEPs, a CCA-based method [39] that maximizes the canonical correlation of a spatially filtered signal and sinusoidal signals is promising. In the CCA-based method, the number of samples corresponds to the signal length. Therefore, if the signal length is short, overfitting may occur, and the extraction accuracy can degrade. A result of SSVEP-based BMI classification in [40] shows that spatially smoothing by GFT lowpass filtering can prevent degradation in extraction accuracy because of short signal length.

Spatial smoothing using GFT is simple. Therefore, we can easily implement the graph-based method combined with any feature extraction and pattern recognition method. For example, similar to CSP and CCA, a spatial filter called xDAWN for extracting event-related potentials can gain robustness through spatial smoothing [41]. Moreover, GFT has been utilized for analyzing functional MRI data [42].

4.4.3 Tangent space mapping from Riemannian manifold

A fundamental limitation of the CSP methods is that they only fit two-class classification problems. Another limitation is that they eliminate the power from the covariance, because the CSP is a generalized eigenvector. To cope with these limitations, the idea of regarding sample covariance matrices (SCMs) as points on a smooth manifold has been suggested. Because the Riemannian manifold is not a linear vector space (Euclidean space), linear algebra cannot be applied. Barachant *et al.* proposed to map points onto a tangent space that is Euclidean [43]. With this tangent space mapping (TSM), any machine-learning algorithms that work in the Euclidean space can be applied to the classification of SCMs. However, the computation of tangent mapping needs a heavy computational load that depends on the size of matrices. Besides, SCMs often contain a nonnegligible error in their estimation. This yields widespread distribution in the manifold, resulting in deteriorated performance in classification, so a regularization such as covariance shrinkage is necessary [44]. To solve this problem, the reduction of the dimension of the electrode space of an EEG by the so-called

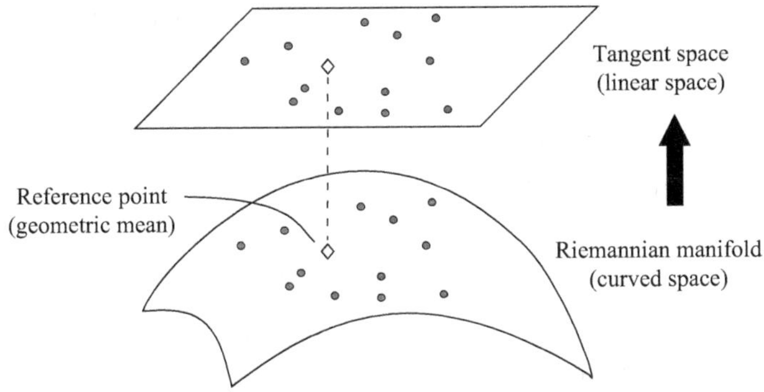

Figure 4.7 Illustration of the tangent space mapping of covariance matrices

GFT [45,46] is useful. The graph is also naturally introduced from the geometrical structure of EEG electrodes, which is static. The dimensionality reduction can also work in a manner similar to regularization.

Let $\boldsymbol{X} \in \mathbb{R}^{M \times T}$ be an M-channel EEG signal, where rows are channels (electrodes) and columns are data points (time instances). In a supervised learning of an EEG, we are given a set of EEG signals together with the corresponding labels, for example, MI of left hand, MI of right foot, and so forth. A dataset can be denoted by $\{(\boldsymbol{X}_i, h_i)\}_{i=0}^{L-1}$, where L is the number of data (trials) and h_i is the label.

A number of approaches to EEG classification during MI use second-order statistics of EEG, that is, the SCMs of $\boldsymbol{X}$ are defined as

$$\boldsymbol{C}_X = \frac{1}{T-1}(\boldsymbol{X} - \boldsymbol{\mu}_X)(\boldsymbol{X} - \boldsymbol{\mu}_X)^T, \tag{4.12}$$

where $\boldsymbol{\mu}_X = \sum_{t=0}^{T-1} \boldsymbol{X}_{:,t}$ and $\boldsymbol{X}_{:,t}$ denotes the tth column of $\boldsymbol{X}$.

The recently proposed method of TSM [43] is a nonlinear transformation to directly classify the SCMs of an EEG. The details for TSM are elaborated in the following: Because a SCM is a positive definite, it is an element of a so-called Riemannian manifold of $n \times n$ symmetric positive definite matrices denoted by $\mathscr{P}(n) = \{\boldsymbol{P} \in S(n) \mid \boldsymbol{u}^T \boldsymbol{P} \boldsymbol{u} > 0, \forall \boldsymbol{u} \in \mathbb{R}^n, \boldsymbol{u} \neq \boldsymbol{0}\}$, where $S(n)$ is a set of symmetric matrices of size $n \times n$. Note that $\mathscr{P}(n)$ is not Euclidean; thus, well-established machine-learning algorithms cannot be employed in $\mathscr{P}(n)$. At a point, $\boldsymbol{Q}$, on $\mathscr{P}(n)$ with an appropriate metric, there exists the tangent space denoted by $T_{\boldsymbol{Q}}\mathscr{P}(n)$, which is Euclidean. The tangent space locally well approximates the Riemannian manifold.

As illustrated in Figure 4.7, the underlying idea behind the TSM method is to map $\boldsymbol{C}_X$ in the Riemannian manifold onto the tangent space where the classification is performed. Specifically, let $\boldsymbol{P}_i \in \mathbb{R}^{n \times n}$ be an SCM and $\bar{\boldsymbol{P}}$ the barycenter of a set of SCMs. Then, a map from $\mathscr{P}(n)$ to $T_{\bar{\boldsymbol{P}}}\mathscr{P}(n)$ can be given as

$$\boldsymbol{s}_i = \text{upper}\left(\log\left(\bar{\boldsymbol{P}}^{-1/2}\boldsymbol{P}_i\bar{\boldsymbol{P}}^{-1/2}\right)\right), \tag{4.13}$$

where the upper $(\cdot)$ operator vectorizes the upper triangular part of a symmetric matrix by applying the weight $\sqrt{2}$ only to the off-diagonal entries. Letting $\boldsymbol{P}_i = \boldsymbol{C}_{\boldsymbol{X}_i}$ in (4.13), we obtain a dataset $\{(\boldsymbol{s}_i, h_i)\}_{i=0}^{L-1}$ in an Euclidean space, for which well-known supervised learning algorithms, such as support vector machine, can be applied. In practice, the choice of the barycenter can affect the performance of the classification of an EEG (see [47]).

Because the EEG signals at neighbor electrodes often exhibit a strong correlation, the corresponding SCM can be ill-conditioned. A well-known and effective approach for solving this problem is regularization, called covariance shrinkage [44]. Another way of regularization is dimensionality reduction of the column space of $\boldsymbol{X}$, which is more appropriate, in terms of TSM discussed in the previous section, because the size of the covariance matrix affects the computational load of tangent mapping.

The dimensionality reduction of the column space is described as $\boldsymbol{Y} = \boldsymbol{V}_r^T\boldsymbol{X}$, where $\boldsymbol{V}_r$ is a projection matrix of size $r \times M$. It should be noted that the size of $\boldsymbol{Y}$ is $r \times T$, where $r < M$. A promising choice of $\boldsymbol{V}_r$ is the lowest r basis vectors of GFT. In the subspace spanned by $\boldsymbol{V}_r$, the SCM of $\boldsymbol{Y}$ can be given as

$$\boldsymbol{C}_{\boldsymbol{Y}} = \frac{1}{T-1}(\boldsymbol{Y} - \mu_{\boldsymbol{Y}})(\boldsymbol{Y} - \mu_{\boldsymbol{Y}})^T = \boldsymbol{V}_r^T \boldsymbol{C}_{\boldsymbol{X}} \boldsymbol{V}_r, \tag{4.14}$$

where $\mu_{\boldsymbol{Y}} = 1/(T-1)\sum_{t=0}^{T-1}(\boldsymbol{Y})_{:,t}$ with $(\boldsymbol{Y})_{:t}$ denoting the tth column of $\boldsymbol{Y}$. An advantage of the graph-based method over the PCA-based dimensionality reduction is that the GFT does not need the population covariance matrix of the signal, which is indeed unknown.

4.4.4 Smoothing on functional brain structures

In the smoothing method in the spatial domain described in Section 4.4.2, graphs represent the geometric distribution of electrodes. However, graphs can represent other structures as well. We have proposed EEG signal processing techniques using functional structures of the brain. Functional structures are often called functional connectivity. One of the measures of the connectivity for two electrodes is a similarity in their observed signals. The similarity is quantified by measures such as the correlation, phase synchronization, and mutual information [26] of the observed signals. Even if the brain regions are far apart, the regions strongly connect if the activities of the regions synchronize [48].

Figure 4.8 shows an example of the functional connectivity between electrodes when performing an oddball paradigm. The functional connectivity can be defined not only between electrodes but also between time segments. We have proposed a way to utilize functional connectivities in spatial and time domains in multilinear signal processing [49] and applied it for robustly extracting event-related potentials [41,50]. In this method, we designed a classifier by the uncorrelated multilinear discriminant analysis [51] for signals that are lowpass-filtered in the graph spectral domain. Figure 4.9 shows the projection vectors for the spatial domain in the multilinear discriminant analysis. As seen in the coefficients, the projection vector designed with

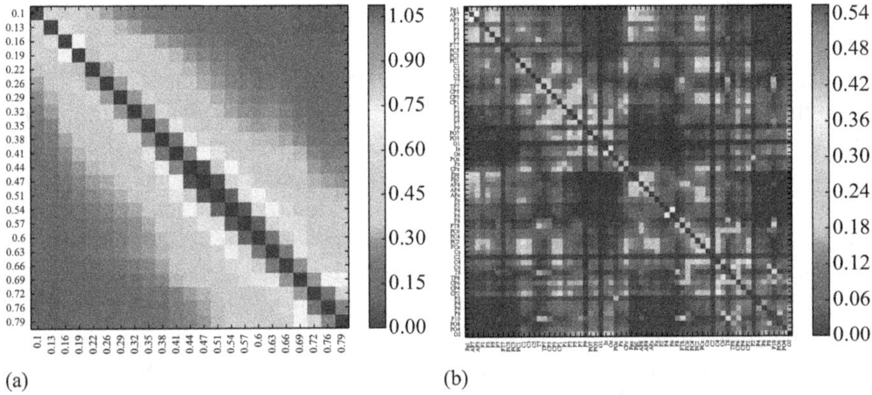

Figure 4.8 Functional connectivity measured by the mutual information between two data points [(a) two time instances or (b) two electrodes] shown in both of the vertical and horizontal axes. This was computed with EEG signals from 12 participants during an oddball paradigm. As the value increases, the nodes connect strongly

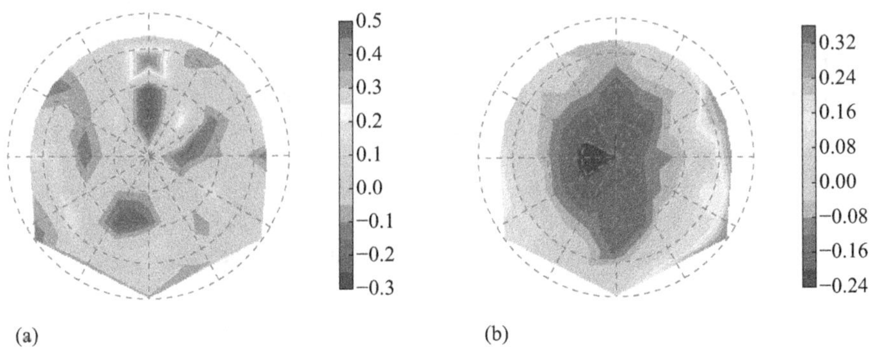

Figure 4.9 Topographical plot of the projection vector given by the linear discriminant analysis. The front and back of the head is located in the upper and lower parts of this figure, respectively. (a) Without the lowpass filter and (b) with the lowpass filter

the lowpass filtering extracts a component around the parietal lobe. This suggests that by using lowpass filtering in the graph spectral domain, the event-related components that are supposed to be observed in the parietal lobe [52] are accurately extracted. It shows that the low-frequency components in the graph spectral domain that represent the smoothness in the connectivity would help with robust extraction.

4.5 Conclusion

This chapter focused on the use of signal structures for feature extraction and pattern recognition of scalp EEG signals. We discussed the methods for incorporating signal structures from the point of view of signal processing in the various domains.

In a source analysis, geometric relations of electrodes, a head model, and volume conductances of the brain's organization are needed to estimate a lead field matrix. Thus, the geometric relations, one of the signal structures, is reflected in the source domain. Some studies reported that by doing feature extraction and pattern recognition in the source domain, the accuracy of the classification in BMIs improves. These studies suggested that incorporating the signal structures can contribute to the robust processing of EEG signals [14]. Therefore, the source analysis is an option that you should be used first, if the required information is present. However, the source localization has the potential to be ill-posedness. If there is information about electrode positions, regularization in the sensor domain can be used. We introduced regularization with the Laplacian matrix and its application with spatial and functional signal structures. Although this regularization is used by combining it with other signal processing, it could be difficult to combine them. If this is the case, the processing in the graph spectral domain can be used. Signals projected onto a graph spectral domain can be processed as general signals. Therefore, not only can conventional signal processing but also advanced signal processing, such as manifolds, be easily combined with signal structures [53]. The projection onto a graph spectral domain can provide robustness for feature extraction and pattern recognition by incorporating signal structures and reducing the dimensionality of signals.

For a source analysis, methods using signal structures measured with the other measurement equipment of brain activity, such as functional MRI and functional near-infrared spectroscopy, have been proposed [16,54]. Information from such multimodal measures can aid in the design of regularization terms and graphs. Moreover, expansion using higher-order statistics used for advanced signal processing, such as blind source separation, is also expected.

References

[1] He B, Ding L. Electrophysiological Mapping and Neuroimaging. In: He B, editor. Neural Engineering. Boston, MA: Springer US; 2013. p. 499–543.

[2] Fazli S, Dahne S, Samek W, Bieszmann F, Muller KR. Learning from more than one data source: Data fusion techniques for sensorimotor rhythm-based brain–computer interfaces. Proceedings of the IEEE. 2015;103(6):891–906.

[3] Lotte F. Signal processing approaches to minimize or suppress calibration time in oscillatory activity-based brain–computer interfaces. Proceedings of the IEEE. 2015;103(6):871–890.

[4] He B, Sohrabpour A. Imaging epileptogenic brain using high density EEG source imaging and MRI. Clinical Neurophysiology. 2016;127(1):5–7.

[5] Holmes CJ, Hoge R, Collins L, Woods R, Toga AW, Evans AC. Enhancement of MR Images Using Registration for Signal Averaging. Journal of Computer Assisted Tomography. 1998;22(2).

[6] Tarkiainen A, Liljeström M, Seppä M, Salmelin R. The 3D topography of MEG source localization accuracy: Effects of conductor model and noise. Clinical Neurophysiology. 2003;114(10):1977–1992.

[7] Sarvas J. Basic mathematical and electromagnetic concepts of the biomagnetic inverse problem. Physics in Medicine and Biology. 1987;32(1):11.

[8] Wolters CH, Anwander A, Tricoche X, Weinstein D, Koch MA, MacLeod RS. Influence of tissue conductivity anisotropy on EEG/MEG field and return current computation in a realistic head model: A simulation and visualization study using high-resolution finite element modeling. NeuroImage. 2006;30(3):813–826.

[9] Hämäläinen MS, Ilmoniemi RJ. Interpreting magnetic fields of the brain: Minimum norm estimates. Medical & Biological Engineering & Computing. 1994;32(1):35–42.

[10] Tikhonov AN, Arsenin VY. Solutions of Ill-Posed Problems. New York: V. H. Winston & Sons; 1977.

[11] Pascual-Marqui RD, Michel CM, Lehmann D. Low resolution electromagnetic tomography: A new method for localizing electrical activity in the brain. International Journal of Psychophysiology. 1994;18(1):49–65.

[12] Congedo M, Lotte F, Lécuyer A. Classification of movement intention by spatially filtered electromagnetic inverse solutions. Physics in Medicine and Biology. 2006;51(8):1971.

[13] Kamousi B, Amini AN, He B. Classification of motor imagery by means of cortical current density estimation and Von Neumann entropy. Journal of Neural Engineering. 2007;4(2):17.

[14] He B, Baxter B, Edelman BJ, Cline CC, Ye WW. Noninvasive brain-computer interfaces based on sensorimotor rhythms. Proceedings of the IEEE. 2015;103(6):1–19.

[15] Edelman BJ, Baxter B, He B. EEG source imaging enhances the decoding of complex right-hand motor imagery tasks. IEEE Transactions on Biomedical Engineering. 2016;63(1):4–14.

[16] Morioka H, Kanemura A, Morimoto S, *et al.* Decoding spatial attention by using cortical currents estimated from electroencephalography with near-infrared spectroscopy prior information. NeuroImage. 2014;90: 128–139. Available from: http://linkinghub.elsevier.com/retrieve/pii/S105381191301255X.

[17] Tomioka R, Müller KR. A regularized discriminative framework for EEG analysis with application to brain–computer interface. NeuroImage. 2010;49(1):415–432.

[18] Tibshirani R. Regression shrinkage and selection via the lasso. Journal of the Royal Statistical Society: Series B (Methodological). 1996;58(1): 267–288.

[19] Zou H, Hastie T. Regularization and variable selection via the elastic net. Journal of the Royal Statistical Society: Series B (Statistical Methodology). 2005;67(2):301–320.

[20] Xi Y, Hasson U, Ramadge PJ, Xiang ZJ. Boosting with spatial regularization. In: Proceedings of Advances in Neural Information Processing Systems 22; 2009. p. 2107–2115.

[21] Lotte F, Guan C. Regularizing common spatial patterns to improve BCI designs: Unified theory and new algorithms. IEEE Transactions on Biomedical Engineering. 2011;58(2):355–362.

[22] Blankertz B, Tomioka R, Lemm S, Kawanabe M, Müller KR. Optimizing spatial filters for robust EEG single-trial analysis. IEEE Signal Processing Magazine. 2008;25(1):41–56.

[23] McFarland DJ, McCane LM, David SV, Wolpaw JR. Spatial filter selection for EEG-based communication. Electroencephalography and Clinical Neurophysiology. 1997;103(3):386–394. Available from: http://www.sciencedirect.com/science/article/pii/S0013469497000222.

[24] Shuman DI, Narang SK, Frossard P, Ortega A, Vandergheynst P. The emerging field of signal processing on graphs: Extending high-dimensional data analysis to networks and other irregular domains. IEEE Signal Processing Magazine. 2013;30(3):83–98.

[25] Higashi H, Cichocki A, Tanaka T. Regularization using geometric information between sensors capturing features from brain signals. In: Proceedings of 2012 IEEE International Conference on Acoustics, Speech and Signal Processing (ICASSP); 2012. p. 721–724.

[26] Bullmore E, Sporns O. Complex brain networks: Graph theoretical analysis of structural and functional systems. Nature Reviews Neuroscience. 2009;10(3):186–198.

[27] Samek W, Kawanabe M, Müller KR. Divergence-based framework for common spatial patterns algorithms. IEEE Reviews in Biomedical Engineering. 2013;7:50–72.

[28] Samek W, Carmen V, Müller KR, Kawanabe M. Stationary common spatial patterns for brain–computer interfacing. Journal of Neural Engineering. 2012;9(2):26013.

[29] Blankertz B, Kawanabe M, Tomioka R, Hohlefeld F, Nikulin V, Müller KR. Invariant common spatial patterns: Alleviating nonstationarities in brain–computer interfacing. Advances in Neural Information Processing Systems. 2008;20:113–120.

[30] Kang H, Nam Y, Choi S. Composite common spatial pattern for subject-to-subject transfer. IEEE Signal Processing Letters. 2009;16(8):683–686.

[31] Lu H, Eng HL, Guan C, Plataniotis KN, Venetsanopoulos AN. Regularized common spatial pattern with aggregation for EEG classification in small-sample setting. IEEE Transactions on Biomedical Engineering. 2010; 57(12):2936–2946.

[32] Suarez E, Viegas MD, Adjouadi M, Barreto A. Relating induced changes in EEG signals to orientation of visual stimuli using the ESI-256 machine. Biomedical Sciences Instrumentation. 2000;36:33–38.

[33] Sandryhaila A, Moura JMF. Discrete signal processing on graphs: Frequency analysis. IEEE Transactions on Signal Processing. 2014;62(12): 3042–3054.

[34] Higashi H, Tanaka T, Tanaka Y. Smoothing of spatial filter by graph Fourier transform for EEG signals. In: Proceedings of 2014 Asia-Pacific Signal and Information Processing Association Annual Summit and Conference; 2014. p. 1–8.

[35] Ramoser H, Muller-Gerking J, Pfurtscheller G. Optimal spatial filtering of single trial EEG during imagined hand movement. IEEE Transactions on Rehabilitation Engineering. 2000;8(4):441–446.

[36] Regan D. Some characteristics of average steady-state and transient responses evoked by modulated light. Electroencephalography and Clinical Neurophysiology. 1966;20(3):238–248.

[37] Middendorf M, McMillan G, Calhoun G, Jones KS. Brain–computer interfaces based on the steady-state visual-evoked response. IEEE Transactions on Rehabilitation Engineering. 2000;8(2):211–214.

[38] Müller MM, Malinowski P, Gruber T, Hillyard SA. Sustained division of the attentional spotlight. Nature. 2003;424(6946):309–312. Available from: http://www.nature.com/nature/journal/v424/n6946/abs/nature01812.html http://www.ncbi.nlm.nih.gov/pubmed/12867981.

[39] Lin Z, Zhang C, Wu W, Gao X. Frequency recognition based on canonical correlation analysis for SSVEP-based BCIs. IEEE Transactions on Biomedical Engineering. 2006;53(12):2610–2614.

[40] Ryu S, Higashi H, Tanaka T, Nakauchi S, Minami T. Spatial smoothing of canonical correlation analysis for steady state visual evoked potential based brain computer interfaces. In: Proceedings of 2016 Annual International Conference of the IEEE Engineering in Medicine and Biology Society (EMBC); 2016. p. 1516–1519.

[41] Higashi H, Rutkowski T, Tanaka T, Tanaka Y. Multilinear discriminant analysis with subspace constraints for single-trial classification of event-related potentials. IEEE Journal of Selected Topics in Signal Processing. 2016;10(7):1295–1305.

[42] Huang W, Goldsberry L, Wymbs NF, Grafton ST, Bassett DS, Ribeiro A. Graph frequency analysis of brain signals. IEEE Journal of Selected Topics in Signal Processing. 2016;10(7):1189–1203.

[43] Barachant A, Bonnet S, Congedo M, Jutten C. Multiclass brain–computer interface classification by Riemannian geometry. IEEE Transactions on Biomedical Engineering. 2012;59(4):920–928.

[44] Bartz D, Müller KR. Covariance shrinkage for autocorrelated data. In: Advances in Neural Information Processing Systems; 2014. p. 1592–1600.

[45] Higashi H, Tanaka T, Tanaka Y. Smoothing of spatial filter by graph Fourier transform for EEG signals. In: Proceedings of The 2014 Asia-Pacific Signal and Information Processing Association Annual Summit and Conference (APSIPA ASC 2014); 2014. p. 1–8.

[46] Shuman DI, Narang SK, Frossard P, Ortega A, Vandergheynst P. The emerging field of signal processing on graphs: Extending high-dimensional data analysis to networks and other irregular domains. Signal Processing Magazine, IEEE. 2013;30(3):83–98.

[47] Uehara T, Tanaka T, Fiori S. Robust averaging of covariance matrices by Riemannian geometry for motor-imagery brain–computer interfacing. In: Advances in Cognitive Neurodynamics (V): Proceedings of the Fifth International Conference on Cognitive Neurodynamics – 2015. 2016; p. 347–353. Available from: http://dx.doi.org/10.1007/978-981-10-0207-6_48.

[48] Biswal B, Zerrin Yetkin F, Haughton VM, Hyde JS. Functional connectivity in the motor cortex of resting human brain using echo-planar MRI. Magnetic Resonance in Medicine. 1995;34(4):537–541.

[49] Cichocki A, Mandic D, Phan AH, *et al.* Tensor decompositions for signal processing applications: From two-way to multiway component analysis. IEEE Signal Processing Magazine. 2014;32(2):145–163.

[50] Higashi H, Rutkowski TM, Tanaka T, Tanaka Y. Subspace-constrained multilinear discriminant analysis for ERP-based brain computer interface classification. In: Proceedings of the 2015 Asia-Pacific Signal and Information Processing Association Annual Summit and Conference; 2015. p. 934–940.

[51] Lu H, Plataniotis KN, Venetsanopoulos AN. Uncorrelated multilinear discriminant analysis with regularization and aggregation for tensor object recognition. IEEE Transactions on Neural Networks. 2009;20(1):103–123.

[52] Sutton S, Braren M, Zubin J, John ER. Evoked-potential correlates of stimulus uncertainty. Science. 1965;150(3700):1187–1188.

[53] Tanaka T, Uehara T, Tanaka Y. Dimensionality reduction of sample covariance matrices by graph Fourier transform for motor imagery brain–machine interface. In: Proceedings of 2016 IEEE Statistical Signal Processing Workshop; 2016. p. 472–475.

[54] Sato M, Yoshioka T, Kajihara S, *et al.* Hierarchical Bayesian estimation for MEG inverse problem. NeuroImage. 2004;23(3):806–826.

Chapter 5

A review on transfer learning approaches in brain–computer interface

Ahmed M. Azab[1], Jake Toth[1], Lyudmila S. Mihaylova[1], and Mahnaz Arvaneh[1]

Abstract

One of the major limitations of brain–computer interface (BCI) is its long calibration time. Typically, a big amount of training data needs to be collected at the beginning of each session in order to tune the parameters of the system for the target user due to between sessions/subjects non-stationarity. To mitigate this limitation, transfer learning can be potentially one useful solution. Transfer learning extracts information from different domains (raw data, features, or classification domain) to compensate the lack of labelled data from the test subject. Within this chapter transfer learning definitions and techniques are fully explained. After that, some of the available transfer learning applications in BCI are explored. Then, a brief discussion about applying transfer learning in the different domains is included. The discussion shows that despite some advances, a successful transfer learning framework for BCI still needs to be developed. Finally, future research directions in this topic are suggested in order to successfully and reliably reduce the calibration time for new subjects and increase the accuracy of the system.

5.1 Introduction

In order to use brain–computer interface (BCI) on a daily basis out of the laboratory, many challenges need to be addressed. One of the main problems is the need for recalibrating the system every new session/subject. Using machine-learning methods for every new session, the BCI system has to learn the user's brain patterns and calibrate the system accordingly. Typically, the calibration could take up to 20–30 minutes for each new session, which is an exhausting and tiring amount of time that the patient has to undergo before the system is fully functional [1,2].

[1]Automatic Control and System Engineering Department, University of Sheffield, United Kingdom

The reasons for having a long calibration in Electroencephalogram (EEG)-based BCI can be as follows: The first reason is the high dimensionality of EEG signals which are very noisy as well. In order to predict the right brain states, features need to be extracted from the training EEG data to train the classifier. The problem here is when there are only a few trials available for training, as it is hard to estimate probability distributions of the features from a few trials of high-dimensional noisy EEG signals where outliers will have tremendously adverse effects. Second, EEG signals are highly non-stationary.

A lot of factors lead to this non-stationarity such as the variations of users' mental and psychological states, miss-concentration and fatigue; also, it may be affected by various measurements circumstances, i.e. changes in the impedance of the electrodes due to sweating. So, the classifier trained on the features extracted from data of the previous sessions usually performs poorly on a new session data. Third, each person has unique brain patterns. The properties of EEG signals typically change when transferring from one subject to another subject. Thus, it is not straightforward to calibrate BCI model for a new subject from EEG data collected from previous subjects.

In order to address the mentioned problem, recent studies try to reduce calibration time based on different methods and algorithms while keeping accuracy in an acceptable range [2–5]. One promising approach to deal with this problem can be transfer learning (TL), where data from different users or sessions can be used to compensate the lack of labelled data [6].

TL is a technique used to improve the accuracy of classifier trained from one domain by transferring useful information from other domains. When there is a limited supply of training data from target domain, the need for TL appears. TL has been successfully applied in different machine-learning applications such as text, image, and human activity classification. For BCI, this is relatively a new field of research that needs to be further explored.

5.2 Transfer learning

Machine-learning methods have granted remarkable success within different engineering research fields. However, most machine-learning methods work well only when data for training and testing purposes is extracted from the same feature space with a fixed distribution. Hence, if any changes happen to this distribution, most statistical models need to be reassembled by collecting new data for training. In many daily life applications, it is expensive and time-consuming to recollect the required data for retraining the model each time we need to use the system. Moreover, in some scenarios, we have access to insufficient labelled data. In such cases, TL between task domains would be a potential solution to reduce the model recalibration efforts.

5.2.1 History of transfer learning

Machine-learning algorithms predict labels of newly coming data by using models that are learned using available labelled (supervised learning) or unlabelled training data (unsupervised learning) [7]. Also, if there are few labelled samples and a large number of unlabelled samples, semi-supervised techniques can be applied [8]. Most of the machine-learning algorithms assume that the labelled and unlabelled data have the same distribution, whereas TL allows the domains, tasks, and distributions used in training and testing to be different, which is more related to real-world situations.

The fundamental motivation for TL in the machine-learning research was firstly discussed in a "Learning-to-Learn" NIPS-95 workshop [9]. Since then, much more attention has been paid to TL. In 2005, the Broad Agency Announcement of Information Processing Technology Office set a formal definition for TL as "the ability of a system to recognise and apply knowledge and skills learned from previous tasks to novel tasks" [10]. Here, the goal of TL is finding usable information in different tasks of different sources and using it to better deal with the target task. Thus, TL is different from multi-task learning where both learnings of the source and target tasks happen at the same time.

TL approaches have been applied efficiently and successfully in many real-world applications, such as learning text data between different domains [11], image classification problem [12], Wi-Fi localisation [13], computer-aided design applications [13], and cross-language classification [14]. Also, TL techniques were applied in some biomedical engineering studies such as human activity, muscle fatigue, drug efficacy, and human activity classification [15].

However, TL solutions were first implemented for multi-language-processing and image-processing classifications, the majority of these TL algorithms can be applied to the different application rather than the one it was implemented for. This property opens the door for TL to be used in other different areas such as attentiveness of drivers, analysing social media reactions and atmospherics data classification [16].

One of the promising applications of TL could be BCI to enhance the overall system accuracy and reduce calibration time. Different studies have tried to apply different TL types in BCI. These studies will be explained in detail in Section 5.4.

5.2.2 Transfer learning definition

In this report, a domain D is defined by its feature space X and its marginal probability distribution $P(X)$, where $X = \{x_1; \ldots; x_n\} \in$ X.

Subsequently, given a specific domain, $D = \{X, P(X)\}$, its task consists of two terms: a label space Y and an objective predictive function $f(.)$ (denoted by $T = \{Y, f(.)\}$), which can be learned using available training data. Thus, for a pair of $\{x_i, y_i\}$, where $x_i \in X$ and $y_i \in Y$, prediction of the labels of new trials is done using $f(.)$.

Generally, when there are two different domains, they have either different feature space, different marginal probability distributions or both. Similarly, two different tasks have either different label space, different predictive function, or both. In this chapter for simplicity, source domain and target domain will be referred as D_S and D_T, respectively.

Definition 5.1. *"Given a D_S, T_S, D_T, and T_T, transfer learning aims to help improve the learning of the target predictive function $f_T(.)$ in D_T using the knowledge in D_S and T_S, where $D_S \neq D_T$ or $T_S \neq T_T$". As $D_S \neq D_T$ means $P_S(X) \neq P_T(X)$ or/and $X_S \neq X_T$. Also, $T_S \neq T_T$ means $Y_S \neq Y_T$ or/and $P_S(Y|X) \neq P_T(Y|X)$ [10].*

When using TL, we need to know which types of knowledge we need to transfer and which types we should not transfer. Moreover, it is important to know how and when to transfer them. To address these questions, different types of TL categories and approaches have been proposed in the literature. We will discuss a number of them in the following sections.

5.2.3 Transfer learning categories

Based on the relationship between source and target domains and tasks, TL can be divided into three main categories. These categories are as follows:

- Inductive TL
- Transductive TL
- Unsupervised TL

5.2.3.1 Inductive transfer learning

The purpose of inductive TL algorithms is to improve estimation of the target predictive function $f_T(.)$ in target domain when the target and source tasks are different ($T_T \neq T_S$). Moreover, it does not matter if the source and target domains are the same or not [10]. It is noted that in the inductive TL, we assume some target domain labels are available to train $f_T(.)$.

Subsequently, depending on the availability of labelled and unlabelled trials from the source domain, two types of the inductive TL are shown here:

- A large amount of source domain labelled data is available. This case is the most common type of the inductive TL. It is noted that the multi-task learning setting also deals with the same situation (i.e. having a large amount of source domain labelled data available). However, in multi-task learning, the learning of both target and source tasks are done at the same time, while in the inductive TL, the target task is learnt based on knowledge transferred from the source task [17].
- There are no labels available in the source domain. Here, it is similar to the self-taught learning [18].

5.2.3.2 Transductive transfer learning

The goal of transductive TL algorithms is to improve estimation of the target predictive function $f_T(.)$ in the target domain when the target and source tasks are same, but the target and source domains are different [10]. It is noted that in the transductive TL we assume no or few labelled trials are available in the target domain, whereas a large amount of labelled trials are available in the source domain.

Subsequently, due to different situations between the source and target domains, transductive TL techniques can be divided into two situations:

- When the feature spaces are different between both the target and source domains, i.e. $X_S \neq X_T$. This is also called heterogeneous TL [19].
- When both the source and target domains have the same features space, $X_S = X_T$, but the features have different marginal probability distributions, $P_S(X) \neq P_T(X)$.

This kind ofTL is related to domain adaptation such as covariate shift method [20]. This is also called homogeneous TL [19].

5.2.3.3 Unsupervised transfer learning

This type of TL tries to solve the learning problem when there are no labelled trials available in both the source and target domains during training. In the unsupervised TL, while both the source and target tasks are different, there is a relation between them. Unsupervised TL algorithms can be applied to problems involving clustering and dimensionality reduction [10].

5.3 Transfer learning approaches

Based on the type of information that needs to be transferred, the TL algorithms can be categorised into four different approaches, explained in the following sub-sections.

5.3.1 Instance-based transfer learning

This approach is based on the assumption that although the entire source domain cannot be used directly, some parts of the source domain data can be reused for learning the target domain. The estimation of the target prediction function is improved by combining the few target labelled data with some instances from the source domain where re-weighted if needed [15]. The well-known techniques using this approach are instance re-weighting and importance sampling [10,21].

5.3.2 Feature-representation transfer learning

This approach focuses on improving the construction of the feature space for the target domain using the data from the source domains, such that the performance of the target task is improved by minimising classification errors.

Depending on the amount of labelled data available in the source domain, the feature-representation TL can be either supervised or unsupervised [10]. The feature-representation TL can also be formulated in two different types, namely asymmetric and symmetric feature-based TL. The former aims to transform the source features of the source domain in a way to be closer to the target domain. The latter tries to discover the underlying representative structures between both domains to find common latent features that have a same marginal distribution across the source and target domain [19].

5.3.3 Classifier-based transfer learning

This approach focuses on improving the construction of the prediction function (classifier) of the target domain using the prediction functions of source domain subjects/sessions. Parameter-based TL assumes that some parameters and prior distributions are shared between the individual functions of the source and target tasks. So, these shared parameters or priors can be transferred to the target prediction function such that the classification errors are reduced. As an example, classifier-based

TL can be done by combining multiple source classifiers (ensemble learners) to form an improved target classifier [22].

5.3.4 Relational-based transfer learning

Different from other approaches, this approach deals with problems that the source and target data are not independent and identically distributed (non-i.i.d.) and can be presented in many relations. So, this approach aims to find the relational patterns between the source and target domains and then transfers the knowledge in the source domain to the target domain based on statistical relational learning techniques [23].

5.4 Transfer learning methods used in BCI

As stated before, BCI applications are obstructed by the long calibration time required at the beginning of each session. TL is a promising approach that can potentially avoid this limitation. TL transfers information from different domains (data, features, or classification domain) to compensate the lack of labelled data from the subject.

Typically in BCI, two types of information have been transferred, i.e. either discriminative or stationary [24]. Transferring discriminative information aims at constructing more discriminative systems (e.g. by focusing on features, classifier, filters). This approach has been successfully applied to scenarios that the available data samples are few to avoid over-fitting and when the source and target domains are very similar. However, this approach may fail when the target and source domains are not very similar. In this case, transferring stationary information which aims at constructing more invariant systems is more successful as it focuses on common information across domains [25].

In this section, we will review the TL methods applied on BCI from point of view of TL approaches mentioned in Section 5.3.

5.4.1 Instance-based transfer learning in BCI

The well-known techniques using this approach in BCI can be categorised under two main divisions: importance sampling, where certain values of the input variables have more impact on the parameter being estimated than others, and instance re-weighting, which aims to weight certain parts of the source domains to be reused in the target domain [10,21].

5.4.1.1 Importance sampling instance-based transfer learning

In [26], a method called Bagged importance-weighted LDA (Bagged IWLDA) based on covariate shift adaptation was proposed to reduce non-stationarities between sessions. Covariate shift adaptation is used to overcome the problem of the supervised learning process which requires a big amount of labelled test samples under the assumption that training and test samples follow the same distribution [27]. But, this basic assumption is mostly violated in real-life applications for BCI. Indeed, covariate shift adaptation is applied under more realistic assumption where the training and testing samples have different distributions, and at the same time, the conditional distribution of output labels are unchanged. So, in [26], a random subset was chosen from the available data to train the classifier. The proposed IWLDA classifier was presented to be an extension of LDA classifier based on the concept of importance sampling as under covariate shift normal LDA is not stable. Importance was calculated as the ratio between the test and training input densities. Then, the proposed classifier was learnt using what was randomly chosen as a training subset. N repetitions of this step were held to compute N number of IWLDA classifiers. In the end, the final predictor was obtained based on the average of these N classifiers.

Another approach based on the principle of covariate shift adaptation was proposed in [28] to reduce non-stationarity between sessions. Marginalised stacked denoising autoencoder was used to calculate the importance weights. The calculated importance weights were used in the learning algorithm to minimise the mismatch between different sessions. The authors assumed, by proposing this algorithm, to overcome the limitation of traditional techniques that were used to calculate the importance weights [i.e. Kernel Mean Matching [27], Kullback–Leibler (KL) Importance Estimation Procedure [27], and Unconstrained Least Squares Importance Fitting [29]]. These methods were used to calculate the importance weights under the assumption that training and testing data must be available, which is not a practical assumption.

The authors of [30] proposed a method of transferring selective instances based on an improved active TL (ATL) algorithm. Active learning (AL) is used to find the most informative samples to be chosen for labelling so that a higher performance learning process can be achieved with less labelling effort. For example, if there are two classes, A and B, with unknown distributions and a few labelled trials and a large amount of unlabelled trials are available from each class. AL task here is how to select the most informative unlabelled samples (say two) to be labelled and added to the training data to enhance the classifier training process. There are different method that can be used for this purpose (i.e. least confident, margin and entropy, query by committee, expected error reduction, variance reduction, and density weighted method [31]). AL was used previously in BCI to select the most informative samples from the unlabelled target domain samples to be labelled and added to the classifier training data [32]. In this research, AL was applied to the source domain labelled samples to choose the samples that were close to target domain labelled samples and could be added to the training domain.

The authors proposed two algorithms, the first algorithm they proposed was called selective instance transfer with AL (SITAL), which aimed to enhance the accuracy of direct TL problem that could lead to a negative transfer. The negative transfer happens when the source and target data have great dissimilarity. So, to find data in source subject's domain that is similar to the target domain data, a similarity finding solution was added with trials that correctly classified using the new subject-specific classifier. This subject-specific classifier was trained using the few trials available in the target domain and was selected for instance transfer. They also proposed another algorithm called selective informative instance transfer with AL (SIITAL). Where AL was used not only to select the most informative samples, after selection of correctly classified trials from other subjects (in SITAL), but also to check the normalised entropy of the selected samples and chose samples with higher entropy from these selected trials for instance transfer.

Results showed that SITAL and SIITAL almost had gained slightly higher classification accuracy compared to baseline approaches for some subjects, but not for all subjects when fewer samples were available for training. These algorithms have some drawbacks which might be due to the class imbalance problem during random selection of instances to be labelled in SITAL and SIITAL. That might be because the criteria of instances selection by informativeness when using SIITAL reduced the number of functioning trials for some subjects.

5.4.1.2 Re-weighting instance-based transfer learning

There is no obvious re-weighting instance-based TL application in BCI, that select some subsets from the source domain data and re-weight them to be used in the target domain, until now based on our knowledge. But there is one application which applied this approach but for the whole available source domain data.

TL in [33] was used to reduce non-stationarity between sessions in BCI. A data space adaptation technique was proposed to linearly transform the EEG data from the target space to the training space in a way to minimise the distribution difference between the two spaces. Two versions of the EEG data space adaptation were proposed using the KL divergence, based on the availability of labels in the testing session: when labelled data were available, it was called the supervised version, and when labelled data were not available, it was called the unsupervised version. The results showed that concerning classification accuracy the proposed algorithm for both versions significantly outperformed the results without adaptation even when applied for subjects with poor BCI performance.

5.4.2 Feature-representation transfer learning in BCI

As mentioned before, feature-representation TL focuses on improving the construction of feature space using some knowledge from the source domains. Multiple BCI TL studies used spatial filters to learn the new feature representation. There are different algorithms to compute spatial features; among them, common spatial pattern (CSP) is the most commonly used algorithm for extracting discriminative features from EEG signals. However, when there are only a few trials available for training, CSP tends to over-fit. So, different improved approaches for CSP were proposed to overcome this limitation. From this point, we can categorise feature-representation TL in BCI into two main subcategories based on the which method is used to extract the information to be transferred. One category deals with approaches using CSP to extract features and the other category include other methods that can be used to extract EEG features.

5.4.2.1 CSP-based feature-representation transfer learning

This category can be implicitly divided also into two subcategories based on how TL is applied. For some applications, modification on how CSP covariance matrix is estimated, and for other applications, the modification can be done within the CSP optimisation function or the algorithm.

In [34], Lotte *et al.* used data from other subjects to improve CSP and linear discriminant analysis (LDA) algorithms. More precisely, it was proposed that using the data from a subset of source subjects could lead to improve the estimation of CSP covariance matrix and the proposed covariance matrix $\widehat{C}_t$ was computed as follows:

$$\widehat{C}_t = (1 - \lambda)C_t + \lambda\left(\frac{1}{|S_t(\Omega)|} \sum_{i \in S_t(\Omega)} C_i\right), \tag{5.1}$$

where C_t denotes the estimated covariance matrix using the target subject's data; Ω is the set of subjects with previously collected trials, $|b|$ is the number of elements in the set B; λ is the regularisation parameter ($0 < \lambda < 1$) which was calculated heuristically, $S_t(\Omega)$ is a subset of subjects from Ω, and C_i is the estimated covariance matrix calculated using data from the subject i. This regularisation aimed to obtain a more stable covariance matrix estimation using covariance matrices of a subset of other available subjects. This subset of available subjects was selected using a sequential algorithm and based on their performance for labelling of the available few trials in the target domain. The results showed that the proposed covariance matrices led to enhance the classification accuracy when few trials were available from the target subject.

The authors of [35] proposed a CSP algorithm for subject-to-subject transfer using a linear combination of covariance matrices of the source and target subjects to estimate a composite covariance matrix. Consequently, the composite CSP could transfer discriminative information from other domains to overcome the CSP limitation when only few samples are available. The composite covariance matrix was calculated using one of the two following proposed methods: Method 1 focused more on covariance matrices calculated using the data from subjects who had large number of trials, where a tuning parameter biased the estimates towards the source domains. Method 2 calculated covariance matrices using subjects' data which were similar to the target domain data, where similarity was calculated using KL-divergence. The general formula to calculate the composite covariance matrix $\widehat{C_c^k}$ for subject k for both methods was as follows:

$$\widehat{C_c^k} = (1 - \lambda)w_{kk}C_c^k + \lambda \sum_{j \neq k}^{K} w_{jk}C_c^j, \tag{5.2}$$

where λ ($0 < \lambda < 1$) is the tuning parameter which controls the importance of the covariance matrix from the new subject related to covariance matrices of other subjects. K is the total number of available subjects. For each $c \in \{+, -\}$, $C_c^{k,orj}$ is the covariance matrix for subject k, orj, and class c. Weights for method 1 were computed as follows:

$$w_{jk} = \begin{cases} \frac{N_c^k}{\sum_{j \neq k}^{K} N_c^j} & \text{for } j = k \\ \frac{N_c^j}{\sum_{j \neq k}^{K} N_c^j} & \text{for } j \neq k \end{cases}, \tag{5.3}$$

where N_c^x is the total number of trials belonging to class c and subject x. Weights for method 2 were calculated as follows:

$$w_{jk} = \begin{cases} 1 & \text{for } j = k \\ a_{jk} & \text{for } j \neq k \end{cases}, \tag{5.4}$$

where a_{jk} are weights for subjects have similar characteristics and can be computed by calculating KL divergence between subjects k and j.

Delvaminck *et al.* modified the CSP objective function by constructing a shared spatial filter between different subjects by dividing the subject's spatial filter w_s into a global part and a subject-specific part [36].

$$w_s = w_0 + v_s, \tag{5.5}$$

where $w_0 \in R^d$ represented the global spatial filter which was shared and learned over all subjects and $v_s \in R^d$ represented the subject-specific part of the filter. The number channels was denoted by d. An optimisation framework was described to couple these two parts using a regularised parameters that were used to make a trade-off between these two parts.

$$\max_{w_0, v_s} \sum_{s=1}^{S} \frac{w_s^T \Sigma_s^{(1)} w_s}{w_s^T \Sigma_s^{(2)} w_s + \lambda_1 ||w_0||^2 + \lambda_2 ||v_s||^2}, \tag{5.6}$$

where $\Sigma_s^{(1)}$ and $\Sigma_s^{(2)}$ are the covariance matrices of the trials for the available two classes 1 and 2, respectively, for subject s. The parameters λ_1 and λ_2 are trade-offs between the global and the specific parts of the filter, by choosing one of them is high and the other one is zero, this leads to force the filter to be specific or more global.

Samek *et al.* [24] proposed an extension for CSP using the same idea described above by dividing CSP into two parts. The proposed algorithm was called stationary subspace CSP (ssCSP) where stationary information across multi-subjects instead of discriminative information was transferred by learning a stationary subspace. At first common invariant information between the available subjects were extracted, for each subject eigenvectors decomposition of the difference between the test and training session covariance matrices was computed. Then for each subject, a predefined number of eigenvectors with the largest absolute values were selected; subsequently, all vectors from all subjects were aggregated in one matrix. After that PCA was used to reduce the dimensionality of this matrix and extract the most common non-stationarities directions. Finally, in order to construct invariant features for the new subject, CSP filters of this subject was regularised towards the orthogonal complement of the most common non-stationary directions extracted in the previous step. The proposed algorithm is promising when there is a significant change between the training and test data, as it was suggested to reduce the shift between the training and test data.

Also, Samek *et al.* introduced a general spatial filter computation framework based on divergence maximisation (divCSP) in [37]. He showed that CSP algorithm could be solved as a divergence maximisation optimisation function. The authors proved that the CSP filters project data to a subspace with the maximum discrepancy, measured by symmetric KL divergence. So, instead of calculating spatial filters using CSP, they obtained another solution based on KL divergence by solving the following regularised objective function which consists of two parts, the CSP term and regularisation term.

$$\ell(V) = (1 - \lambda)D_{KL}(V^T\Sigma_1 V)||(V^T\Sigma_2 V) - \lambda\Delta, \tag{5.7}$$

where λ is the regularisation parameter which is used to make a trade-off between the two parts, and it was obtained here by cross-validation. D_{KL} is the symmetric KL divergence between the whitened covariance matrices of data from the two classes for the new subject. Σ_1 and Σ_2 are covariance matrices of data from the two classes for the new subject. Δ is the regularisation term and can be computed by

$$\Delta = -\frac{1}{K}\sum_{k=1}^{K} D_{KL}(V^T\Sigma_1^k V)||(V^T\Sigma_2^k V), \tag{5.8}$$

where K is the number of the available subjects. Using this approach information from other subjects were used. Whereby, introducing regularisation into the optimisation function solution, it led to the design of a novel spatial filtering algorithm. Thus, by jointly optimising the divergence problems of different users, a subject-independent feature space could be extracted.

5.4.2.2 Non CSP-based feature-representation transfer learning

There are some other feature-representation-transfer applications where information can be transferred using different strategies, such as transfer stationary information using principal component analysis (PCA)-based covariate shift adaptation as in [38]. The authors aimed to minimise the non-stationaries effect by proposing a new covariate shift adaptation method based on PCA. The most important non-stationary principal components were extracted and normalised by shifting a window over the data to reduce the effect of non-stationarity. Each feature normalisation was done individually rather than normalising the combination of all features. This method is beneficial when the number of dimensions is more than the number of trials and enhances the accuracy of CSP-based methods that work with stationary information. There are different applications using covariate shift for TL were proposed in [39,40].

Another method also was presented called a non-negative matrix factorisation (NMF). This method was shown to be useful in capturing discriminative information without using the concept of cross-validation in motor imagery EEG tasks [41]. However, direct application of NMF to EEG data of different subjects captures only intra-subject variations. In [42], the authors applied NMF in a way called group non-negative matrix factorisation (GNMF) where discriminative information was transferred from multiple subjects. Given EEG data measured from several subjects under the same conditions, the goal was to estimate common task-related bases in a linear model capturing intra-subject variations and at the same time inter-subject variations.

5.4.3 Classifier-based transfer learning in BCI

Classifier-based TL in BCI can be divided into two subcategories: domain adaptation and ensemble learning of classifiers. In domain adaptation, source domains classifiers are reused by adjusting classifier parameters concerning the target domain. Commonality between the source and target domains are compulsory to apply this type of TL. This domain adaptation is commonly used to transfer the discriminative and stationary information between sessions. In the ensemble learning of classifiers, different classifiers trained from different source domains are combined to acquire better classification accuracy of the target domain samples [6].

5.4.3.1 Domain adaptation in classifiers

In order to solve the problem of EEG non-stationarity between sessions, Bamdadian *et al.* proposed an algorithm named adaptive extreme learning machine (ELM) [43]. At first, the ELM algorithm was trained using previous sessions EEG data trials, and then the trained ELM classifier was used to label test session data. Then, these labelled data trials from test session were added to the training set to be used in the training process of the final ELM classifier.

A domain adaptation algorithm called domain adaptation support vector machine was proposed in [44]. This algorithm was detailed in three steps: (1) using source domain to initialise the discriminative function; (2) replacing samples from source domain with samples from the target domain to adjust the discriminative function; (3) final discriminative function was learned using only data from the target domain. Samples replacement were determined based on different settings that can be found in [44].

5.4.3.2 Ensemble learning of classifier

An example of ensemble learning of the classifier was proposed in [45] as a framework for subject transfer. This framework was consisted of three parts: (1) two sparse filter sets called robust filter bank and adaptive filter bank were learnt for each subject using the subject's CSP filter bank; (2) two classifiers models were trained for each subject based on these two filter banks; and finally (3) a two-level ensemble strategy was applied to integrate all classifiers from the robust ensemble models and adaptive ensemble models into one robust ensemble learner and one adaptive ensemble learner. Then, at the second level these two learners were combined into one final ensemble classifier using a tuning parameter for controlling the balance between adaptiveness and robustness.

When the test sample x_i was to be classified, the robust models of all subjects $M_{rj}(j = 1, \ldots, K + 1)$ were used to construct a robust ensemble learner as follows:

$$RE(x_i) = \sum_{j}^{K+1} W_{rj} \times M_{rj}(x_i), \tag{5.9}$$

where $RE(x_i)$ denotes the robust ensemble result of test sample x_i, $M_{rj}(x_i)$ is the result of the robust model of subject j, and W_{rj} is the weight of the model M_{rj}. The adaptive ensemble learner was computed using the following equation:

$$AE(x_i) = \sum_{j}^{K+1} W_{aj} \times M_{aj}(x_i), \tag{5.10}$$

where $AE(x_i)$ denotes the adaptive ensemble result of test sample x_i and $M_{aj}(x_i)$ is the result of the adaptive model of subject j. $M_{aj}(j = 1, \ldots, K + 1)$ are the adaptive models of all subjects.

A dynamic ensemble strategy based on the prediction consistency with the neighbourhood structure of the test example surrounded by subject j samples was applied to assign different weights for distinct test samples. Then, the final ensemble learner was defined as follows:

$$E(x_i) = (1 - \lambda)RE(x_i) + \lambda AE(x_i), \tag{5.11}$$

where $\lambda \in [0, 1]$ represents the tuning parameter which was calculated by cross validation.

The authors of [32] proposed a novel application of TL for online calibration of a single-trial error related potential (ERP) classifier. First, labelled training trials from the new subject only were used to train an SVM classifier. Then, data available from each other subject was combined with the few labelled trials available from the new subject to train an SVM classifier for each subject. After that, the final classifier C_{new} was constructed by combining classifiers from all these subjects as follows:

$$C_{new} = C_i + \sum_{m=1}^{M} C_m \times a_m, \tag{5.12}$$

where the subject-specific classifier of the new subject, C_i, had a unit weight, and the weight of the classifier of each subject m, C_m was assumed to be the average of the cross-validation accuracy a_m which is how accurately the few available trials of the testing subject were labelled. This iteration was repeated ten times, and each time two new labelled trials from the target domain were added to the training domain. Selection of these two trials was done either randomly or using AL. The authors attempted to enhance the classification accuracy by integrating AL with TL. They selected some unlabelled samples from the target domain to be labelled and added to the training domain and they named this algorithm ATL. The data instances that had the greatest amount of uncertainty were selected to be the most informative, as these samples had the most disagreement within the trained classifiers [32]. Results of the proposed methods, when compared with two baseline approaches using SVM classifier showed that TL and ATL almost outperformed baseline approaches when there were few trials available for training from the new subject. ATL mostly outperformed TL, and this could be because during TL the random selection of trials may lead to class trials imbalance when there are only a few labelled trials available.

In [46], the same authors again proposed another ensemble classifier approach named weighted adaptation regularisation (wAR), which used data from other subjects to reduce the amount of labelled data required in the offline single-trial classification of ERPs. The proposed model explicitly handles class-imbalance problems which are common in many real-world BCI applications. They also aimed to reduce the computational cost of wAR by proposing a source domain selection (SDS) approach which selects the closest source domains (i.e. existing subjects) to the target domain. Thus, SDS was performed to select the closest source domains, and then wAR was applied on selected source domain separately to obtain the best classifier parameters for that specific source domain. The final classification was a weighted average of these individual classifiers, with the weight being calculated based on the training accuracy of the corresponding wAR.

Besides, an adaptive accuracy-weighted ensemble (AAWE) approach was proposed in [47] to allow tracking of non-stationarities in EEG signals using data from other subjects. AAWE combines different individual classifiers, and each classifier trained using data recorded from each individual subject, the weight given to each classifier was initialised based on the accuracy of classifying calibration data for the new subject. After that, the weights were updated using ensemble learning within feedback phase, when there were no true class labels available in the classification domain.

Another approach used multi-task techniques to transfer information between session or subjects was proposed in [48]. In this algorithm, a parametric probabilistic approach that used shared priors was employed to calculate the classification parameters of a new session/subject by defining the relation between this session/subject parameters and the shared priors of available sessions/subjects. These shared parameters were used to compute the classifier parameters of a new subject when there were only a few trials available from this subject.

This algorithm works as follows: $s = \{1, \ldots, S\}$ is the multiple subjects or recording sessions with n_s trials. The class label of a new trial can be predicted by

$$y_s^{i+1} = \text{sign}(w_s^T x_s^{i+1}), \tag{5.13}$$

where w_s is the classification parameter being used to predict the class label for subject/session s trials, and x_{i+1} denotes the feature vector extracted from new trial of subject s. y_s^{i+1} presents the classes, for example the left or right hand movement motor imagery is performed in trial i at session/subject s is presented by $y_s^i \varepsilon \{-1, 1\}$. So using the available data sets and labels, the objective is to determine the best w_s which lead to the best labels prediction of the trials for each subject/session such that the number of errors in this data set D_s is reduced.

The authors claimed that for a BCI problem, each subject/session is treated as one task, where the shared structure (μ, Σ) can be presented respectively by the mean vector and covariance matrix of W where $W = w_1, \ldots, w_S$. So, the goal of this model is to calculate these shared parameters from all the tasks jointly in a way that these

w_s reduce the error and also are close together, and this can be achieved by solving the following optimisation problem:

$$\min LP(W; D_s, \mu, \Sigma, \lambda) = \min\left(\frac{1}{\lambda}\right)\sum_s ||(X_s w_s - y_s)||^2 + \sum_s \Omega(w_s; \mu, \Sigma) + C, \tag{5.14}$$

where the first term of this optimisation problem is the sum of the losses from each task, and by minimising it all sessions are ensured to be well fitted together. The divergence of each task model from shared structures is controlled using the second term. Finally, by solving this optimisation problem with respect to W and holding (μ, Σ) fixed this yields the following equation to update w_s:

$$w_s = \left(\left(\frac{1}{\lambda}\right)\Sigma X_s^T X_s + I\right)\left(\left(\frac{1}{\lambda}\right)\Sigma X_s^T y_s + \mu\right) \tag{5.15}$$

For fixed W, solving the optimisation problem yields to identify the update equations of μ and Σ, which also how the multi-task algorithm works until finding the new subject classification parameter [17].

In [49], a method for unsupervised TL named spectral transfer using information geometry was proposed. This process aimed to rank and combine unlabelled predictions from individual different subjects' ensemble classifiers. The authors claimed that the proposed method significantly outperformed the existing techniques of classifying ERPs when few trials are available for training.

The within-session and subjects differences can be understood as geometric transformations of the covariance matrices using the Riemannian framework. Riemannian geometry presents an optimum method for looking over the TL problem because of the affine invariance property of the Riemannian distance and mean.

In [50], the authors aimed to make EEG data of different subjects/session comparable by the affine transformation of the spatial covariance matrices of the EEG signals of every session/subject. The authors assumed that covariance matrices shifts concerning a reference (resting) state could be happened due to different source configurations and electrode positions. Using a reference covariance matrix, the covariance matrices of every session/subject were placed with respect to the reference, so that only the displacement with respect to the reference state was observed when there was a new task. For every session, there was a reference matrix estimation, but different subjects. Then, a congruent transformation was performed using the available data and this reference matrix. Although there were different reference matrices within sessions and subjects, but if the reference matrix was chosen accurately, different sessions/subjects data could be compared.

The proposed procedure was tested in a classification problem, where data from different sessions (subjects) were used to estimate the class parameters that needed to classify new trials. But this approach improved the accuracy of classification for subjects performing BCI with excellent and medium accuracy.

5.4.4 Unsupervised transfer learning

Unsupervised adaptive TL for was proposed in [51]. This approach provided robust class separation in feature space of the target subject by learning SSVEP templates for this subject which led to enhanced classification accuracy. By using an extended version of Canonical correlation analysis (CCA) called Adaptive Combined-CCA (Adaptive-C3A) that used a set of reference signals consisting of sinuses and cosines at the fundamental and harmonic frequencies of the SSVEP stimuli to define linear multivariate projections in EEG data. After that a simple matching classifier template was selected to predict the target class label by allocating the frequency label to the EEG segment which best coincided with the corresponding template frequency.

There is a problem with unsupervised adaptation procedures that is how likely the potential of trials which are atypical given the class structure learned during transfer to be adapted. In order to solve this problem, the Adaptive-C3A choose high confidence trials only for adaptation according to the ratio between the best and second-best predicted classes.

5.5 Challenges and discussion

Although several studies focused on TL in BCI, there are still many open questions that need further investigation. These include what to transfer, when to transfer and how to transfer. Current studies focused on transferring either discriminative or nonstationary information. However, how to identify them are not reliably investigated across all subjects. Future studies are needed to better identify these subsets of information. Moreover, finding the subsets of information that satisfy both being stationary and discriminative would lead to better results. In addition, exploring information that more specifically reflect the mental activity performed by the user might better address the problem. Last but not the least, identifying different clusters of previous subjects and using only a subset that is similar to the target subject might be useful in improving TL in BCI. Below, we will further discuss the limitations of different TL approaches in BCI.

5.5.1 Instance-based transfer learning in BCI

Instance-based TL is still a developing research area; different aspects need to be further investigated. For example, data from some subjects that are similar to the target subject can be chosen instead of using all the available data from all subjects. Selection of subjects and trials can be made based on the few trials available from the target subject. There is a need for algorithms that can accurately identify which parts of information should be transferred and what is the most suitable approach to transfer them, and how to re-weight these selected data if required. In addition, TL algorithms should be able to properly deal with the unbalanced class trials, as in real scenarios, the user should not be stressed to perform equal balanced class actions.

5.5.2 Feature-representation transfer learning in BCI

From what was described previously it is shown that existing traditional CSP-based methods calculate covariance matrices on a subject-specific basis. When there are only a few trials for training available, the performance of CSP methods on a subject-specific basis is degraded as the estimated covariance matrices are over-fitted. Different modifications were applied to traditional CSP algorithm to overcome this limitation as stated before. However, there are still some subjects that may not gain from these modifications. Moreover, finding the optimum regularisation parameter is still a challenge. For many cases, the regularisation parameter is calculated using cross validations over a number pre-defined values which cost a long computational time. For some other cases, the regularisation parameter is calculated empirically which is not optimum.

Unlike current approach that attempt to make the CSP filters (covariance matrices) of the current subject similar to the CSP filters (covariance matrices) of other subjects, another possible approach could be focusing on CSP filters of different subjects and choosing the most informative ones.

5.5.3 Classifier-based transfer learning in BCI

Since classifier-based TL is just focused on construction of the classifier, it might not be useful for subjects who have non-separable features, as changing the parameters of the classifiers does not add any separability to the feature space. Thus, classifier-based TL might be better to be coupled with either instance-based or feature-based TL approaches in order to be useful for all subjects including those with initially poor BCI performance.

5.6 Summary

Within this chapter, a general introduction to the TL in BCI was described. Moreover, TL definitions and techniques were explained. Finally, some of the available TL applications in the BCI were explored to better identify the suitable approaches that can be used to reduce calibration time and at the same time increase the accuracy of the BCI-based system. These approaches could be summarised as follows:

- TL algorithms on classification domain that reuse classifiers learned from other domains to aid better classification of new data.
- TL algorithms on feature space to improve feature extraction/selection. The investigated algorithms explore the common information across subjects/sessions to find more robust features that can enhance the model trained by a small training data.
- TL algorithms that can be applied on raw EEG directly to mine and reuse certain parts of data from other subjects/session for training a better model for a new subject.

Finally, the challenges and limitations of the available TL approaches in BCI were discussed and some possible future research directions were suggested.

References

[1] Millán JdR, Rupp R, Müller-Putz GR, *et al.* Combining brain–computer interfaces and assistive technologies: State-of-the-art and challenges. Frontiers in Neuroscience. 2010;4:161.

[2] Krauledat M, Schröder M, Blankertz B, *et al.* Reducing calibration time for brain–computer interfaces: A clustering approach. In: Advances in Neural Information Processing Systems; 2007. p. 753–760.

[3] Tu W, Sun S. A subject transfer framework for EEG classification. Neurocomputing. 2012;82:109–116. Available from: http://dx.doi.org/10.1016/j.neucom.2011.10.024.

[4] Fazli S, Popescu F, Danóczy M, *et al.* Subject-independent mental state classification in single trials. Neural Networks. 2009;22(9):1305–1312. Available from: http://dx.doi.org/10.1016/j.neunet.2009.06.003.

[5] Lotte F, Guan C. Learning from other subjects helps reducing brain–computer interface calibration time. In: ICASSP, IEEE International Conference on Acoustics, Speech and Signal Processing – Proceedings. 2010; p. 614–617.

[6] Wang P, Lu J, Zhang B, *et al.* A review on transfer learning for brain–computer interface classification. In: 2015 5th International Conference on Information Science and Technology, ICIST 2015. 2015; p. 315–322.

[7] Baralis E, Chiusano S, Garza P. A lazy approach to associative classification. IEEE Transactions on Knowledge and Data Engineering. 2008;20(2):156–171.

[8] Zhu X. Semi-supervised learning literature survey. Computer Science, Citeseer: University of Wisconsin-Madison. 2005;2(3):4.

[9] Thrun S, Pratt L. Learning to learn: Introduction and overview. In: Learning to learn. Boston, MA: Springer; 1998. p. 3–17.

[10] Pan SJ, Yang Q. A survey on transfer learning. IEEE Transactions on Knowledge and Data Engineering. 2010;22(10):1345–1359.

[11] Aggarwal CC, Zhai C. Mining text data. Springer Science & Business Media; 2012.

[12] Quattoni A, Collins M, Darrell T. Transfer learning for image classification with sparse prototype representations. In: Computer Vision and Pattern Recognition, 2008. CVPR 2008. IEEE Conference on. IEEE; 2008. p. 1–8.

[13] Pan SJ, Zheng VW, Yang Q, *et al.* Transfer learning for wifi-based indoor localization. In: Association for the Advancement of Artificial Intelligence (AAAI) Workshop; 2008. p. 6.

[14] Lu J, Behbood V, Hao P, *et al.* Transfer learning using computational intelligence: a survey. Knowledge-Based Systems. 2015;80:14–23.

[15] Weiss K, Khoshgoftaar TM, Wang D. A survey of transfer learning. vol. 3. Springer International Publishing; 2016. Available from: http://journalofbigdata.springeropen.com/articles/10.1186/s40537-016-0043-6.

[16] Dai W, Yang Q, Xue GR, *et al.* Boosting for transfer learning. Proceedings of the 24th International Conference on Machine Learning. ACM; 2007. p. 193–200.

[17] Alamgir M, Grosse-Wentrup M, Altun Y. Multitask learning for brain–computer interfaces. In: Proceedings of the Thirteenth International Conference on Artificial Intelligence and Statistics (AISTATS-10). vol. 9; 2010. p. 17–24.

[18] Raina R, Battle A, Lee H, *et al.* Self-taught learning: transfer learning from unlabeled data. In: Proceedings of the 24th International Conference on Machine Learning; 2007. p. 759–766.

[19] Weiss K, Khoshgoftaar TM, Wang D. A survey of transfer learning. Journal of Big Data. 2016;3(1):9.

[20] Shimodaira H. Improving predictive inference under covariate shift by weighting the log-likelihood function. Journal of Statistical Planning and Inference. 2000;90(2):227–244.

[21] Liao X, Xue Y, Carin L. Logistic regression with an auxiliary data source. In: Proceedings of the 22nd International Conference on Machine Learning. ACM; 2005. p. 505–512.

[22] Evgeniou T, Pontil M. Regularized multi–task learning. In: Proceedings of the Tenth ACM SIGKDD International Conference on Knowledge Discovery and Data Mining. ACM; 2004. p. 109–117.

[23] Mihalkova L, Mooney RJ. Transfer Learning by Mapping with Minimal Target Data; In Proceedings of the AAAI-08 workshop on transfer learning for complex tasks, 2008. Association for the Advancement of Artificial Intelligence; 2008. Available from http://www.aaai.org.

[24] Samek W, Meinecke FC, Müller KR. Transferring subspaces between subjects in brain–computer interfacing. IEEE Transactions on Biomedical Engineering. 2013;60(8):2289–2298.

[25] Wang P, Lu J, Zhang B, *et al.* A review on transfer learning for brain–computer interface classification. In: Information Science and Technology (ICIST), 2015 5th International Conference on. IEEE; 2015. p. 315–322.

[26] Li Y, Kambara H, Koike Y, *et al.* Application of covariate shift adaptation techniques in brain–computer interfaces. IEEE Transactions on Biomedical Engineering. 2010;57(6):1318–1324.

[27] Sugiyama M, Nakajima S, Kashima H, *et al.* Direct importance estimation with model selection and its application to covariate shift adaptation. In: Advances in Neural Information Processing System Foundation NIPS; 2008. p. 1433–1440.

[28] Abid F, Hassan A, Abid A, *et al.* Transfer learning for electroencephalogram signals. In: International Conference on Computer and Electrical Engineering, ICCEE. vol. 7; 2017. p. 143–152.

[29] Kanamori T, Hido S, Sugiyama M. A least-squares approach to direct importance estimation. Journal of Machine Learning Research. 2009;10(Jul): 1391–1445.

[30] Hossain I, Khosravi A, Nahavandhi S. Active transfer learning and selective instance transfer with active learning for motor imagery based BCI.

In: Neural Networks (IJCNN), 2016 International Joint Conference on. IEEE; 2016. p. 4048–4055.

[31] Settles B. Active learning literature survey. University of Wisconsin, Madison. 2010;52(55–66):11.

[32] Wu D, Lance B, Lawhern V. Transfer learning and active transfer learning for reducing calibration data in single-trial classification of visually-evoked potentials. In: Systems, Man and Cybernetics (SMC), 2014 IEEE International Conference on. IEEE; 2014. p. 2801–2807.

[33] Arvaneh M, Guan C, Ang KK, *et al.* EEG data space adaptation to reduce intersession nonstationarity in brain–computer interface. Neural Computation. 2013;25(8):2146–2171.

[34] Lotte F, Guan C. Learning from other subjects helps reducing brain–computer interface calibration time. In: Acoustics Speech and Signal Processing (ICASSP), 2010 IEEE International Conference on. IEEE; 2010. p. 614–617.

[35] Kang H, Nam Y, Choi S. Composite common spatial pattern for subject-to-subject transfer. IEEE Signal Processing Letters. 2009;16(8):683–686.

[36] Devlaminck D, Wyns B, Grosse-Wentrup M, *et al.* Multisubject learning for common spatial patterns in motor-imagery BCI. Computational Intelligence and Neuroscience. 2011;2011:8.

[37] Samek W, Kawanabe M, Muller KR. Divergence-based framework for common spatial patterns algorithms. IEEE Reviews in Biomedical Engineering. 2014;7:50–72.

[38] Spüler M, Rosenstiel W, Bogdan M. Principal component based covariate shift adaption to reduce non-stationarity in a MEG-based brain–computer interface. EURASIP Journal on Advances in Signal Processing. 2012;2012(1):129.

[39] Sugiyama M, Krauledat M, MÃžller KR. Covariate shift adaptation by importance weighted cross validation. Journal of Machine Learning Research. 2007;8(May):985–1005.

[40] Raza H, Cecotti H, Li Y, *et al.* Adaptive learning with covariate shift-detection for motor imagery-based brain–computer interface. Soft Computing. 2016;20(8):3085–3096.

[41] Lee H, Cichocki A, Choi S. Nonnegative matrix factorization for motor imagery EEG classification. In: Artificial Neural Networks–ICANN 2006. 2006; p. 250–259.

[42] Lee H, Choi S. Group nonnegative matrix factorization for EEG classification. In: Artificial Intelligence and Statistics; 2009. p. 320–327.

[43] Bamdadian A, Guan C, Ang KK, *et al.* Improving session-to-session transfer performance of motor imagery-based BCI using adaptive extreme learning machine. In: Engineering in Medicine and Biology Society (EMBC), 2013 35th Annual International Conference of the IEEE. IEEE; 2013. p. 2188–2191.

[44] Bruzzone L, Marconcini M. Domain adaptation problems: A DASVM classification technique and a circular validation strategy. IEEE Transactions on Pattern Analysis and Machine Intelligence. 2010;32(5):770–787.

[45] Tu W, Sun S. A subject transfer framework for EEG classification. Neurocomputing. 2012;82:109–116.

[46] Wu D, Lawhern VJ, Lance BJ. Reducing offline BCI calibration effort using weighted adaptation regularization with source domain selection. In: Systems, Man, and Cybernetics (SMC), 2015 IEEE International Conference on. IEEE; 2015. p. 3209–3216.

[47] Dalhoumi S, Dray G, Montmain J, *et al.* An adaptive accuracy-weighted ensemble for inter-subjects classification in brain–computer interfacing. In: Neural Engineering (NER), 2015 7th International IEEE/EMBS Conference on. IEEE; 2015. p. 126–129.

[48] Jayaram V, Alamgir M, Altun Y, *et al.* Transfer Learning in Brain–Computer Interfaces. 2015. p. 1–20. Available from: http://arxiv.org/abs/1512.00296{\%}0Ahttp://dx.doi.org/10.1109/MCI.2015.2501545.

[49] Waytowich NR, Lawhern VJ, Bohannon AW, *et al.* Spectral transfer learning using information geometry for a user-independent brain–computer interface. Frontiers in Neuroscience. 2016;10:430.

[50] Zanini P, Congedo M, Jutten C, *et al.* Transfer learning: a Riemannian geometry framework with applications to brain–computer Interfaces. IEEE Transactions on Biomedical Engineering. 2017;65:1107–1116.

[51] Waytowich NR, Faller J, Garcia JO, *et al.* Unsupervised adaptive transfer learning for Steady-State Visual Evoked Potential brain–computer interfaces. In: IEEE International Conference on Systems, Man, and Cybernetics (SMC), 2016.

Chapter 6
Unsupervised learning for brain–computer interfaces based on event-related potentials

Pieter-Jan Kindermans[1], David Hübner[2], Thibault Verhoeven[3], Klaus-Robert Müller[1,4,5], and Michael Tangermann[2]

Abstract

Machine learning has become a core component in brain–computer interfaces (BCIs). Unfortunately, the use of machine learning typically requires the collection of subject specific labelled data. This process is time-consuming and not productive from a user's point of view, as during calibration the user has to follow given instructions and cannot make own decisions. Only after calibration, the user is able to use the BCI freely. In this chapter, we describe how the supervised calibration process can be circumvented by unsupervised learning in which the decoder is trained while the user is utilising the system. We discuss three variations. First, expectation–maximisation (EM)-based training, which works wells empirically but can sometimes be unstable. Learning from label proportions (LLP)-based training, which is guaranteed to converge to the optimal solution, but learns more slowly. Third, a hybrid approach combining the stability of LLP with the speed of learning of EM in a highly efficient and effective approach that can readily replace supervised decoders for event-related potential BCI.

6.1 Introduction

In the early years of brain–computer interfaces (BCIs), users had to learn how to master the device, while the computer's behaviour was pre-programmed and fixed. It often took an enormous effort for a user to learn how to generate the required brain signatures such that the computer could recognise them, which could take up to several

[1]Department of Machine Learning, Technical University of Berlin, Germany
[2]Department of Computer Science, University of Freiburg, Germany
[3]Electronics and Information Systems, Ghent University, Belgium
[4]Department of Brain and Cognitive Engineering, Korea University, Korea
[5]Max Planck Institute for Informatics, Saarbrücken, Germany

weeks [1]. The introduction of machine learning methods into the field of BCI made it possible to move from user-centred learning to computer-centred learning [2–8]. In modern BCIs, mostly the computer adapts to the user, while user learning is not focused on by many publications. These BCI systems are capable to learn the specific brain signature for each new user which allows most healthy subjects to control the system within hours or less [5,9].

Learning to decode brain signals, though, is a rather difficult machine learning problem due to a combination of multiple challenges. First of all, the signal-to-noise ratio in the electroencephalogram (EEG), which typically is exploited to assess brain activity, is very low [10]. Despite its high temporal resolution, the spatial resolution of the EEG is limited. As a result, the recorded signal reflects the superposed brain activity of many neural ensembles that are active simultaneously. Consequently, we obtain a mixed measurement of all active brain processes. On top of that, EEG brain activity differs greatly not only between users but even within a single user [11]. Changing statistics of brain signals over time [12,13] make the decoding problem specifically hard.

To obtain optimal decoding performance, a BCI can be calibrated for each individual user. But, even with a subject-specific decoder, performance drops when the BCI's decoder is not re-calibrated regularly, see for instance [14]. Traditionally, the calibration is done using a calibration recording at the beginning of a session. Here, the user is instructed to perform a predefined mental task, e.g., pay attention to a specific stimulus or imagine a specific movement. This procedure constructs a labelled data set containing EEG segments together with the correct labels. The EEG segments and the labels are the ingredients to train a supervised machine learning algorithm, and only after this training, the BCI system can be used productively on the actual desired application [7,15].

As the user cannot utilise the BCI application until this calibration is over, the calibration time spent is not productive. This can be extremely problematic, if the user has a limited attention span, e.g., for patients who want to communicate with a BCI. The BCI research community is aware of this problem. To mitigate it, various improvements on the machine learning side have been proposed with the aim to either shorten the calibration period or completely abandon it. These improvements are realised, e.g., by unsupervised, semi-supervised and transfer learning methods. In transfer learning, information is transferred within the same subject [11] or from other subjects [16–20]. In semi-supervised learning, a pre-trained classifier is obtained from transfer learning and then adapted with the unlabelled data from a new subject. Different strategies have been proposed in this context, e.g., adaptation based on predicted labels [21], based on co-training of two classifiers [22], based on error-related potentials [23–26] or based on unsupervised global mean and covariance adaptation [27].

We will focus on unsupervised learning, which does not require labelled data at all and thus bears the hope to abolish the (re-)calibration periods. Comparing unsupervised learning with supervised learning, it becomes obvious that an explicit unproductive calibration period is no longer required. Second, an adaptive unsupervised method can directly cope with changes in the statistics of the data. As it is able to continuously adapt to changing data, the decoding performance may remain stable in cases, where it would drop over time using a supervised non-adaptive decoder [14]. The reason for this important difference is found, when we take a look at the productive phase of BCI use, i.e., when the user works with the intended application. Brain signal data obtained during this online use is not labelled (except for special cases), such that this data becomes worthless for a supervised machine learning model, since it cannot be used to improve the model over time. An unsupervised model, to the contrary, can still make use of the unlabelled data to improve decoding with longer and longer periods of online use. Unsupervised adaptation or learning was described as one of the key technologies for successful BCI control [28,29].

Adaptive methods and semi-supervised methods were investigated for both systems based on motor tasks (e.g., motor imagery, attempted motor execution) and systems based on event-related potentials (ERPs). In the remaining part of this chapter, we will focus on ERP paradigms. The reason for this decision is that the constraints introduced by ERP paradigms allow us to formulate three unsupervised decoding approaches: (1) expectation–maximisation (**EM**)-based decoding [14,30,31], (2) decoders based on learning from label proportions (**LLP**) [32] and (3) an analytical mixture of both approaches (**MIX**) [33,34]. We start by presenting a short overview of the different methods and their properties.

A decoder based on EM has the following properties:

- EM decoding can be applied to most known ERP paradigms.
- Given a sufficient amount of data, EM can compete with supervised classification.
- During longer experimental sessions, EM decoders can outperform a non-adaptive supervised model.
- The quality of the EM solution has high variance when data is limited.
- EM works well empirically, but quality guarantees cannot be provided.
- EM works best when the number of classes is low and the number of iterations is high.

An LLP-based decoder has the following properties:

- LLP decoding can only be applied to a modified ERP paradigm.
- Given a sufficient amount of stationary data, LLP will converge to the optimal classifier.
- LLP decoding can adapt to changes in the data distribution.
- The LLP solution is deterministic, unique and stable.
- LLP is more stable but on average learns slower than a comparable EM-based decoder.

Finally, we can perform an analytical mixing of LLP and EM. This method called MIX has the following properties:

- MIX can be applied to BCI where LLP can be applied.
- MIX is very stable with respect to the initialisation.
- MIX on average learns faster than any of the single methods (EM and LLP).

- Empirical results indicate that MIX can be used as a direct replacement for a traditional supervised decoder.

In the next section, we introduce ERP-based BCI systems, corresponding paradigms and naming conventions. Afterwards, we discuss the three decoding methods in a section each. This is followed by a discussion of the empirical evaluation and a conclusion.

6.2 Event-related potential based brain–computer interfaces

Developing an unsupervised approach to BCI decoding is possible using constraints embedded into the BCI paradigm. Hence, without at least a basic understanding of how ERP paradigms work, it is not possible to build a reliable unsupervised decoder. Therefore, we will detail the experimental components first, beginning with the most important, the ERP component of the EEG.

An ERP is a transient brain response time-locked to an event. It can be observed not only in non-invasive EEG recordings but also in recordings of the electrocorticogram and magnetoencephalogram. 'Time-locked' means that there is a fixed delay between the event and the appearance of an ERP brain response. Many of these responses (also called 'components') have been described in the literature and under various experimental conditions. Their amplitude, timing and spatial distribution are influenced by different user tasks and vary with attention. In BCI the best known ERP component is the P300 (P3). The P300 is a positive deflection, around 300 ms (or more, depending on the exact user task) after an attended stimulus is presented [35,36]. Typically, P300 components are observed in oddball paradigms, where the user has to focus his attention on a rare target stimulus that is embedded into a stream of frequent non-target stimuli. The P300 is observed for various stimulus modalities (visual, auditory or even haptic). Even though it is a comparably strong ERP component, the P300 is not the only one, which is exploited by BCI systems to decode attended from non-attended stimuli. However, other ERP components, which may be observed with different latency, amplitude, polarity and spatial distribution on the scalp than the P300, seem to be much more sensitive to the chosen user task and parameters of the stimulus presentation.

To utilise the oddball paradigm for BCI, Farwell and Donchin proposed to couple stimuli to actions. To execute an action, e.g. to write a letter or ring for a caregiver, the user has to focus on the stimulus associated with that desired action. The presentation of the attended stimulus will elicit so-called target ERP responses, which should be different from non-target responses. Since the computer knows at what time which stimulus has been presented to the user, and as the ERP response is time-locked, the desired action can be inferred by determining which stimulus elicited an ERP target response and which stimuli did not. Once the user's intention was decoded, the action could be executed by the computer system or an attached physical device. In the specific case of the matrix speller introduced in the 1980s by Farwell *et al.* [37], a 6×6 grid (matrix) of symbols/letters was presented on a computer screen. Stimuli consisted of visual brightness intensifications of individual rows and columns of this symbol grid. Hence, this paradigm realises two oddball paradigms running concurrently, one for the rows and the other for the columns. Each has a 1 over 6 probability of highlighting the desired stimulus. By determining which row and which column resulted in a target ERP response, the desired symbol can be determined. Over time, different intensification structures which are not restricted to rows and columns became popular, because they yield improved performance, e.g., the checkerboard paradigm by Townsend *et al.* [38]. Nevertheless, the basic principle of all those grid spellers remains the same.

As the signal-to-noise ratio of the EEG is so low, a reliable decoding of target from non-target stimuli requires that all stimuli are presented multiple times. A series of stimulus presentations where each stimulus is presented once is often called an iteration, and we adopted this terminology in our chapter. Several (typically 5–15) iterations form a single trial, where a single trial corresponds to the selection/execution of a single action. The presentation of a single stimulus, the smallest time unit in these paradigms, leads to an EEG response, which often is referred to as an EEG epoch. It typically comprises a signal window of a few hundred milliseconds around the stimulus onset.

In summary, an ERP speller works by assigning commands to stimuli. The computer presents stimuli to the user and determines if a target or non-target ERP response has appeared. This way the attended stimulus and consequently the desired action can be inferred and executed. From the viewpoint of machine learning, the classifying problem consists of determining, if an observed EEG epoch belongs to the class of the attended (target) stimuli or to the class of un-attended (non-target) stimuli. Traditionally, supervised classification approaches are utilised to solve this problem, for which labelled data is required and must be recorded in a preceding calibration measurement [15]. However, we can also build a mathematical model of the application to enable reliable unsupervised learning, as discussed next.

6.3 Decoding based on expectation maximisation

The unsupervised EM-based decoder describes the ERP paradigm probabilistically. To learn how to decode the data, it optimises the parameters such that the likelihood of the observations is maximised. In this model, the unknown latent variable is the user's intention. In the description of the model below, we will present a minor variation, which is simpler to implement than the original model from Kindermans *et al.* [30].

6.3.1 The probabilistic model for ERP BCI

Let us start by formalising the BCI paradigm. Denote with $\boldsymbol{x}_{t,i,j}$ the feature vector recorded during trial t in stimulus iteration i after stimulus presentation j. In the original grid speller by Farwell and Donchin, an iteration i comprises 12 stimulus presentations from which 6 stimulus presentations highlight rows and 6 highlight columns. We assume that there are C different actions to choose from, which are all equally likely to be the desired one c_t. Our model assumes that the EEG can be projected by the vector $\boldsymbol{w}$ into a single dimension. This projection will be centred around $y_{t,i,j}(c_t)$. The value of $y_{t,i,j}(c_t)$ is T_+ when it is a target stimulus, and T_- if it is a non-target stimulus. We will be using the values $T_+ := N/N_+$ and $T_- := N/N_-$. Selecting these targets has the advantage that when centred data is used, the resulting weight vector is the same as a linear discriminant analysis model (see e.g. [33]).

We assume that the projections have a per class variance of β^{-1}.

$$
\begin{aligned}
p(c_t) &= \frac{1}{C} \\
p\left(\boldsymbol{x}_{t,i,j}^T \boldsymbol{w} | c_t, \beta\right) &= \mathcal{N}\left(\boldsymbol{x}_{t,i,j}^T \boldsymbol{w} | y_{t,i,j}(c_t), \beta^{-1}\right) \\
y_{t,i,j}(c_t) &= \begin{cases} T_+ & \text{if } c_t \in \text{stimulus } j \text{ of iteration } i \\ T_- & \text{otherwise} \end{cases}
\end{aligned}
$$

In this model, the constraints on the possible predictions from the repetitive structure of the paradigm are contained in the function $y_{t,i,j}$. When the desired action c_t is known, the values of $y_{t,i,j}$ are known. Since there are only C possible values for each c_t the possible assignations of all the $y_{t,i,j}$ are also limited to C in trial t. A naive approach would have an exponential number of possible assignations of the different $y_{t,i,j}$. The goal of the decoder is to infer c_t. Inference in the model is based on Bayes's rule where the predicted symbol is the most likely one.

In the following formula, X_t denotes all feature vectors belonging to trial t.

$$
p\left(c_t | X_t, \boldsymbol{w}, \beta\right) = \frac{p\left(c_t\right) p\left(X_t^T \boldsymbol{w} | c_t, \beta\right)}{\sum_{c_t} p\left(c_t\right) p\left(X_t^T \boldsymbol{w} | c_t, \beta\right)}.
$$

Finally, the decoded symbol will be the one with the highest likelihood.

6.3.2 Training the model

The unsupervised training procedure for this model is the EM algorithm [39]. In the EM algorithm, we treat the attended stimulus c_t as latent variable and the weight vector $\boldsymbol{w}$ and the class-wise variance of the projection β^{-1} as parameters of the model.

The EM algorithm estimates how likely each stimulus is to be the attended one and subsequently modifies the model such that it maximises its belief. To obtain the update equations, we compute the gradient with respect to the parameters and set it equal to zero. We will only discuss the resulting updates. First, the estimated weight vector $\hat{\boldsymbol{w}}$ becomes.

$$
\hat{\boldsymbol{w}} = \left(XX^T\right)^{-1} X \sum_{\boldsymbol{c}} p\left(\boldsymbol{c} | X, \boldsymbol{w}, \beta\right) \boldsymbol{y}\left(\boldsymbol{c}\right).
$$

If the data is centred around zero, which we will ensure in our experiments by standardising the data, then XX^T is equal to N times the covariance matrix of the data

$$
N \Sigma_x = XX^T.
$$

Furthermore, the other term $X\sum_{c} p\left(\boldsymbol{c}|X,\boldsymbol{w},\beta\right)\boldsymbol{y}\left(\boldsymbol{c}\right)$ can be simplified too. This term is equal to N times the difference between the target and non-target means

$$X\sum_{c} p\left(\boldsymbol{c}|X,\boldsymbol{w},\beta\right)\boldsymbol{y}\left(\boldsymbol{c}\right) = N(\hat{\boldsymbol{\mu}}_{+}^{EM} - \hat{\boldsymbol{\mu}}_{-}^{EM}).$$

If we combine these two terms, we see that

$$\hat{\boldsymbol{w}} = \Sigma_{x}^{-1}(\hat{\boldsymbol{\mu}}_{+}^{EM} - \hat{\boldsymbol{\mu}}_{-}^{EM})$$

Clearly, this equation has the same functional form as the equation for the weight vector in an LDA model. We introduce regularisation to the model by estimating Σ_x with Ledoit–Wolf analytical regularisation [40].

The variance of the projections is updated to equal the expected mean squared error between the target projections and the actual projection:

$$\hat{\beta}^{-1} = \left\langle \sum_{c_t} p\left(c_t|X,\boldsymbol{w},\beta\right)\left(\boldsymbol{x}_{t,i,j}^{T}\boldsymbol{w} - y_{t,i,j}(c_t)\right)^2 \right\rangle_{t,i,j}.$$

In the formula above $\langle\ldots\rangle_{t,i,j}$ denotes taking the average over the indices.

The EM algorithm optimises the likelihood of the data. Hence, there is no guarantee that the classifier will actually learn to solve the correct classification task. The underlying classification problem however is part of the probabilistic model, and we have observed that there is a high correlation between the data log likelihood and the classification accuracy. As a result, the data log likelihood can be used to select the best classifier from a group of unsupervised models.

Such a group of models is often needed. Due to local minima, a randomly initialised classifier can have very low accuracies. For this reason, we typically use groups of initialisations to maximise the probability of finding a well-performing classifier. The number of initialisations depends on the data quality, e.g., for a visual ERP paradigm, fewer initialisations are needed, compared to an auditory paradigm which typically has a lower signal-to-noise ratio. Using this trick we observed that the EM decoder can be used reliably across different ERP paradigms. However, there is no theoretical guarantee that the optimal solution will be found by EM.

To summarise, the model can be updated as follows given a previous parameter setting $\hat{\boldsymbol{w}}, \hat{\beta}$.

1. Use the current parameter setting to estimate the difference between the class means

$$\hat{\boldsymbol{\mu}}_{+}^{EM} - \hat{\boldsymbol{\mu}}_{-}^{EM} = \frac{1}{N}\sum_{c} p\left(\boldsymbol{c}|X,\boldsymbol{w},\beta\right)\boldsymbol{y}\left(\boldsymbol{c}\right).$$

2. Compute the new weight vector:

$$\hat{\boldsymbol{w}} = \Sigma_{X}^{-1}(\hat{\boldsymbol{\mu}}_{+}^{EM} - \hat{\boldsymbol{\mu}}_{-}^{EM})$$

with Σ_X the estimate of the data covariance (optionally using Ledoit–Wolf shrinkage for regularisation [15]).

3. Update β.

$$\hat{\beta}^{-1} = \left\langle \sum_{c_t} p\left(c_t|X, \boldsymbol{w}, \beta\right) \left(\boldsymbol{x}_{t,i,j}^T \boldsymbol{w} - y_{t,i,j}(c_t)\right)^2 \right\rangle_{t,i,j}.$$

6.4 Decoding based on learning from label proportions

While the EM algorithm described above works well empirically, it does not always converge to a good solution. To obtain a theoretical guarantee on the performance we can resort to the LLP idea and the Mean-Map algorithm [41]. But, there is a price to pay for this guarantee, as the experimental paradigm has to be modified slightly. In this section, we will first explain the background to LLP and then discuss how it can be applied to BCI.

6.4.1 Learning from label proportions

In the section on EM-based decoding, we have shown that we only need label information to estimate the means of the classes. The labels of the individual data points are not really needed. This property holds for a subset of supervised classification approaches where we optimise a (surrogate) loss function that depends on the data $\boldsymbol{x}_i$ and the labels $y_i \in \{-1, 1\}$, where the index $i = 1, \ldots, N$ denotes the sample. This subset of losses are the *symmetric proper scoring losses*, including the logistic loss and the square loss, which can be re-written in a form that depends only on the class means, the class proportions and the input data [42]. Please observe that this formulation of the loss does *not* depend on the individual labels. Consider for example the squared error:

$$\sum_{i=1}^{N} \left(\boldsymbol{w}^T \boldsymbol{x}_i - y_i\right)^2 = \sum_{i=1}^{N} \left((\boldsymbol{w}^T \boldsymbol{x}_i)^2 + 1\right) - 2\boldsymbol{w}^T \left(\sum_{i_+} \boldsymbol{x}_{i_+} - \sum_{i_-} \boldsymbol{x}_{i_-}\right).$$

It is clear that the first term $\sum_{i=1}^{N} \left((\boldsymbol{w}^T \boldsymbol{x}_i)^2 + 1\right)$ does not depend on label information. The second term $2\boldsymbol{w}^T \left(\sum_{i_+} \boldsymbol{x}_{i_+} - \sum_{i_-} \boldsymbol{x}_{i_-}\right)$ does, but it can be re-formulated in terms of the class means and the number of data points per class:

$$2\boldsymbol{w}^T \left(\sum_{i_+} \boldsymbol{x}_{i_+} - \sum_{i_-} \boldsymbol{x}_{i_-}\right) = 2\boldsymbol{w}^T \left(N_+ \hat{\boldsymbol{\mu}}_+ - N_- \hat{\boldsymbol{\mu}}_-\right).$$

In the formula above, $\hat{\boldsymbol{\mu}}_+$ is the estimated average feature vector of the positive class and N_+ is the number of data points for this class. This implies that the loss can be computed when only the class means are known. This in turn implies that the loss can be optimised knowing only these means. Please observe that the optimal weight vectors for the linear discriminant analysis introduced before also depends on the class means only, and not on the individual labels. One can see this clearly when this weight vector is computed as

$$w = \Sigma_X^{-1}(\mu_+ - \mu_-). \tag{6.1}$$

To benefit from the dependency on the class means, we must find an approach where estimating the class means is easier or less expensive than obtaining all the single labels. Previously, we have used EM for this, but a more reliable approach can be taken using the Mean-Map algorithm. Mean-Map can be applied when we have G groups of data where these class-wise proportions per group are given by Π. The expected value of the feature vectors in the groups $\mu_1, \ldots, \mu_G$ can be expressed in terms of the class means μ_+, μ_- as

$$\begin{bmatrix} \mu_1 \\ \vdots \\ \mu_G \end{bmatrix} = \Pi \begin{bmatrix} \mu_+ \\ \mu_- \end{bmatrix}, \quad \Pi = \begin{bmatrix} \pi_+^1 & \pi_-^1 \\ \vdots & \vdots \\ \pi_+^G & \pi_+^G \end{bmatrix}.$$

An empirical estimate of the group means $\mu_1, \ldots, \mu_G$ can be obtained directly by taking the group-wise average. The system of linear equations above can also be inverted to give an estimate of the class means:

$$\begin{bmatrix} \hat{\mu}_+ \\ \hat{\mu}_- \end{bmatrix} = \left(\Pi^T \Pi\right)^{-1} \Pi^T \begin{bmatrix} \mu_1 \\ \vdots \\ \mu_G \end{bmatrix}.$$

The idea behind the Mean-Map algorithm lies in the fact that the empirical class means $\hat{\mu}_+, \hat{\mu}_-$ estimated in this way are used in the objective function. When the system of linear equations can be solved, these will converge to the true class means μ_+, μ_-. As a result, our approximated loss function will converge to the supervised loss function and the optimised model will converge to the ideal solution.

In the next section, we will discuss how the ERP-based BCI can be adapted to allow for LLP-based unsupervised decoding.

6.4.2 A modified ERP paradigm

The standard ERP paradigm uses, e.g., a 6×6 grid on the screen, and in each iteration each row and each column is highlighted once. As a result, there are two groups of events, row and column highlighting events. However, since both have a 1–5 target to non-target ratio, LLP cannot be applied. LLP needs differing target to non-target ratios between the groups of events. A possible solution therefore would be to use a distinct non-square grid shape, e.g., a 4×9 grid. In this case, there is a 1–3 target to non-target ratio in the first group and a 1–8 ratio in the second group. This would support LLP training, but it would violate one of the key assumptions, as there would only be four row intensifications which highlight many more symbols each than the column intensifications would. Thus, this non-square grid design bears the risk, which target responses for rows could differ from the column target responses. In this case, our class-wise means of the resulting ERP epochs would differ between the groups and LLP could not be applied.

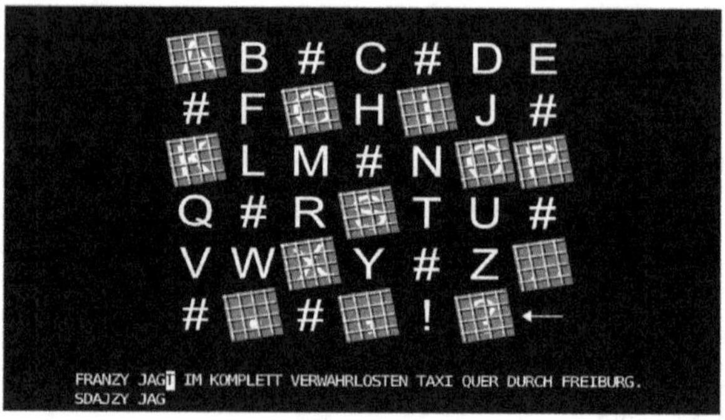

Figure 6.1 Example of an intensification in the LLP paradigm. The hash symbols are included to balance the saliency of the stimulus events between the two sequences

Thus, we propose a slightly more complex setup, which ensures that the class-wise data distribution is as similar as possible between the two groups of EEG epochs. Using the advanced stimulus presentation paradigm by Verhoeven *et al.* [43], we can also have target/non-target ratios that differ maximally between groups, which speed up convergence of LLP to the true class means.

In the approach by Hübner *et al.* [32], we use a bigger 6×7 grid containing 42 symbols. The symbol set is composed of 26 letters, punctuation and white-space symbols and 10 hash symbols (#) as shown in Figure 6.1. The hash symbols serve as uninformative blanks and always take the role of non-targets. These blanks are included such that the stimuli's brightness can be kept constant between the two groups of stimuli. One stimulus sequence contained 2 targets per 18 stimuli, the other sequence contained 3 targets per 8 stimuli. The resulting 2×2 matrix Π contains $\pi_+^1 = 3/8, \pi_-^1 = 5/8$ and $\pi_+^2 = 2/18, \pi_-^2 = 16/18$. This matrix can be inverted analytically. A schematic of this approach is given in Figure 6.2.

6.4.3 Training of the LLP model

Training an LLP decoder is rather straightforward. The classifier can be updated continuously using all data by following these steps:

1. Compute the group-wise empirical means $\hat{\boldsymbol{\mu}}_1, \hat{\boldsymbol{\mu}}_2$.
2. Estimate the class-wise means:

$$\begin{bmatrix} \hat{\boldsymbol{\mu}}_+^{LLP} \\ \hat{\boldsymbol{\mu}}_-^{LLP} \end{bmatrix} = \begin{bmatrix} 3.37 & -2.37 \\ -0.42 & 1.42 \end{bmatrix} \begin{bmatrix} \hat{\boldsymbol{\mu}}_1 \\ \hat{\boldsymbol{\mu}}_2 \end{bmatrix}.$$

3. Estimate the weight vector:

$$\hat{\boldsymbol{w}} = \Sigma_X^{-1}(\hat{\boldsymbol{\mu}}_+^{LLP} - \hat{\boldsymbol{\mu}}_-^{LLP})$$

with Σ_X being the estimate of the data covariance (optionally using Ledoit–Wolf shrinkage for regularisation).

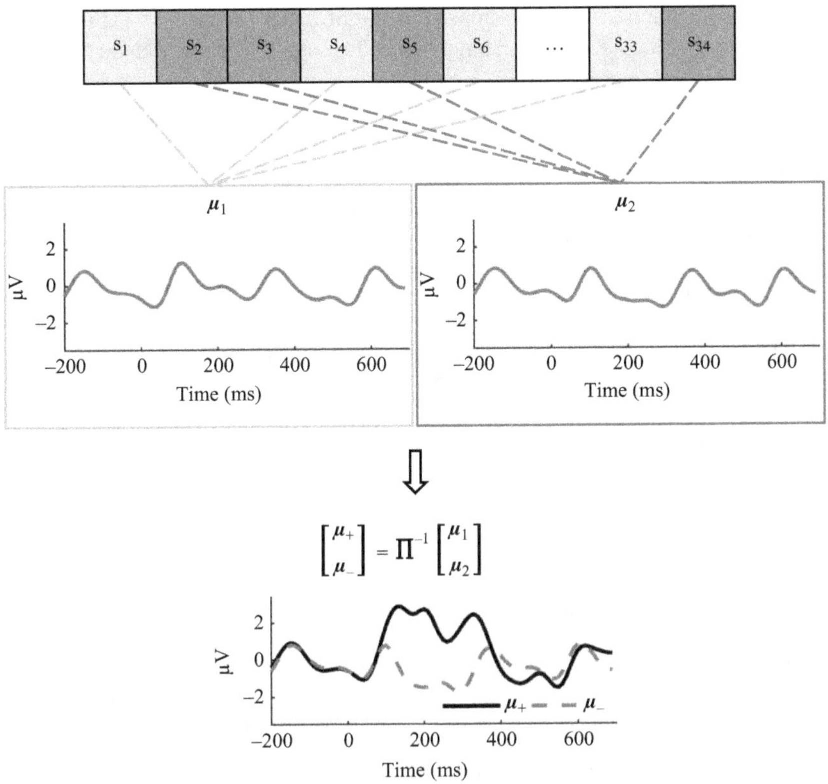

Figure 6.2 LLP decoding works by interleaving two oddball sequences with unknown targets but known target to non-target ratios. From the average response per sequence ($\boldsymbol{\mu}_1, \boldsymbol{\mu}_2$), the average target $\boldsymbol{\mu}_+$ and the average non-target response $\boldsymbol{\mu}_-$ can be estimated when the proportion Π are known for the individual sequences

Since we have control over the stimulus presentation sequence, we know which group the different ERP responses belong to. Thus, we can compute the average stimulus response within a group and use the second of the three steps above to estimate the class means. These can then be plugged into an LDA or least squares model to obtain the resulting weight vector $\boldsymbol{w}$. For each trial, the weight vector is then applied to each individual stimulus. Whenever a character was highlighted, its score is summed up and the character with the highest score is predicted (except when it is a # symbol).

6.5 Combining EM and LLP decoders analytically

The LLP and the EM solution for the weight vector $\boldsymbol{w}$ only differ in how they estimate the class-means. The estimation of LLP is guaranteed to converge to the correct solution, but in practice, the EM approach might learn faster. It is clear that both approaches to the mean estimation have their advantages. For this reason, Verhoeven *et al.* [33] proposed to combine these methods analytically by

$$\hat{\boldsymbol{\mu}}(\gamma) = (1-\gamma)\hat{\boldsymbol{\mu}}^{EM} + \gamma\hat{\boldsymbol{\mu}}^{LLP}.$$

Here, the mixing parameter γ is chosen to trade off the influence between LLP and EM. It is optimised to minimise the mean squared error between the true mean $\boldsymbol{\mu}$ and the mixture of the two estimators.

$$\gamma^* = \arg\min_{\gamma} E\left[\|\boldsymbol{\mu} - \hat{\boldsymbol{\mu}}(\gamma)\|^2\right] \tag{6.2}$$

Under the simplifying assumption that $E[\hat{\boldsymbol{\mu}}^{EM}] = \mu$, an optimal solution for γ can be computed as follows:

$$\gamma^* = \frac{1}{2}\left(\frac{\sum_d Var\left[\hat{\boldsymbol{\mu}}_d^{EM}\right] - \sum_d Var\left[\hat{\boldsymbol{\mu}}_d^{LLP}\right]}{\|\hat{\boldsymbol{\mu}}^{EM} - \hat{\boldsymbol{\mu}}^{LLP}\|^2} + 1\right), \tag{6.3}$$

with d indicating the feature dimension.

6.5.1 Training the MIX model

The training procedure of the decoder encompasses the basic mean estimation from the pure LLP and EM models, followed by analytically mixing these. Putting everything together, we obtain the following procedure for performing an update of the model:

1. Estimate the LLP means $\hat{\boldsymbol{\mu}}_+^{LLP}, \hat{\boldsymbol{\mu}}_-^{EM}$.
2. Estimate the EM means $\hat{\boldsymbol{\mu}}_+^{EM}, \hat{\boldsymbol{\mu}}_-^{EM}$.
3. Combine the means optimally using the following formula with γ computed as described above:

$$\hat{\boldsymbol{\mu}}_+(\gamma_+) = (1-\gamma_+)\hat{\boldsymbol{\mu}}_+^{EM} + \gamma_+\hat{\boldsymbol{\mu}}_+^{LLP} \tag{6.4}$$

$$\hat{\boldsymbol{\mu}}_-(\gamma_-) = (1-\gamma_-)\hat{\boldsymbol{\mu}}_-^{EM} + \gamma_-\hat{\boldsymbol{\mu}}_-^{LLP} \tag{6.5}$$

4. Compute the weight vector:

$$\hat{\boldsymbol{w}}_{u+1} = \Sigma_X^{-1}(\hat{\boldsymbol{\mu}}_+ - \hat{\boldsymbol{\mu}}_-)$$

with Σ_X denoting the estimate of the data covariance (optionally using Ledoit–Wolf shrinkage for regularisation [15]).

6.6 Experimental setup

6.6.1 Data

In our experiments, we will compare all methods on the same data obtained during an LLP decoding experiment. The data is available at the Zenodo database (DOI: http://doi.org/10.5281/zenodo.192684) and is described in detail in [32]. The study was approved by the ethics committee of the University Medical Center Freiburg and subjects were requested written informed consent prior to the start of the experiment. EEG signals were recorded at 1 kHz sampling rate by a BrainAmp EEG system using 31 passive Ag/AgCl electrodes (EasyCap) placed according to the extended 10–20 system. The data was referenced against the nose and the ground was at channel position AFz. The data set contains recordings from 13 subjects (five female, average age: 26 years).

During an online spelling task, each of the subjects spelled a 63-symbol sentence three times with the visual ERP speller. The sentence was kept fixed throughout the experiment. The following parameter settings were used for the paradigm. The stimulus duration was 100 ms and the interstimulus interval was 150 ms. Per trial, 68 stimuli were recorded. The visual stimulation realised the optimised visual highlighting effect reported in [44] and combined a coloured grid overlay effect with rotation and brightness enhancement.

This stimulus presentations comprised two sequences with differing target to non-target ratios to allow for using LLP. The sequence ratios were designed as described in Section 6.5. These two sequences are interleaved randomly to reduce the chance of observing differing target or non-target class means for the different sequences. For efficiency reasons, each sequence presented to the user was randomly selected from a set of 100 pre-computed sequences. During these sequences, every symbol (apart from the uninformative hash symbols) was highlighted 16 times.

6.6.2 Data processing

In our experiments, we use the pre-recorded data to mimic a spelling run for all the different methods. The data is processed trial by trial and at the end of each trial a prediction has to be made.

Before feeding the data to a decoder it is band-pass filtered between 0.5 and 8 Hz using a third order Chebyshev Type II filter. Each epoch, which describes the effect of a single stimulus was limited in dimensionality by downsampling its data to 100 Hz. Then the ERP responses in the time window [−200, 700] ms relative to the stimulus presentation were used for feature extraction. For each epoch, the mean value of the [−200, 0] ms interval is subtracted to baseline the data. In contrast to the work in [32] the samples are averaged in every interval of length 100 ms, resulting in 9 features per channel or a total of 279 features per ERP epoch.

After the per-stimulus pre-processing, we take all the available data points, centre and whiten them before updating the classifier and making the predictions for the current trial with the respective methods.

6.6.3 Methods and hyperparameters

In our experiments, we compare the EM, LLP, MIX and a supervised decoding model abbreviated as SUP. The methods only differ in how the means are estimated. EM uses the EM approach, LLP the learning from label proportions approach, MIX uses an analytical mixture of EM and LLP, while SUP makes use of label information of the first ten symbols spelled in order to estimate the class means. This last method SUP is included as a reference. It marks the ideal case and will allow to judge the quality of the unsupervised approaches. To evaluate the single epoch classification accuracy we use the area under the receiver operating curve (AUC), which shows the true positive rate of the decoder versus the false positive rate at various classification threshold values. The AUC measures how well target stimuli can be discriminated from non-target stimuli. It is a more refined measure of classifier performance compared to the symbol spelling accuracy.

6.7 Results

In Figure 6.3, we show the AUC values after the 20th symbol has been spelled during the experiments by the four methods. For EM, LLP and MIX the decoder is trained without any label information at all and on the complete set of EEG epochs recorded during these first 20 trials. It is tested on the same set of EEG epochs. When working with these three unsupervised decoders, any label information is only used for the performance evaluation. For the SUP decoder, the situation is different. Here, the responses from the first ten trials and the corresponding labels are used to train the decoder in a supervised way. For SUP, the reported AUC values are derived from the classification of the test data, consisting of the second block of ten trials.

Two main observations can be made in Figure 6.3. The first one is that the EM and LLP decoders obtain comparable AUC levels on data from 20 symbols. The second observation is astonishing: the unsupervised MIX approach can on an average compete with the optimal supervised decoder SUP. In all except for two subjects, MIX performs on level or even better to SUP. This is a very remarkable achievement for an unsupervised decoding approach!

Investigating this closer, Figure 6.3 reveals that the MIX method outperforms the two methods it is composed of (EM and LLP) for every single subject. As this

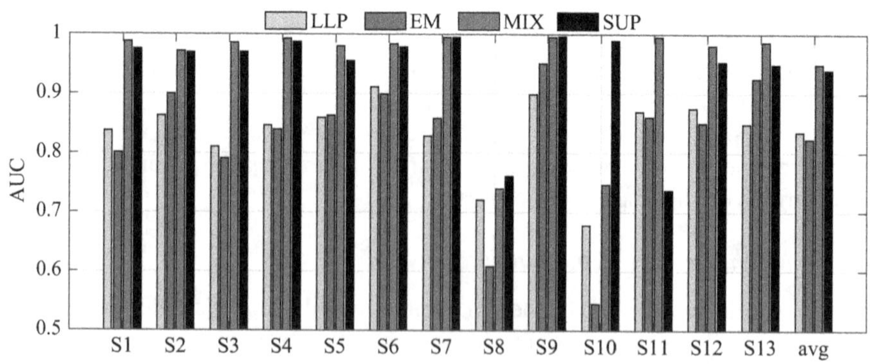

Figure 6.3 Comparison of the single epoch accuracy after the spelling of 20 symbols

mini-ensemble MIX for many trials/symbols is able to compensate for specific errors committed by either of the two single methods, it can be concluded that the learning methods EM and LLP either exploit different information contained in the data, or that the learning processes of EM and LLP must differ from each other. As all methods are provided with the exact same features, the latter reason is the more probable one.

To investigate the learning processes of the three unsupervised methods, we must move away from looking at the final accuracy and rather look at how the methods behave over time and with growing data. Figure 6.4 illustrates the evolution of the spelling accuracies obtained over the course of the complete sentence. As online spelling was performed only for LLP, we approximated the development of the EM and MIX method by simulations and provide the results of ten simulation runs performed under randomly initialised parameters. For each method, the development of the median AUC level is shown. In addition, the area between the 10% and 90% percentile is shaded. For each trial, the p-value of a Wilcoxon signed-rank test, comparing the EM result to the MIX result, is reported. The thin black line indicates the threshold for statistical significance ($p = 0.05$). The $\times$ symbol indicates the point where there is no more variation in one of the two methods, making the statistical comparison inapplicable. The result is illustrated in Figure 6.4(a) and (b) for the first spelling block recorded in subject S1 and S13, respectively. The trends observed here are shared across most subjects. Figure 6.4(c) shows the grand average result over all the 13 subjects and their three recorded blocks of data.

We make three observations. First, at early time points of the spelling when data is still scarce, the LLP decoder performs better than EM. A comparison of the shaded areas shows that the EM decoder varies strongly between re-initialisations. It is highly dependent on the random initialisation, making it less robust. Once a sufficient amount of data has been collected, however, the EM jumps to a high AUC level. This behaviour confirms the complementary characteristics of the two decoding methods. Second, the MIX method exploits this complementary learning behaviour of its two building blocks LLP and EM and obtains a classification performance at least as good as the better one of the two. Finally, the mixing of the EM decoder with the robust LLP decoder makes the MIX result far less dependent on the random initialisation of parameters. This is illustrated by the reduced variation (indicated by the shaded area) for MIX in Figure 6.4(a) and (b).

In Figure 6.5, we selected a single prototypical subject (S1) to visualise the actual mean estimation performance for LLP, EM and MIX. The top graph shows the chronological development of AUC levels obtained with the different decoders. For EM and MIX, the best and worst result are illustrated. Below the graph, the true and the estimated mean target responses after the spelling of the 22nd symbol are illustrated by scalp maps for the different estimation methods. Scalp maps in the first row demonstrate the supervised estimates obtained with label information. The LLP estimate approximates this ground truth but still has room for improvement. For EM, the estimation again shows the dependency upon the random initialisation of parameters. The resulting variance is overcome by the mixing of the EM estimator with LLP. Once again, the MIX method shows remarkable performance.

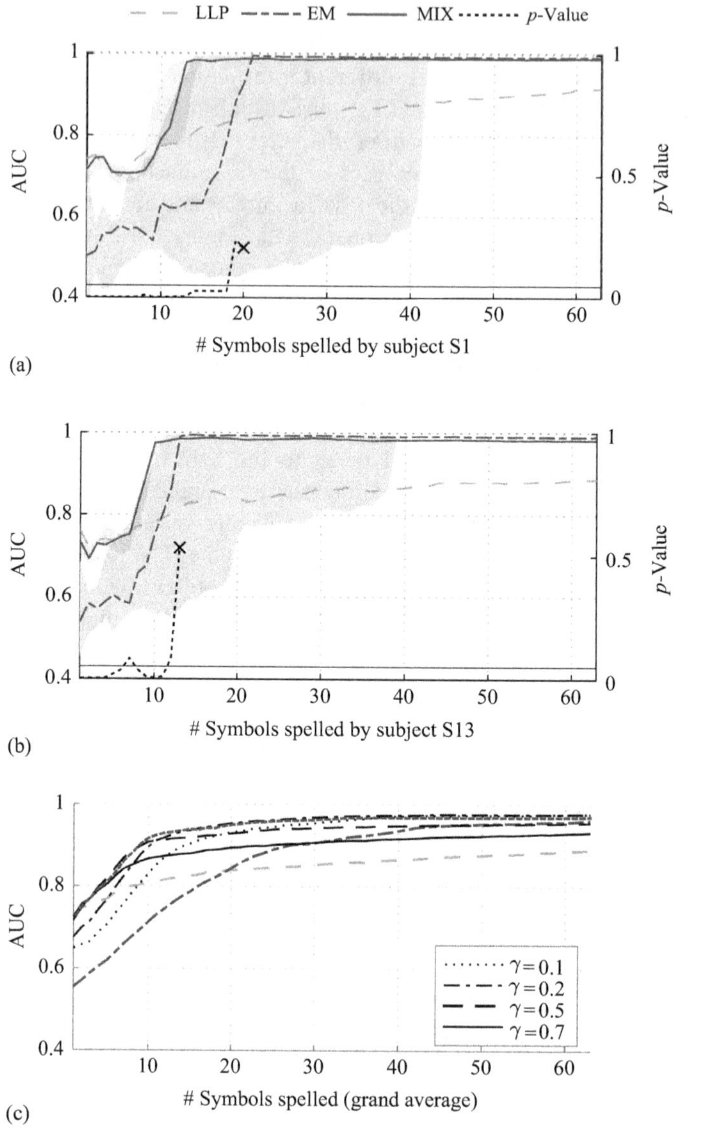

Figure 6.4 Evolution of the single epoch accuracy for two selected subjects (plots a and b) and the grand average (plot c) during the spelling of a complete sentence of 63 symbols with the three unsupervised decoding methods. In addition, we also show the performance of a mixture for predefined values of γ

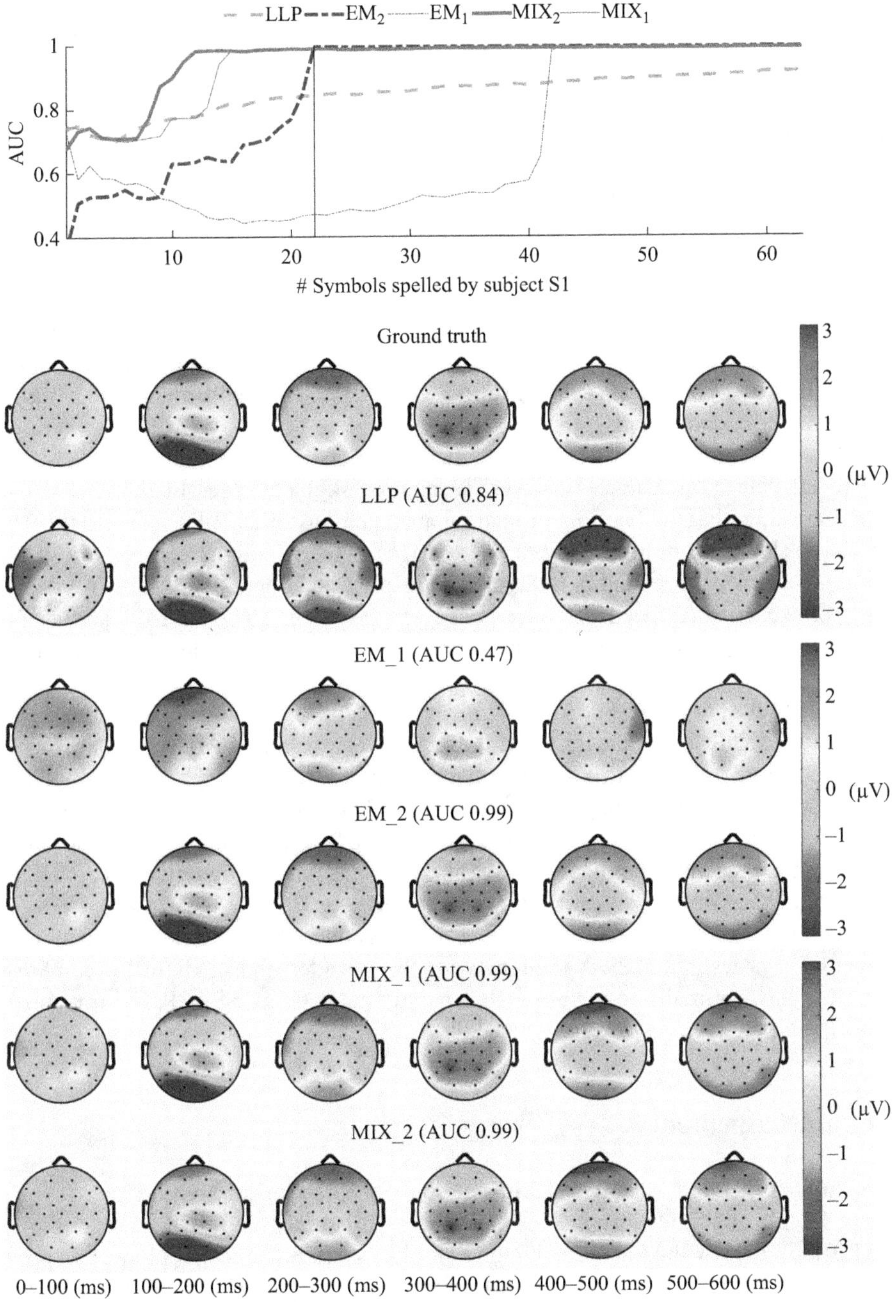

Figure 6.5 Subject S1: Mean target response patterns estimated at six different time intervals post stimulus. Besides the ground truth (top row), the estimate of the LLP (second row), two estimates of the EM method (worst in row three, best in row four) and two estimates of the MIX method (worst in row five, best in row six) are depicted

6.8 Conclusion

Unsupervised learning has become a viable alternative to supervised decoding in ERP paradigms. For the example of ERP spellers, we presented two basic approaches and a mixed approach. The basic unique approaches, EM and LLP, on an average perform comparably but have their individual strengths and weaknesses.

The EM decoder can be applied to most existing ERP paradigms without any change, works well empirically but comes without guarantees. Its key idea is to model the application directly: by encoding the structure of the BCI paradigm, the decoding problem becomes much easier and label information is no longer required. To increase the rather low reliability of the basic EM approach, the use of several EM model instances in parallel is advised, which of course comes at increased computational costs. However, even parallel initialisations cannot guarantee convergence to a high-quality solution in the end.

Our second approach LLP is guaranteed to obtain the optimal decoder when sufficient data is available. However, it requires slight changes of the ERP–BCI paradigm such that the learning problem is made easier for LLP. For the visual ERP speller we have exemplified how simple changes of the paradigm allow LLP to estimate the average target and non-target ERP responses without using explicit labels. As a drawback, the speed of convergence can be rather slow for LLP, and especially for subjects with strong signal-to-noise ratio the EM decoder can be a better choice.

Finally, we combined both approaches in an analytically optimal manner to profit from the individual advantages of LLP and EM. The resulting MIX decoder learns efficiently **and** reliably. Remarkably, it performs much better than the individual parts. Decoding performance of this unsupervised decoder is at about the same level as that of a supervised adaptive decoder would be. Consequently, we argue that using modern machine learning models the calibration sessions can be reliably skipped in ERP paradigms without a drawback in performance.

Future work should focus on taking the key concepts of this work: modelling structure, co-designing the paradigm with the decoder and analytically combining decoders optimally to different BCI paradigms such as SSVEP and imagined movement.

Acknowledgements

The authors DH and MT are thankful for support by BrainLinks-BrainTools, Cluster of Excellence funded by the German Research Foundation (DFG, grant number EXC1086). PJK by the European Union's Horizon 2020 research and innovation programme under the Marie Sklodowska-Curie grant agreement No 657679; TV by the Special Research Fund from Ghent University and KRM by the BK21 programme funded by Korean National Research Foundation grant No. 2012-005741.

References

[1] Kübler A, Neumann N, Kaiser J, Kotchoubey B, Hinterberger T, Birbaumer NP. Brain–computer communication: self-regulation of slow cortical potentials for verbal communication. Archives of Physical Medicine and Rehabilitation. 2001;82(11):1533–1539.

[2] Blankertz B, Curio G, Müller KR. Classifying single trial EEG: towards brain computer interfacing. In: Advances in neural information processing systems; 2002. p. 157–164.

[3] Schröder M, Bogdan M, Rosenstiel W, Hinterberger T, Birbaumer N. Automated EEG feature selection for brain computer interfaces. In: Proceedings of the first int. IEEE EMBS conf. on neural engineering; 2003. p. 626–629.

[4] Schröder M, Lal TN, Hinterberger T, *et al.* Robust EEG channel selection across subjects for brain computer interfaces. EURASIP Journal of Applied Signal Processing. 2005;19:3103–3112.

[5] Blankertz B, Dornhege G, Krauledat M, Müller KR, Curio G. The non-invasive Berlin brain–computer interface: fast acquisition of effective performance in untrained subjects. NeuroImage. 2007;37(2):539–550.

[6] Blankertz B, Tomioka R, Lemm S, Kawanabe M, Müller KR. Optimizing spatial filters for robust EEG single-trial analysis. IEEE Signal processing magazine. 2008;25(1):41–56.

[7] Lemm S, Blankertz B, Dickhaus T, Müller KR. Introduction to machine learning for brain imaging. NeuroImage. 2011;56(2):387–399.

[8] Dornhege G, Blankertz B, Curio G, Müller KR. Increase information transfer rates in BCI by CSP extension to multi-class. In: Advances in neural information processing systems; 2004. p. 733–740.

[9] Höhne J, Holz E, Staiger-Sälzer P, Müller KR, Kübler A, Tangermann M. Motor imagery for severely motor-impaired patients: evidence for brain–computer interfacing as superior control solution. PLoS ONE. 2014;9(8): e104854.

[10] Wolpaw JR, Birbaumer N, McFarland DJ, Pfurtscheller G, Vaughan TM. Brain–computer interfaces for communication and control. Clinical Neurophysiology. 2002;113(6):767–791.

[11] Krauledat M, Tangermann M, Blankertz B, Müller KR. Towards zero training for brain–computer interfacing. PLoS ONE. 2008;3(8):e2967.

[12] Shenoy P, Krauledat M, Blankertz B, Rao RP, Müller KR. Towards adaptive classification for BCI. Journal of Neural Engineering. 2006;3(1):R13.

[13] Sugiyama M, Krauledat M, Müller KR. Covariate shift adaptation by importance weighted cross validation. Journal of Machine Learning Research. 2007;8:1027–1061.

[14] Kindermans PJ, Schreuder M, Schrauwen B, Müller KR, Tangermann M. True zero-training brain–computer interfacing – an online study. PLoS ONE. 2014;9(7):e102504.

[15] Blankertz B, Lemm S, Treder M, Haufe S, Müller KR. Single-trial analysis and classification of ERP components, a tutorial. NeuroImage. 2011;56(2): 814–825.

[16] Vidaurre C, Krämer N, Blankertz B, Schlögl A. Time domain parameters as a feature for EEG-based brain–computer interfaces. Neural Networks. 2009;22(9):1313–1319.

[17] Fazli S, Popescu F, Danóczy M, Blankertz B, Müller KR, Grozea C. Subject-independent mental state classification in single trials. Neural Networks: The Official Journal of the International Neural Network Society. 2009 Jun;22(9):1305–1312.

[18] Lotte F. Signal processing approaches to minimize or suppress calibration time in oscillatory activity-based brain–computer interfaces. Proceedings of the IEEE. 2015;103(6):871–890.

[19] Fazli S, Dähne S, Samek W, Bießmann F, Müller KR. Learning from more than one data source: data fusion techniques for sensorimotor rhythm-based brain–computer interfaces. Proceedings of the IEEE. 2015;103(6):891–906.

[20] Jayaram V, Alamgir M, Altun Y, Schölkopf B, Grosse-Wentrup M. Transfer learning in brain–computer interfaces. IEEE Computational Intelligence Magazine. 2016;11(1):20–31.

[21] Lu S, Guan C, Zhang H. Unsupervised brain computer interface based on intersubject information and online adaptation. IEEE Transactions on Neural Systems and Rehabilitation Engineering. 2009;17(2):135–145.

[22] Panicker RC, Puthusserypady S, Sun Y. Adaptation in P300 brain–computer interfaces: A two-classifier cotraining approach. IEEE Transactions on Biomedical Engineering. 2010;57(12):2927–2935.

[23] Chavarriaga R, Ferrez PW, Millán JdR. To err is human: learning from error potentials in brain–computer interfaces. In: Advances in cognitive neurodynamics ICCN 2007; 2008. p. 777–782.

[24] Spüler M, Rosenstiel W, Bogdan M. Online adaptation of a c-VEP brain–computer interface (BCI) based on error-related potentials and unsupervised learning. PLoS ONE. 2012;7(12):e51077.

[25] Grizou J, Iturrate I, Montesano L, Oudeyer PY, Lopes M. Calibration-free BCI based control. In: AAAI; 2014. p. 1213–1220.

[26] Zeyl T, Yin E, Keightley M, Chau T. Partially supervised P300 speller adaptation for eventual stimulus timing optimization: target confidence is superior to error-related potential score as an uncertain label. Journal of Neural Engineering. 2016;13(2):026008.

[27] Vidaurre C, Kawanabe M, von Bünau P, Blankertz B, Müller KR. Toward unsupervised adaptation of LDA for brain–computer interfaces. IEEE Transactions on Biomedical Engineering. 2011;58(3):587–597.

[28] Millán JdR, Rupp R, Müller-Putz GR, *et al.* Combining brain–computer interfaces and assistive technologies: state-of-the-art and challenges. Frontiers in Neuroscience. 2010;4:161.

[29] Nicolas-Alonso LF, Gomez-Gil J. Brain computer interfaces, a review. Sensors. 2012;12(2):1211–1279.

[30] Kindermans PJ, Verstraeten D, Schrauwen B. A Bayesian model for exploiting application constraints to enable unsupervised training of a P300-based BCI. PLoS ONE. 2012;7(4):e33758.

[31] Kindermans PJ, Tangermann M, Müller KR, Schrauwen B. Integrating dynamic stopping, transfer learning and language models in an adaptive zero-training ERP speller. Journal of Neural Engineering. 2014;11(3):035005.

[32] Hübner D, Verhoeven T, Schmid K, Müller KR, Tangermann M, Kindermans PJ. Learning from label proportions in brain–computer interfaces: online unsupervised learning with guarantees. PLoS ONE. 2017;12(4):e0175856.

[33] Verhoeven T, Hübner D, Tangermann M, Müller KR, Dambre J, Kindermans PJ. Improving zero-training brain–computer interfaces by mixing model estimators. Journal of Neural Engineering. 2017;14(3):036021.

[34] Hübner D, Verhoeven T, Müller KR, Kindermans PJ, Tangermann M. Unsupervised learning for brain–computer interfaces based on event-related potentials: review and online comparison. IEEE Computational Intelligence Magazine. 2018;13(2):66–77.

[35] Polich J. Updating P300: an integrative theory of P3a and P3b. Clinical Neurophysiology. 2007;118(10):2128–2148.

[36] Fazel-Rezai R, Allison BZ, Guger C, Sellers EW, Kleih SC, Kübler A. P300 brain computer interface: current challenges and emerging trends. Frontiers in Neuroengineering. 2012;5:14.

[37] Farwell LA, Donchin E. Talking off the top of your head: toward a mental prosthesis utilizing event-related brain potentials. Electroencephalography and Clinical Neurophysiology. 1988;70(6):510–523.

[38] Townsend G, LaPallo B, Boulay C, *et al.* A novel P300-based brain–computer interface stimulus presentation paradigm: moving beyond rows and columns. Clinical Neurophysiology. 2010;121(7):1109–1120.

[39] Dempster AP, Laird NM, Rubin DB. Maximum likelihood from incomplete data via the EM algorithm. Journal of the Royal Statistical Society Series B (Methodological). 1977;39(1):1–38.

[40] Ledoit O, Wolf M. A well-conditioned estimator for large-dimensional covariance matrices. Journal of Multivariate Analysis. 2004;88(2):365–411.

[41] Quadrianto N, Smola AJ, Caetano TS, Le QV. Estimating labels from label proportions. Journal of Machine Learning Research. 2009 Oct;10:2349–2374.

[42] Patrini G, Nock R, Caetano T, Rivera P. (Almost) No Label No Cry. In: Ghahramani Z, Welling M, Cortes C, Lawrence ND, Weinberger KQ, editors. Montréal, Canada: Advances in Neural Information Processing Systems 27. Curran Associates, Inc.; 2014. p. 190–198. Available from http://papers.nips.cc/book/advances-in-neural-information-processing-systems-27-2014.

[43] Verhoeven T, Buteneers P, Wiersema J, Dambre J, Kindermans P. Towards a symbiotic brain–computer interface: exploring the application–decoder interaction. Journal of Neural Engineering. 2015;12(6):066027.

[44] Tangermann M, Schreuder M, Dähne S, *et al.* Optimized stimulation events for a visual ERP BCI. International Journal of Bioelectromagnetism. 2011;13(3):119–120.

Chapter 7

Covariate shift detection-based nonstationary adaptation in motor-imagery-based brain–computer interface

Haider Raza[1] and Dheeraj Rathee[2]

Abstract

Nonstationary learning refers to the process that can learn patterns from data, adapt to shifts, and improve performance of the system with its experience while operating in the nonstationary environments (NSEs). Covariate shift (CS) presents a major challenge during data processing within NSEs wherein the input-data distribution shifts during transitioning from training to testing phase. CS is one of the fundamental issues in electroencephalogram (EEG)-based brain–computer interface (BCI) systems and can be often observed during multiple trials of EEG data recorded over different sessions. Thus, conventional learning algorithms struggle to accommodate these CSs in streaming EEG data resulting in low performance (in terms of classification accuracy) of motor imagery (MI)-related BCI systems. This chapter aims to introduce a novel framework for nonstationary adaptation in MI-related BCI system based on CS detection applied to the temporal and spatial filtered features extracted from raw EEG signals. The chapter collectively provides an efficient method for accounting nonstationarity in EEG data during learning in NSEs.

7.1 Introduction

In electroencephalogram (EEG)-based brain–computer interface (BCI) systems, the majority of the learning algorithms assume, either implicitly or explicitly, that the EEG data have statistically stationary/fixed distributions over different sessions and/or runs of the recording [1]. However, such an assumption is simply not true as the EEG data obtained over different sessions and/or runs possess nonstationary characteristics [2,3]. In real-world BCI applications, nonstationarity is a ubiquitous phenomenon,

[1]Institute for Analytics and Data Science, School of Computer Science and Electronics Engineering, University of Essex, United Kingdom
[2]Intelligent Systems Research Centre, School of Computing, Engineering & Intelligent Systems, Ulster University, United Kingdom

especially when the system interacts with the dynamic and evolving environments, e.g., driving robotic wheelchair or playing virtual games. Hence, developing machine-learning models that are optimized for nonstationary environments (NSEs) is of great need. Machine-learning methods are generally categorized into unsupervised, semi-supervised, and supervised learning methods, whereas adaption to nonstationarity is a common practice to all these three categories. Furthermore, the nonstationary learning (NSL) methods are majorly grouped into passive and active approaches. In the passive approach to NSL, it is assumed that the input distribution should be continuously shifting over time. Thus, passive approach-based NSL methods adapt to new data distributions in continuous fashion, i.e., the model is updated for every new input observation or new batch of input observations from the streaming data. In contrast, an active approach-based NSL method uses a shift detection test to detect the presence of shifts in the streaming data, and correspondingly, based upon the time of detected shift, an adaptive action is initiated.

A BCI system aims to provide an alternative means of communication or rehabilitation for the physically challenged population so as to allow them to express their intentions using brain signal modulations without explicit muscular exertion [4]. Noninvasive EEG-based BCI systems acquire neural signals at scalp level, analyze them to evaluate specific features of EEG activity that are related to voluntary imagery/execution tasks, and finally utilize the outcomes as control signals that are further relayed to an output device [5]. Such a system operates typically in two phases, namely, the training phase and the evaluation (testing) phase [6]. The EEG signals are acquired through a multichannel EEG amplifier, and a preprocessing step is performed to reduce noise and enhances the signal-to-noise ratio. In the next step, the discriminable features are extracted from the artifact-cleaned signals using feature extraction techniques, such as spatial filtering [e.g., common spatial pattern (CSP), Laplacian filtering, current source density] [7–9]. Further, the optimal features are employed for training the classifier model. With an EEG-based BCI system that operates online in real-time nonstationary/changing environments, it is required to consider the input features that are invariant to shifts of the data or the learning approaches that can track the changes repeating over time. However, due to the nonstationary nature of the brain response characteristics in the EEG signal, it may be difficult to classify the EEG patterns reliably using a conventional classification algorithms used in motor imagery (MI)-related BCI systems [1,6]. The nonstationarities of the EEG signals may be caused by various reasons, such as changes in the user attention level, fatigue, imagery preferences, and/or placement and aging of electrodes [2,10–11]. These nonstationarities cause notable variations or shifts in the EEG signals both during trial-to-trial and session-to-session transfers [11,12]. As a result, these variations may often appear as covariate shifts (CSs), wherein the input data distributions differ significantly between training and testing phases, while the conditional distribution remains the same [13–16]. The traditional BCI systems are built upon a conservative approach to NSL, i.e., passive approach, where the learning system is updated continuously, assuming that the environment is constantly shifting. Furthermore, both single and ensemble of classifier(s)-based learning methods were developed for these passive approaches to improve the MI detection performance. In contrast, an active approach provides a more intuitive option for NSL involving a CS detection (CSD) test to capture the presence of CSs in the streaming EEG features followed by an adaptive action based on the correctly detected CSs.

The aim of this chapter is to present a CSD-related nonstationary adaptation (CSD-NSA) algorithm for EEG-based BCI systems. Different from the existing methods, the CSD-NSA algorithm is an active approach to learning in NSEs, wherein a CSD test is applied to initiate adaptation by adding new information from the testing data to the existing training data and retrain the updated classifier. This active approach aims at better managing computational resources, by adapting the classifier only when it is necessary, i.e., once the data from a novel distribution have to be processed. Specifically, for the detection of CSs in the EEG features, we considered an exponential weighted moving average (EWMA) model-based CSD test that reacts to the CSs in nonstationary conditions [15]. At the point of shift detection, the classifier can be adapted either in supervised or unsupervised manner. To assess the performance of the proposed CSD-NSA algorithms (i.e., supervised and unsupervised), experimental evaluations have been performed on publicly available MI-related EEG dataset.

7.2 Background

7.2.1 Covariate shift in EEG signals

CS is a case where the conditional probability distribution remains the same, i.e., $(P_{train}(y|x) = P_{test}(y|x))$, whereas the input data distribution shifts, i.e., $(P_{train}(x) \neq P_{test}(x))$, while transitioning from the training to testing stage. Figure 7.1 illustrates a typical example of CS wherein scatter plots between first best and second best features for the ten overlapping frequency bands (FBs) i.e., [8–12], [10–14],…, [26–30] Hz in the train and test data of the subject A07 (BCI competition-IV dataset 2A) are presented. For each plot, two classes were illustrated in crosses and dots, class 1 and class 2, respectively. The ellipse for each class shows the feature space. The light color dashed lines represent the training hyperplane and bold dashed black line represent the testing hyperplane.

7.2.2 Adaptive learning methods in EEG-based BCI

To enhance the performance of MI-related BCI systems, a large variety of signal processing methods have been proposed to extract features in the temporal and spatial domains to manage the nonstationarity in EEG signals. In the temporal domain, band-power and bandpass-based filtering methods have been commonly used [12,16], whereas in the spatial domain, common averaging, current source density, and CSP-based features were widely explored for the detection of MI-related responses [7,17,18]. The issue of low classification accuracy with the existing BCI systems has been one of the main concerns in their rather low uptake among people with severe physical disability [17].

To tackle the issue of low performance of a BCI system due to the CSs, several adaptive learning algorithms have been proposed to devise adaptive BCI systems

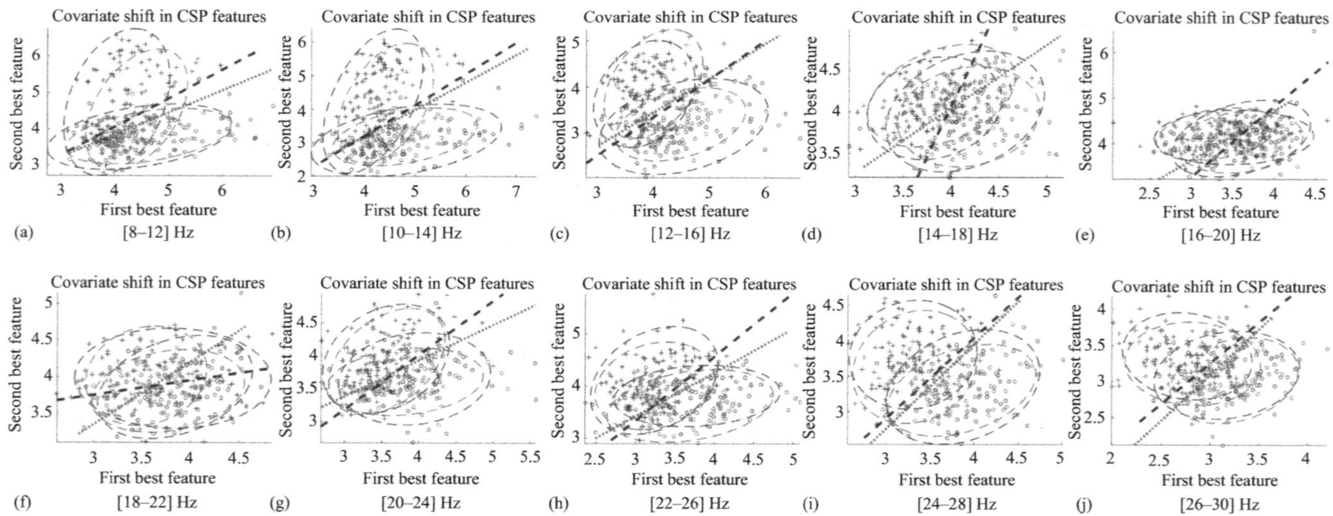

Figure 7.1 Covariate shift between the training and test distributions of subject A07 of BCI competition-IV dataset 2A. (a)–(j) For each plot, two classes were illustrated in crosses and dots, class 1 and class 2, respectively. The ellipse for each class shows the feature space. The light color dashed lines represent the training hyperplane and bold dashed black line represent the testing hyperplane

with encouraging results, such as Vidaurre *et al.* [19] have developed an adaptive classifier using an adaptive estimation of information matrix. Next, Shenoy *et al.* [1] have provided quantified systematic evidence of statistical differences in data recorded during multiple sessions and various adaptive schemes to enhance the BCI performance. Additionally, a CS minimization method was proposed for the NSA to reduce the feature set overlap and unbalance for different classes in the feature set domain [13]. More interestingly, Li *et al.* [10] have contributed a CS adaptation (CSA) method for the BCI system based on a density ratio estimation technique. Density ratio-based CSA is an unsupervised adaptation method that can adapt to the testing session without knowing the true labels. However, there exists a limitation that requires all the testing unlabeled data before starting the testing phase to estimate the importance for the nonstationarity adaptation, which makes this approach impractical in the real-time BCI systems for both communication and rehabilitation purposes.

Furthermore, since the last decade, ensemble-based machine-learning methods have become popular for NSL, where a set of classifiers is coupled to provide an overall decision. In the EEG-based BCI systems, ensemble learning-based methods have been evaluated to improve the classification performance (e.g., bagging, boosting, and random subspace [20]). Impressively, a dynamically weighted ensemble classification (DWEC) method has been proposed, wherein an ensemble of multiple classifiers is trained on cluster features to handle the issue of nonstationarity adaptation [21]. The DWEC method partitions the EEG data using clustering analysis and multiple classifiers were trained using different partitioned datasets. The final decision of the ensemble was then obtained by appropriately weighting the classification decisions of individual classifiers. In a recent study, the ensemble of CSP patches has shown a potential for improving the performance of online MI-related BCI system [22]. Both single and ensemble of classifier(s)-based approaches were developed based on the passive mechanism to improve the MI detection performance. In contrast, an active approach-based NSL method in BCI system could offer a possible solution that involves an active shift detection mechanism, which leads to an efficient adaptation to nonstationarity in the streaming EEG features.

7.3 Covariate shift detection-based nonstationary adaptation (CSD-NSA) algorithm

7.3.1 *Problem formulation*

Given a set of training samples $X^{Tr} = \{x_i^{tr}, y_i^{tr}\}$, where $i \in \{1 \dots n\}$ is the number of training data observations, $x_i^{tr} \in \mathbb{R}^d$ (d denotes the input data dimensionality) is a set of input training features drawn from a probability distribution with density $P_{tr}(x)$, and $y_i^{tr} \in \{C_1, C_2\}$ is a set of training labels, where $y_i^{tr} = C_1$, if x_i^{tr} belongs to class ω_1, and $y_i^{tr} = C_2$, if x_i^{tr} belongs to class ω_2. Here, we have assumed that the input data distribution remains stationary during the training phase. Furthermore, given an unlabeled test input data samples $X^{Ts} = \{x_i^{ts}\}$, where $i \in \{1 \dots m\}$ is the number of test data observations, $x_i^{ts} \in \mathbb{R}^d$ is a set of test input features, drawn independently from a probability distribution with density $P_{ts}(x)$. Note that $P_{tr}(x) \neq P_{ts}(x)$ in general, and thus, the input distributions may be different during the training and test phases, leading to CSs.

7.3.2 Covariate shift detection (CSD) test

The CSD test is an unsupervised method for detecting nonstationary changes in the unlabeled testing data (X^{Ts}) during the testing phase [23]. For this study, we estimated the CSD parameters [i.e., smoothing constant (λ) and control limit multiplier (L)] during the training phase with an assumption that the training data distribution is different from testing data distribution ($P_{tr}(x) \neq P_{ts}(x)$). Next, during the testing phase, an exponentially weighted moving average (EWMA) model was implemented for the detection of the CSs in the incoming data-stream (trial-based estimation). The EWMA is a type of infinite impulse response filter that applies weighting factors, which decrease exponentially. The weight of each older observation decreases exponentially, however, never reaching zero value. The weighting factor is one of the strengths of the EWMA model. The EWMA control chart overtakes other control charts because it pools the present and the past data together in such a way that even small shifts in the time-series can be detected more easily and quickly. Thus, the incoming observations are continuously examined to provide a 1-step-ahead prediction (1-SAP). Next, the 1-SAP error was plotted on the control chart and if the estimated error falls outside the control limits (L), the point was said to be a point of CS. The EWMA model can be presented as

$$z_{(i)} = \lambda x_{(i)} + (1 - \lambda) z_{(i-1)} \tag{7.1}$$

where $z_{(i)}$ is an EWMA statistics of the current trial, λ is a smoothing constant, which is selected based on minimizing 1-SAP error on the training dataset (X^{Tr}). The selection of the value of λ is a key issue in the CSD test. In particular, for the autocorrelated time series data, it is suggested to select a value of λ that minimizes the sum of the squares of the 1-SAP errors [24]. In the current approach, the value of λ was obtained by testing different values of λ in the range of $[0 : 0.01 : 1]$ on the training dataset. In particular, the CSD test further consists of two stages. In the first stage, it detects the CS in the first principal component extracted from a principal component analysis (PCA)-based EEG feature vector of each trial using the EWMA model and issue a CS warning (CSW). In case of positive CSW outcome, a CS validation (CSV) process is performed in the second stage in order to reduce the number of false alarms. During CSV stage, a multivariate two-sample Hotelling's T-Square statistical hypothesis test was performed to compare the two distinct samples with an equal number of observations generated before and at the CSW time point [25]. If the test rejects the null hypothesis, the existence of CS was confirmed during the CSV stage; otherwise, it was considered as a false alarm [23]. This CSD test has been successfully applied in our previous studies for the detection of CSs in EEG features for different CS adaptive methods [15].

Algorithm 1: Supervised CSD-NSA algorithm

Input : $X^{Tr} = \{x_i^{tr}, y_i^{tr}\}$, where $i \in \{1 \dots n\}$
: $X^{Ts} = \{x_i^{ts}\}$ where $i \in \{1 \dots m\}$
Output : Y^{Ts}

TRAINING:
$Tr(f_0, X^{Tr})$
Set $s = 0$, where s counts the number of covariate shift detection (CSD)
TEST:
Start evaluation using testing dataset X^{Ts}
Set $i = 1$
$\hat{y}_i = f(x_i)$
for $i = 2$ to m **do**
  $p \leftarrow CSD(x_i^{ts})$
  if ($p < 0.05$) **then**
    $s = s + 1$
    $X^{New} = \{(x_v^{ts}, y_v^{ts})\}_{v=1:l}$
    $X^{Tr} = (X^{Tr} \cup X^{New})$
    $Train(f_s, X^{Tr})$
  end if
  $\hat{y}_i = f(x_i)$
end for
return Y^{Ts}

7.3.3 Supervised CSD–NSA algorithm

The Supervised Covariate Shift Detection-Nonstationary Adaptation (S-CSD-NSA) algorithm combines the aforementioned CSD test and a supervised adaptation (SA) method as described in Algorithm 1. In the training phase, a support vector machine (SVM) classifier f_s was trained using the dataset X^{Tr}. The total number of detected and validated shifts was denoted by s, whereas in the training phase, s was set equal to 0, and the index of the current example is denoted by i. The next step involves the classification of the first trial of the testing data using classifier (f_s). Next, for each new unlabeled trial, the features were monitored to detect a CS using the CSD test. If the test provides positive outcome, then a CS was confirmed in the features of the current trial; otherwise, the trial was classified with the existing classifier f_s. In case of confirmed CS, the value of s was incremented by 1. The next step is to select and store the correctly predicted trials in a buffer $X^{New} = \{(x_v^{ts}, y_v^{ts})\}_{v=1:l}$, where l is the number of correctly predicted trials before the ith trial in which a CS was most recently detected. Afterward, the training dataset X^{Tr} was merged with X^{New} to update the training dataset. Also, a new classifier f_s was trained on the updated dataset and discard the old classifier f_{s-1}. As the initial classifier was obtained from the training phase, wherein, the initial value of s was set to zero. This procedure of single-trial EEG classification was repeated for each new incoming trial until all the m trials were classified in the testing phase.

7.3.4 Unsupervised CSD-NSA algorithm

The Unsupervised Covariate Shift Detection-Nonstationary Adaptation (U-CSD-NSA) algorithm combines the aforementioned CSD test and an unsupervised adaptation method using a transductive–inductive classifier as described in the Algorithm 2. The idea of the proposed U-CSD-NSA algorithm is to adapt to the nonstationary changes by using both the training dataset and the new knowledge obtained in unsupervised mode from the testing phase. The transductive classifier was only used for adding new information to the existing training dataset and the inductive classifier was used for predicting the BCI outputs, after being retrained each time the CS was detected. It is thus a learning approach wherein the transductive and the inductive learning approaches were combined to update the training dataset and to adapt to the evolution of CS over the time period in the feature set of the testing phase. The transductive learning was implemented using a probabilistic weighted k-nearest neighbor (PWkNN) method (i.e., instance-based learning). The output from the transductive method was used to determine if a trial and its corresponding estimated label can be added to the training dataset and subsequently, the learning model was updated. Transductive learning combines induction and deduction in a single step and is related to the field of semisupervised learning (SSL), which uses both labeled and unlabeled data during learning [26]. By eliminating the need to construct a global model, transductive learning offers prospect to achieve higher accuracy. However, in order to make use of unlabeled data, it is necessary to assume some structure to the underlying distribution of the data. Additionally, it is essential that the SSL approach must satisfy at least one of the following assumptions such as smoothness, cluster, or manifold assumption [26, 27]. U-CSA-NDA algorithm makes use of the smoothness assumption (i.e., the points which are closest to each other are more likely to share the same label) to implement a transductive learning algorithm. The second classifier, i.e., a linear SVM classifier (f), was inductive and its outputs were used to determine the single-trial BCI outputs.

7.3.4.1 Probabilistic k-nearest neighbor

Probability theory plays a vital role in solving several of pattern recognition problems, as majority of these problems can be solved using a density estimation technique [27]. The task involved in this is to model a probability density function $P(x)$ of a random variable X, given a training input data X_{Tr}. There are two approaches for the density estimation, namely, parametric and nonparametric. One of the key limitations of the parametric approach is that it assumes a precise practical form for the distribution which may be incompatible for a specific application. A nonparametric density estimation is an alternative approach that estimates density function without applying any assumption about underlying data distribution. Here, we considered a nonparametric approach based on k-nearest neighbors (kNN); being a transductive learning methods, it uses the test data point to determine a decision. In the kNN algorithm, we considered a small sphere centered at the point x where the density $P(x)$ should be estimated. We allowed the radius of the sphere to grow until it contains k data points, and the estimate of the density is then given by

$$P(x) = \frac{k}{(N \cdot V)} \tag{7.2}$$

Algorithm 2: Unsupervised CSD-NSA algorithm

Input : $X^{Tr} = \{x_i^{tr}, y_i^{tr}\}$, where $i \in \{1 \dots n\}$
: $X^{Ts} = \{x_i^{ts}\}$ where $i \in \{1 \dots m\}$
Output : Y^{Ts}

TRAINING:
$Tr(f_0, X^{Tr})$
Set $s = 0$, where s counts the number of covariate shift detection (CSD)
Set λ by minimizing 1-step-ahead-prediction error on the training dataset (X^{Tr})
TEST:
Start evaluation using testing dataset X^{Ts}
Set $i = 1, sl(0) = 1$
$\hat{y}_i = f(x_i)$
for $i = 2$ to m **do**
  $p \leftarrow CSD(x_i^{Ts}, \lambda)$
  if ($p < 0.05$) **then**
    $s = s + 1$
    $sl(s) = i$
    $X^{New} = \{x_v^{ts}\}_{v=sl(s-1):sl(s)}$
    $(CR, \hat{y}) = PWkNN(X_{Tr}, X_{New}, CR_\alpha, k)$
    $X^{Tr} = (X^{Tr} \cup X^{New})$
    $Train(f_s, X^{Tr})$
  end if
  $\hat{y}_i = f(x_i)$
end for
return Y^{Ts}

where V is set to the volume of the sphere, and N is the total number of points. The parameter k governs the degree of smoothing. The technique of kNN density estimation may be extended to the classification task in which the kNN density estimation is obtained for each class and the Bayes' theorem is used to perform a classification task. Now, let us suppose that we have a dataset comprising N_{ω_i} points in the class ω_i within the set of classes ω, where $i \in 1, 2$, so that $\sum_{\omega_i} N_{\omega_i} = N$. If we wish to classify a new data point x, we draw a sphere centered on x containing precisely k points irrespective of their classes. Now, if this sphere has a volume V and contains $k_{(\omega_i)}$ from class ω_i as an estimate of the density associated with each class or likelihood can be obtained by

$$P(x|\omega_i) = \frac{k_{\omega_i}}{N_{\omega_i} \cdot V} \tag{7.3}$$

Similarly, the unconditional density is given by $P(x) = k/(N \cdot V)$. The class prior probability is given by

$$P(\omega_i) = \frac{N_{\omega_i}}{N} \tag{7.4}$$

Using the Bayes' theorem, we can obtain the posterior probability of the class membership:

$$P(\omega_i|x) = \frac{P(x|\omega_i)P(\omega_i)}{P(x)} = \frac{k_{\omega_i}}{k} \tag{7.5}$$

Further, we wish to minimize the probability of misclassification; this can be achieved by assigning the test point x to the class ω_i having the largest posterior probability, i.e., corresponding to the largest value of k_{ω_i}/k. Thus, to classify a new point, identify the k-nearest points from the training dataset and then assign the new point to the set having the largest number of representatives. This posterior probability is also known as the Bayesian belief or confidence ratio (CR). However, the overall estimate obtained by the kNN method may not be satisfactory because the resulting density is not a true probability density since its integral over all the samples space diverges [28]. Another drawback is that it considers only k points to build the density and each neighbor has an equal weight. An extension to the above kNN method is to assign the weight to each sample that depends on its distance to x. A radial basis function (RBF) kernel was used to obtain these weights. Using RBF Kernel, the nearest points have weights with the higher value than furthest points. A PWkNN approach based on an RBF kernel is thus proposed to devise the transductive classifier with $RBF_{(p,q)}$, which is given as

$$RBF_{(p,q)} = exp\left(-\frac{d^2_{(p,q)}}{2\sigma^2}\right) \tag{7.6}$$

where $d_{(p,q)}$ is the Euclidean distance from the unlabeled data point x_p to the labeled data point x_q is computed as given below:

$$d_{(p,q)} = \sqrt{\sum_{i=1}^{m} (x_p(i) - x_q(i))^2} \tag{7.7}$$

and $x(i)$ is the ith feature of x and m is the number of features. For binary detection, the confidence ratio of CR_{ω_i} of the class ω_i, for a data point x_p, is defined by

$$CR_{\omega_1} = \frac{\sum_{j=1}^{K} RBF_{(p,j)} \cdot (l_j == \omega_1)}{\sum_{j=1}^{K} RBF_{(p,j)}} \tag{7.8}$$

$$CR_{\omega_2} = 1 - CR_{\omega_1} \tag{7.9}$$

where $1 \leq j \leq k$ corresponds to the jth nearest neighbor of x_p. The outputs of PWkNN include the overall confidence of the decision, given by

$$CR = max(CR_{\omega_1}, CR_{\omega_2}) \tag{7.10}$$

and the output class $\widehat{y}$ is equals to 1 if x_p is assigned to ω_1 otherwise equals to 0.

7.4 Experimental validation of the CSD-NSA algorithms

7.4.1 EEG dataset

The BCI Competition-IV dataset 2*A* [29] comprised EEG signals acquired from nine healthy participants, namely [$A01 - A09$], during two sessions on separate days using a cue-based MI paradigm. Each session consists of 6 runs where each run comprised 48 trials (12 trials for each class). Thus, the complete study involved 576 trials from both sessions of the dataset. The total trial length is 7.5 s with variable intertrial duration. The data were acquired from 25 channels (22 EEG channels along with three monopolar EOG channels) with a sampling frequency of 250 Hz and bandpass filtered between 0.5 and 100 Hz (notch filter at 50 Hz). Reference and ground were placed at the left and right mastoid, respectively. Among the 22 EEG channels, 10 channels, which are responsible for capturing most of the MI-related activations, were selected for this study (i.e., channels: $C3$, $FC3$, $CP3$, $C5$, $C1$, $C4$, $FC4$, $CP4$, $C2$, and $C6$). The dataset consist of four different MI tasks: left hand (class 1), right hand (class 2), both feet (class 3), and tongue (class 4). Only the classes corresponding to the left hand and right hand were considered in the present study. The MI data from the session-I were used for training phase and the MI data from the session-II was used for testing phase.

7.4.2 Signal processing and feature extraction

Figure 7.2 illustrates the complete signal processing pipeline implemented in this study for CSD-based NSA in MI-related EEG patterns. The following steps have been executed: bandpass filtering, feature extraction (log variance of CSP), detection and validation of CSs, NSA in supervised or unsupervised manner, and finally the binary classification.

In the signal processing and feature extraction stage, a set of bandpass filters was used to decompose the EEG signals into different FBs by employing an eighth order, zero-phase forward and reverse bandpass Butterworth filter. A total of 10 bandpass filters (i.e., filter bank) with overlapping bandwidths, including [8–12], [10–14], [12–16], [14–18], [16–20], [18–22], [20–24], [22–26], [24–28], and [26–30] Hz, were used for temporal filtering of the data. Next, spatial filtering using CSP algorithm was performed to maximize the divergence of bandpass filtered signals under one class and minimize the divergence for the other class. In MI-related BCI systems, both physical and imaginary movements cause a growth of bounded neural rhythmic activity known as event-related synchronization/desynchronization (ERD/ERS). The CSP algorithm has been widely implemented for estimating spatial patterns for detecting ERD/ERS [21]. Each combination of the bandpass filter and extreme left and right components of the CSP filter provides the discriminative features that are specific to a particular frequency range. Next to CSP filtering, the discriminating features were extracted using a time window of 3 s after the cue onset so as to continue our further analysis on the MI-related features only. Finally, the obtained features from all FBs were merged to create the set of input features for a linear SVM classifier.

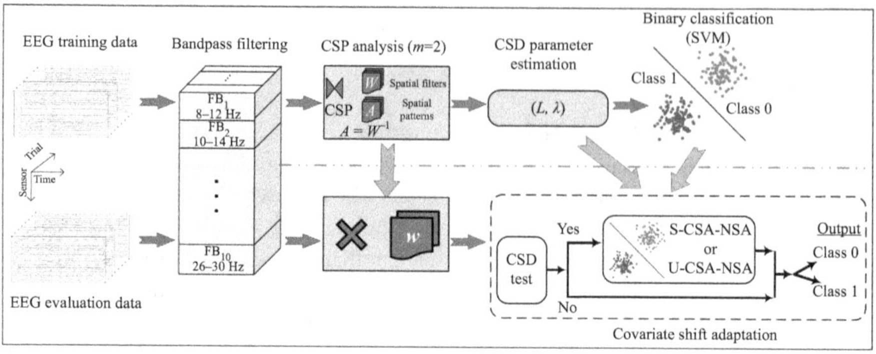

Figure 7.2 Schematic diagram of signal processing pipeline for implementation of CSD-NSA for MI-BCI system. The analysis consists of two phases. During the training phase, the features were extracted in the signal processing block from the filter-bank band-passed EEG signals using CSP algorithm and a classifier was trained on the labeled observation (i.e., XTrain). Also, two parameters [i.e., smoothing constant (λ) and control limit multiplier (L)] were estimated/set during this phase. In the testing phase, a similar signal processing approach was applied initially and CSP features were monitored by the CSA block. In the CSA block, the CSD test detects the presence of CSs and initiates adaptation using supervised or unsupervised adaptation. Finally, the classification was performed with the trained classifier model

7.4.3 Feature selection and parameter estimation

The parameters for the CSD test (i.e., λ and L) were computed during the training stage. The smoothing constant (λ) was selected by minimizing the sum of squared 1-SAP errors. The control limit multiplier L was set equal to 2. The value of L is needed to be carefully selected because it has a major impact in the performance of the CSD test. A small value of L makes the system more sensitive to minor shifts in EEG features. Moreover, the shifts in the feature set generated due to the noise were smoothed by proper selection of λ. In the testing phase, the CSD test was applied on the multivariate inputs features of EEG data using the estimated values of parameters. Due to the high dimensionality of the EEG features, the PCA algorithm was used to reduce the dimensionality to a single component [30]. Next, the CSD test was applied to the PCA output features for detecting CSs at the first stage of the CSD test. Based on the positive outcome of the CSD test in stage I, stage II gets activated. In stage II, a window of 3 s of CSP features after the cue onset in the current trial was extracted to use as a first sample and a window of averaged CSP features from the previous data trials was used as the second sample to execute the multivariate two-sample Hotelling's T-Square statistical hypothesis test. If the p-value of the Hotelling's T-Square test is less than 0.05, a CSD was confirmed in the current trial and an

Table 7.1 Performance of the covariate shift detection-based nonstationary adaptation (CSD-NSA) algorithm

Subjects	Baseline	S-CSD-NSA	U-CSD-NSA
A01	87.50	91.75	93.06
A02	58.33	58.33	59.03
A03	84.72	91.75	95.14
A04	63.89	68.33	71.53
A05	67.55	68.33	71.53
A06	62.50	62.50	62.50
A07	70.83	71.53	71.53
A08	86.11	91.75	91.75
A09	86.11	89.58	89.58
Mean	74.17	77.09	78.41
SD	11.84	13.92	14.00
p-Value		0.0156	0.0078

adaptive action was initiated. In U-CSD-NSA algorithm, the value of CR_{Thres} was set equal to 0.70, which means if the probability of the classification is more than 0.70, then only the example was added to the training dataset.

7.4.4 Empirical results

The performance of the CSD-NSA algorithm has been evaluated with active approach for CSA. With a single classifier, active approach was employed with both SA and unsupervised adaptation (i.e., S-CSD-NSA and U-CSD-NSA, respectively). In the S-CSD-NSA algorithm, the adaptation was achieved after detection of each shift, wherein the data from the correctly predicted trials were merged with the existing training dataset to enrich the data distribution, and the classifier was retrained on the updated dataset. Likewise, U-CSD-NSA algorithm has been implemented using the PWkNN method for the unsupervised adaptation, wherein the training dataset was updated at the instances of shift detection only.

We have compared the classification accuracies (in %) obtained by the two proposed algorithms with the baseline method (i.e., CSP features) for a binary classification task of MI-related BCI system. Table 7.1 provides the classification accuracies (in %) for the nine healthy participants along with the mean and SD. The average binary classification accuracy (mean±SD) for the baseline method, S-CSD-NSA, and U-CSD-NSA are 74.17 ± 11.84, 77.09 ± 13.92, and 78.41 ± 14.00, respectively. Thus, both CSD-NSA algorithms enhance the discriminability of the MI features as compared to the nonadaptive system. Furthermore, the Wilcoxon signed rank statistical test provided significant p-values for both methods, i.e., 0.0156 for S-CSD-NSA and 0.0078 for U-CSD-NSA in comparison with the baseline method.

7.5 Discussion and future prospects

Learning in NSEs provides a challenging and inspiring area of research in the fields of machine learning and computational intelligence and secured escalating interest of the researchers globally because of its increasing prevalence in real-world applications involving streaming and big data. EEG-based BCI systems involve learning of features from brain-generated electrical potentials which in turn provide a highly dynamic and complex environment. In such applications, using traditional approaches that either ignore the underlying shifts or their passive approach-based handling are inevitably bound to low performance. Thus, an active approach-based shift detection and subsequent adaptive measure is necessary to achieve high classification accuracies. In this chapter, we presented two algorithms for supervised and unsupervised adaptive learning in NSEs based on CSD and NSA frameworks. The results, with a benchmark MI-related BCI dataset, showed statistically significant improvement of the system performance.

The EWMA-CSD test is a good option for detecting CSs because it circumvents false detections resulting from noise or spurious shifts through much more intense smoothing of the EEG signal. A central issue in the CSD test is the selection of the value for λ and L. For the autocorrelated data, it is suggested to select a value of λ that minimizes the sum of the squares of the 1-SAP errors. Moreover, for L, considering smaller value ($L = 2$) results in concentrating on trivial shifts, such as temporary disturbance in user concentration and muscular artifacts emerging during trial-to-trial transfer. Conversely, the long-term CS may be handled by fixing a large value ($L = 3$). The value of parameter CR_{Thres} resolves importance of the current information. If the value of CR is above this threshold, then it is beneficial to be into the existing knowledge base (KB) otherwise rejected. The rejected instance may belong to a shifted distribution, but it is not providing higher confidence to be merged into the KB. Thus, the values of CR_{Thres} and L are needed to be cautiously selected in order to get greater classification accuracy, and these hyperparameters can be subject specific as well [31–33].

There are, however, a few limitations to be considered while using the active approach-based NSL algorithm for BCI systems and can be taken into account in future studies. First, the CSD test has been applied on the combined CSP features of multiple FBs, which creates a high-dimensional input vector and may affect the robustness of the CSD process. Furthermore, several other feature types have been utilized in MI-BCI systems and NSA for these feature sets may warrant further investigation of the proposed methods [8]. Second, we have used a supervised approach (i.e., S-CSD-NSA) for adaptation, which can be of limited usage in the case of real-time BCI-based applications such as for rehabilitation or communication, where the labels are not available during the testing phase. However, the unsupervised adaptation method to learn with the upcoming shifts in unlabeled streaming EEG data. Third, the proposed system has been implemented on the feature set of two sessions only, whereas practical BCI applications may have more number of sessions recorded on different days, and this condition can make the adaptive learning task further challenging. Moreover, single classifier based learning can be replaced by ensemble-based learning by recruiting several classifiers. In that case, a recurrent concept handling method may be needed to dynamically replace the old classifier with the updated classifier in the ensemble, and when the concept reappears, the old classifier will be reactivated. Fourth, datasets are assumed to be labeled when presented to a supervised algorithm or unlabeled for an unsupervised one. However, the data streams may contain a mixture of labeled and unlabeled data. Thus, one interesting prospect of further research is to examine the collective concept of supervised and unsupervised learning in NSEs. Fifth, one of the central issues with current BCI technology is the need to increase the number of classes for optimizing the practical application. Also, hybrid BCI systems can provide further flexibility to the system [34]. However, these enhancements generate data with different characteristics, such as multidimensionality, multiscale, and multilabel. Thus, future-learning algorithms should include new modeling and adaptive strategies to be able to cope with such data and conditions.

References

[1] Shenoy P, Krauledat M, Blankertz B, *et al.* Towards adaptive classification for BCI. Journal of Neural Engineering. 2006 Mar;3(1):13–23.

[2] Arvaneh M, Guan C, Ang KK, *et al.* Optimizing spatial filters by minimizing within-class dissimilarities in electroencephalogram-based brain-computer interface. IEEE Transactions on Neural Networks and Learning Systems. 2013;24(4):610–619.

[3] Arvaneh M, Guan C, Quek C. EEG data space adaptation to reduce intersession nonstationary in brain–computer interface. Journal of Neural Computation. 2013;25:1–26.

[4] Wolpaw JR, McFarland DJ, Neat GW, *et al.* An EEG-based brain–computer interface for cursor control. Electroencephalography and Clinical Neurophysiology. 1991 Mar;78(3):252–259.

[5] Wolpaw JR, Birbaumer N, McFarland DJ, *et al.* Brain–computer interfaces for communication and control. Clinical Neurophysiology: Official Journal of the International Federation of Clinical Neurophysiology. 2002 Jun;113(6): 767–791.

[6] Lotte F, Congedo M, Lécuyer A, *et al.* A review of classification algorithms for EEG-based brain–computer interfaces. Journal of Neural Engineering. 2007 Jun;4(2):1–13.

[7] Ramoser H, Müller-Gerking J, Pfurtscheller G. Optimal spatial filtering of single trial EEG during imagined hand movement. IEEE Transactions on Rehabilitation Engineering. 2000 Dec;8(4):441–446.

[8] Rathee D, Raza H, Prasad G, *et al.* Current source density estimation enhances the performance of motor-imagery related brain–computer interface. IEEE Transactions on Neural Systems and Rehabilitation Engineering. 2017;25(12):2461–2471.

[9] Rathee D, Cecotti H, Prasad G. Single-trial effective brain connectivity patterns enhance discriminability of mental imagery tasks. Journal of Neural Engineering. 2017;14(5):056005.

[10] Li Y, Kambara H, Koike Y, *et al.* Application of covariate shift adaptation techniques in brain–computer interfaces. IEEE Transactions on Biomedical Engineering. 2010 Jun;57(6):1318–1324.

[11] Raza H, Cecotti H, Li Y, *et al.* Learning with covariate shift-detection and adaptation in non-stationary environments: application to brain–computer interface. In: International Joint Conference on Neural Networks (IJCNN); 2015. p. 1–8.

[12] Blankertz B, Tomioka R, Lemm S, *et al.* Optimizing spatial filters for robust EEG single-trial analysis. IEEE Signal Processing Magazine. 2008;25(1): 41–56.

[13] Satti A, Guan C, Coyle D, *et al.* A covariate shift minimization method to alleviate non-stationarity effects for an adaptive brain–computer interface. In: Proceedings—International Conference on Pattern Recognition. IEEE; 2010. p. 105–108.

[14] Sugiyama M, Krauledat M, Müller KR. Covariate shift adaptation by importance weighted cross validation. Journal of Machine Learning Research. 2007;8:985–1005.

[15] Raza H, Prasad G, Li Y. EWMA based two-stage dataset shift-detection in non-stationary environments. In: IFIP Advances in Information and Communication Technology. vol. 412. Berlin Heidelberg: Springer; 2013. p. 625–635.

[16] Shahid S, Prasad G. Bispectrum-based feature extraction technique for devising a practical brain–computer interface. Journal of Neural Engineering. 2011 Apr;8(2):025014.

[17] Suk HI, Lee SW. A novel Bayesian framework for discriminative feature extraction in brain–computer interfaces. IEEE Transactions on Pattern Analysis and Machine Intelligence. 2013;35(2):286–299.

[18] Raza H, Cecotti H, Prasad G. Optimising frequency band selection with forward-addition and backward-elimination algorithms in EEG-based brain–computer interfaces. In: 2015 International Joint Conference on Neural Networks (IJCNN). IEEE; 2015. p. 1–7.

[19] Vidaurre C, Schlöogl A, Cabeza R, *et al.* A fully on-line adaptive BCI. IEEE Transactions on Biomedical Engineering. 2006;53(6):1214–1219.

[20] Sun S, Zhang C, Zhang D. An experimental evaluation of ensemble methods for EEG signal classification. Pattern Recognition Letters. 2007;28(15): 2157–2163.

[21] Liyanage SR, Guan C, Zhang H, *et al.* Dynamically weighted ensemble classification for non-stationary EEG processing. Journal of Neural Engineering. 2013 Apr;10(3):036007.

[22] Sannelli C, Vidaurre C, Müller KR, *et al.* Ensembles of adaptive spatial filters increase BCI performance: an online evaluation. Journal of Neural Engineering. 2016 May;13(4):046003.

[23] Raza H, Prasad G, Li Y. EWMA model based shift-detection methods for detecting covariate shifts in non-stationary environments. Pattern Recognition.

2015;48(3):659–669. Available from: http://dx.doi.org/10.1016/j.patcog.2014.07.028.

[24] Montgomery D, Mastrangelo C. Some statistical process control methods for autocorrelated data. Journal of Quality Technology. 1991;23(3):179–204.

[25] Kuncheva L, Faithfull W. PCA feature extraction for change detection in multidimensional unlabeled data. IEEE Transactions on Neural Networks and Learning Systems. 2014;25(1):69–80.

[26] Raza H, Prasad G, Li Y, *et al.* Toward transductive learning classifiers for non-stationary EEG. In: Engineering in Medicine and Biology Society (EMBC), 2014 35th Annual International Conference of the IEEE; 2014. p. 1.

[27] Sugiyama M, Kanamori T, Suzuki T, *et al.* Density-difference estimation. Neural Computation. 2013 Oct;25(10):2734–2775.

[28] Bishop CM. Pattern Recognition and Machine Learning. New York: Springer; 2006.

[29] Tangermann M, Müller KR, Aertsen A, *et al.* Review of the BCI competition IV. Frontiers in Neuroscience. 2012 Jan;6:55.

[30] Kolter JZ, Maloof MA. Dynamic weighted majority: a new ensemble method for tracking concept drift. In: Third IEEE International Conference on Data Mining. IEEE Comput. Soc; 2003. p. 123–130.

[31] Raza H, Cecotti H, Prasad G. A combination of transductive and inductive learning for handling non-stationarities in motor imagery classification. In: International Joint Conference on Neural Networks (IJCNN); 2016. p. 763–770.

[32] Chowdhury A, Raza H, Dutta A, *et al.* EEG-EMG based hybrid brain–computer interface for triggering hand exoskeleton for neuro-rehabilitation. In: Proceedings of the Advances in Robotics; 2017. p. 45:1–45:6.

[33] Chowdhury A, Raza H, Meena YK, *et al.* Online covariate shift detection based adaptive brain–computer interface to trigger hand exoskeleton feedback for neuro-rehabilitation. IEEE Transactions on Cognitive and Developmental Systems. 2017;(99):1. Available from https://doi.org/10.1109/TCDS.2017.2787040"10.1109/TCDS.2017.2787040.

[34] O'Doherty D, Meena YK, Raza H, *et al.* Exploring gaze-motor imagery hybrid brain–computer interface design. In: 2014 IEEE International Conference on Bioinformatics and Biomedicine (BIBM). IEEE; 2014. p. 335–339.

Chapter 8
A BCI challenge for the signal-processing community: considering the user in the loop

Fabien Lotte[1], Camille Jeunet[2,3], Jelena Mladenovic[1,5], Bernard N'Kaoua[4], and Léa Pillette[1]

Abstract

Electroencephalography (EEG)-based brain–computer interfaces (BCIs) have proven promising for a wide range of applications, from communication and control for motor impaired users to gaming targeted at the general public, real-time mental state monitoring and stroke rehabilitation, to name a few. Despite this promising potential, BCIs are still scarcely used outside laboratories for practical applications. The main reason preventing EEG-based BCIs from being widely used is arguably their poor usability, which is notably due to their low robustness and reliability. To operate a BCI, the user has to encode commands in his/her EEG signals, typically using mental imagery tasks, such as imagining hand movement or mental calculation. The execution of these tasks leads to specific EEG patterns, which the machine has to decode by using signal processing and machine learning. So far, to address the reliability issue of BCI, most research efforts have been focused on command decoding only. However, if the user is unable to encode commands in her EEG patterns, no signal-processing algorithm would be able to decode them. Therefore, we argue in this chapter that BCI design is not only a decoding challenge (i.e., translating EEG signals into control commands) but also a human–computer interaction challenge, which aims at ensuring the user can control the BCI.

Interestingly enough, there are a number of open challenges to take the user into account, for which signal-processing and machine-learning methods could provide solutions. These challenges notably concerns (1) the modeling of the user and (2) understanding and improving how and what the user is learning. More precisely, the BCI community should first work on user modeling, i.e., modeling and updating the

[1]Team Potioc, Inria/LaBRI (CNRS/Bordeaux INP/University of Bordeaux), France
[2]CNBI, EPFL, Switzerland
[3]Team Hybrid, University of Rennes, Inria, IRISA, CNRS, France
[4]Team Handicap, Activité, Cognition, Santé, University of Bordeaux, France
[5]Team Dycog, Inserm Lyon, CRNL, France

user's states and skills overtime from his/her EEG signals, behavior, BCI performances and possibly other sensors. The community should also identify new performance metrics—beyond classification accuracy—that could better describe users' skills at BCI control. Second, the BCI community has to understand how and what the user learns to control the BCI. This includes thoroughly identifying the features to be extracted and the classifier to be used to ensure the user's understanding of the feedback resulting from them, as well as how to present this feedback. Being able to update machine-learning parameters in a specific manner and a precise moment to favor learning without confusing the user with the ever-changeable feedback is another challenge. Finally, it is necessary to gain a clearer understanding of the reasons why mental commands are sometimes correctly decoded and sometimes not; what makes people sometimes fail at BCI control, in order to be able to guide them to do better. Overall, this chapter identifies a number of open and important challenges for the BCI community, at the user level, to which experts in machine learning and signal processing could contribute.

8.1 Introduction

Electroencephalography (EEG)-based brain–computer interfaces (BCIs) have proven promising for a wide range of applications, from communication and control for severely motor-impaired users, to gaming targeted at the general public, real-time mental state monitoring and stroke rehabilitation, to name a few [1,2]. Despite this promising potential, BCIs are still scarcely used outside laboratories for practical applications. The main reason preventing EEG-based BCIs from being widely used is arguably their poor usability, which is notably due to their low robustness and reliability. To operate a BCI, the user has to encode commands in his/her EEG signals, typically using mental imagery (MI) tasks, such as imagining hand movement or mental calculations. The execution of these tasks leads to specific EEG patterns, which the machine has to decode by using signal processing and machine learning. So far, to address the reliability issue of BCI, most research efforts have been focused on command decoding only. This present book contains numerous examples of advanced machine-learning and signal-processing techniques to robustly decode EEG signals, despite their low spatial resolution, their noisy and nonstationary nature. Such algorithms contributed a lot to make BCI systems more efficient and effective, and thus more usable.

However, if users are unable to encode commands in their EEG patterns, no signal-processing or machine-learning algorithm would be able to decode them. Therefore, we argue in this chapter that BCI design is not only a decoding challenge (i.e., translating EEG signals into control commands) but also a human–computer interaction challenge, which aims at ensuring the user can control the BCI. Indeed, BCI control has been shown to be a skill, which needs to be learned and mastered [3,4]. Recent research results have actually shown that the way BCI users are currently trained was suboptimal, both theoretically [5,6] and practically [7]. Moreover, the user is known to be one of the main causes of EEG signals variability in BCI, due to his/her change in mood, fatigue, attention, etc. [8,9].

Therefore, there are a number of open challenges to take the user into account during BCI design and training, for which signal-processing and machine-learning methods could provide solutions. These challenges notably concern (1) the modeling of the user and (2) understanding and improving how and what the user is learning.

More precisely, the BCI community should first work on user modeling, i.e., modeling and updating the user's mental states and skills overtime from their EEG signals, behavior, BCI performances and possibly other sensors. This would enable us to design individualized BCI, tailored for each user, and thus maximally efficient for each user. The community should also identify new performance metrics—beyond classification accuracy—that could better describe users' skills at BCI control.

Second, the BCI community has to understand how and what the user learns to control the BCI. This includes thoroughly identifying the features to be extracted and the classifier to be used to ensure the user's understanding of the feedback resulting from them, as well as how to present this feedback. Being able to update machine-learning parameters in a specific manner and a precise moment to favor learning without confusing the user with the ever-changeable feedback is another challenge. Finally, it is necessary to gain a clearer understanding of the reasons why mental commands are sometimes correctly decoded and sometimes not; what makes people sometimes fail at BCI control, in order to be able to guide them to do better.

Altogether, solving these challenges could have a substantial impact in improving BCI efficiency, effectiveness and user-experience, i.e., BCI usability. Therefore, this chapter aims at identifying and describing these various open and important challenges for the BCI community, at the user level, to which experts in machine learning and signal processing could contribute. It is organized as follows: Section 8.2 addresses challenges in BCI user modeling, while Section 8.3 targets the understanding and improvement of BCI user learning. For each section, we identify the corresponding challenges, the possible impact of solving them and first research directions to do so. Finally, the chapter summarizes these open challenges and possible solutions in Section 8.4.

8.2 Modeling the user

In order to be fully able to take the user into account into BCI design and training, the ideal solution would be to have a full model of the users, and in particular of the users' traits, e.g., cognitive abilities or personality, and states, e.g., current attention level or BCI skills at that stage of training. Signal-processing and machine-learning tools and research can contribute to these aspects by developing algorithms to estimate the users' mental states (e.g., workload) from EEG and other physiological signals, by estimating how well users can self-modulate their EEG signals, i.e., their BCI skills, and by dynamically modeling, using machine learning, all these aspects together. We detail these points below.

8.2.1 Estimating and tracking the user's mental states from multimodal sensors

Increase in the number of available low-cost sensors [10] and development in machine learning enables real-time assessment of some cognitive, affective and motivational processes influencing learning, such as attention for instance. Numerous types of applications are already taking advantage of these pieces of information, such as health [11], sport [12] or intelligent tutoring systems (ITS) [13]. Such states could thus be relevant to improve BCI learning as well.

Among the cognitive states influencing learning, attention deserves a particular care since it is necessary for memorization to occur [14]. It is a key factor in several models of instructional design, e.g., in the Attention, Relevance, Confidence, Satisfaction (ARCS) model where A stands for Attention [15]. Attention levels can be estimated in several ways. Based on the resource theory of Wickens, task performance is linked to the amount of attentional resources needed [16]. Therefore, performances may be used as a rough estimate of the level of attentional resources the user dedicates to the task. However, this metric also reflects several other mental processes and should thus be considered with care. Moreover, attention is a broad term that encompasses several types of concepts [17,18]. For example, focused attention refers to the amount of information that can be processed at a given time, whereas vigilance refers to the ability to pay attention to the apparition of an infrequent stimulus over a long period of time. Each type of attention has particular ways to be monitored, for example, vigilance can be detected using blood-flow velocity measured by transcranial Doppler sonography [19]. Focused visual attention, which refers to the selection of visual information to process, can be assessed by measuring eye movements [20]. While physiological sensors provide information about the physiological reactions associated with processes taking place in the central nervous system, neuroimaging has the advantage of recording information directly from the source [21]. EEG recordings enable to discriminate some types of attention with various levels of reliability given the method used. For instance, alpha band (7.5–12.5 Hz) can be used for the discrimination of several levels of attention [22], while the amplitude of event-related potentials (ERPs) are modulated by visual selective attention [23]. While specific experiments need to be carried out to specify the exact nature of the type(s) of attention involved in BCI training, a relationship between gamma power (30–70 Hz) in attentional network and mu rhythm-based BCI performance has already been shown by Grosse-Wentrup *et al.* [24,25]. Such linear correlation suggests the implication of focused attention and working memory (WM) [25] in BCI learning.

The WM load or workload is another cognitive factor of influence for learning [26,27]. It is related to the difficulty of the task, depends on the user's available resources and to the quantity of information given to the user. An optimal amount of load is reached when the user is challenged enough not to get bored and not too much compared with his abilities [28]. Behavioral measures of workload include accuracy and response time, when physiological measures comprise eye-movements [29], eye blinks [30], pupil dilatation [31] or galvanic skin response [32]. However, as most behavioral measures, these measures change due to WM load, but not only, making them unreliable to measure uniquely WM load. EEG is a more reliable measure of workload [33]. Gevins *et al.* [34] showed that WM load could be monitored using theta (4–7 Hz), alpha (8–12 Hz) and beta (13–30 Hz) bands from EEG data. Low amount of workload could be discriminated from high amount of workload in 27 s long epochs of EEG with a 98% accuracy using Joseph–Viglione's neural network algorithm [35,36]. Interestingly, they also obtained significant classification accuracies when training their network using data from another day (i.e., 95%), another person (i.e., 83%) and another task (i.e., 94%) than the data used for classification. Several experiments have since reported online (i.e., real time) classification rate ranging from 70% to 99% to distinguish between two types of workload [37,38]. The results depend greatly on the length of the signal epoch used: the longer the epoch, the better the performance [38,39]. The importance of monitoring WM in BCI applications is all the more important because BCI illiteracy is associated with high theta waves [40] which is an indicator of cognitive overload [41]. Finally, another brain-imaging modality can be used to estimate mental workload: functional Near-Infrared Spectroscopy (fNIRS). Indeed, it was shown that hemodynamic activity in the prefrontal cortex, as measured using fNIRS, could be used to discriminate various workload levels [42–44].

Learners' state assessment has mostly focused on cognitive components, such as the ones presented above, because learning has often been considered as information processing. However, affects also play a central role in learning [45]. For example, Isen [46] has shown that positive affective states facilitate problem solving. Emotions are often inferred using contextual data, performances and models describing the succession of affective states the learner goes through while learning. The model of Kort *et al.* [47] is an example of such model. Though physiological signals can also be used such as electromyogram, electrocardiogram, skin conductive resistance and blood volume pressure [48,49]. Arroyo *et al.* [50] developed a system composed of four different types of physiological sensors. Their results show that the facial recognition system was the most efficient and could predict more than 60% of the variance of the four emotional states. Several classification methods have been tried to classify EEG data and deduce the emotional state of the subject. Methods such as multilayer perceptron [51], K-nearest neighbor, linear discriminant analysis (LDA), fuzzy K-means or fuzzy C means were explored [52,53], using as input alpha, beta and gamma frequency bands power. The results are promising and vary around 75% accuracy for two to five types of emotions. Note, however, that the use of gamma band power features probably means that the classifiers were also using EMG activity due to different facial expressions. For emotion monitoring as well, fNIRS can prove useful. For instance, in [54], fNIRS was shown to be able to distinguish two classes of affective stimuli with different valence levels with average classification accuracies around 65%. Recognizing emotion represents a challenge because most of the studies rely on the assumption that people are accurate in recognizing their emotional state and that the emotional cues used have a similar and the intended effect on subject. Moreover, many brain structures involved into emotion are deep in the brain, e.g., the amygdala, and as such activity from these areas are often very weak or even invisible in EEG and fNIRS.

Motivation is interrelated with emotions [55,56]. It is often approximated using the performances [37]. Several EEG characteristics are modulated by the level of motivation. For example, this is the case for the delta rhythm (0.5–4 Hz) which could originate from the brain reward system [57]. Motivation is also known to modulate the amplitude of the P300 ERP and therefore increases performance with ERP-based BCI [58]. Both motivation and emotions play a major role in biofeedback learning [59–63] and in BCI performances [64,65].

Cognitive, affective and motivational states have a great impact on learning outcome and machine learning plays a key role in monitoring them. However, some challenges remain to be overcome, such as detecting and removing artifacts in real time. For example, facial expressions often occur due to change in mental states and may create artifacts polluting EEG data and for which real time removal still represents an issue. Limitations also arise from the number of different states we are able to differentiate. The quantity of data to train the classifier increases with the number of classes to differentiate. Future studies should also focus on the reliability and stability of the classification within and across individuals [66]. Indeed, classification accuracy, particularly online one, still needs to be improved. Furthermore, calibration of classifiers is often needed for each new subject or session which is time consuming and might impede the use of such technology on a larger scale. Finally, while several emotional states can be recognized from user's behavior, there is usually very limited overt behavior, e.g., movements or speech, during BCI use. Thus, future studies should try to differentiate more diverse emotional states, e.g., frustration, directly from EEG and physiological data.

8.2.2 Quantifying users' skills

As mentioned above, part of the user modeling consists in measuring the users' skills at BCI control. Performance measurement in BCI is an active research topic, and various metrics were proposed [67,68]. However, so far, the performance considered and measured was that of the whole BCI system. Therefore, such performance metrics reflected the combined performances of the signal-processing pipeline, the sensors, the user, the BCI interface and application, etc. Standard performance metrics used cannot quantify specifically and uniquely the BCI users' skills, i.e., how well the user can self-modulate their brain activity to control the BCI. This would be necessary to estimate how well the user is doing, where are their strengths and weaknesses, in order to provide optimal instructions, feedback, application interface and training exercises.

We recently proposed some new metrics to go in that direction, i.e., to estimate specifically users' skills at BCI control, independently of a given classifier [69]. In particular, we proposed to quantify the users' skills at BCI control, by estimating their EEG patterns distinctiveness between commands, and their stability. We notably used Riemannian geometry to quantify how far apart from each other the EEG patterns of each command are, as represented using EEG spatial covariance matrices, and how variable over trials these patterns are. We showed that such metrics could reveal clear user learning effects, i.e., improvements of the metrics over training runs, when classical metrics such as online classification accuracy often failed to do so [69].

This work thus stressed the need for new and dedicated measures of user skills and learning. The metrics we proposed are however only a first attempt at doing so, with more refined and specific metrics being still needed. For instance, our metrics can mostly quantify control over spatial EEG activity (EEG being represented using spatial covariance matrices). We also need metrics to quantify how much control the user has over their spectral EEG activity, as well as over their EEG temporal dynamics. Notably, it would seem useful to be able to quantify how fast, how long and how precisely a user can self-modulate their EEG activity, i.e., produce a specific EEG pattern at a given time and for a given duration. Moreover, such new metrics should be able to estimate successful voluntary self-regulation of EEG signals amidst noise and natural EEG variabilities and independently of a given EEG classifier. We also need metrics that are specific for a given mental task, to quantify how well the user can master this mental command, but also a single holistic measure summarizing their control abilities over multiple mental tasks (i.e., multiclass metrics), to easily compare users and give them adapted training and BCI systems. The signal-processing and machine-learning community should thus address all these open and difficult research problems by developing new tools to quantify the multiple aspects of BCI control skills.

8.2.3 Creating a dynamic model of the users' states and skills

8.2.3.1 A conceptual model of mental imagery BCI performance

In order to reach a better understanding of the user-training process, a model of the factors impacting MI-BCI skill acquisition is required. In other words, we need to understand which users' traits and states impact BCI performance, how these factors do interact and how to influence them through the experimental design or specific cognitive training procedures. We call such a model a cognitive model. Busemeyer and Diederich describe cognitive models as models which aim to scientifically explain one or more cognitive processes or how these processes interact [70]. Three main features characterize cognitive models: (1) their goal: they aim at explaining cognitive processes scientifically, (2) their format: they are described in a formal language, (3) their background: they are derived from basic principles of cognition [70]. Cognitive models guarantee the production of logically valid predictions, they allow precise quantitative predictions to be made and they enable generalization [70].

In the context of BCIs, developing a cognitive model is a huge challenge due to the complexity and imperfection of BCI systems. Indeed, BCIs suffer from many limitations, independent from human-learning aspects, that could explain users' modest performance. For instance, the sensors are often very sensitive to noise and do not enable the recording of high-quality brain signals, while the signal-processing algorithms sometimes fail to recognize the encoded mental command. But it is also a huge challenge due to the lack of literature on the topic and to the complexity and cost associated with BCI experiments that are necessary to increase the quantity of experimental data required to implement a complete and precise model [71].

Still, a cognitive model would enable us to reach a better understanding of the MI-BCI user-training process, and consequently to design adapted and adaptive training protocols. Additionally, it would enable BCI scientists to guide neurophysiological analyses by targeting the cognitive and neurophysiological processes involved in the task. Finally, it would make it possible to design classifiers robust to variabilities, i.e., able to adapt to the neurophysiological correlates of the factors included in the model. To summarize, building such a model, by gathering the research done by the whole BCI community, could potentially lead to substantial improvements in MI-BCI reliability and acceptability.

Different steps are required to build a cognitive model [70]. First, it requires a formal description of the cognitive process(es)/factors to be described based on conceptual theories. Next, since the conceptual theories are most likely incomplete, ad hoc assumptions should be made to complete the formal description of the targeted factors. Third, the parameters of the model, e.g., the probabilities associated with each factors included in the model, should be determined. Then, the predictions made by the model should be compared to empirical data. Finally, this process should be iterated to constrain and improve the relevance of the model.

By gathering the results of our experimental studies and of a review of the literature, we proposed a first formal description of the factors influencing MI-BCI performance [71]. We grouped these factors into three categories [72]. The first category is "task-specific," i.e., it includes factors related to the BCI paradigm considered. Here, as we focused on MI-based BCIs, this category gathers factors related to spatial abilities (SA), i.e., the ability to produce, transform and manipulate mental images [73]. Both the second and third categories include "task-unspecific" factors, or, in other words, factors that could potentially impact performance whatever the paradigm considered. More precisely, the second category includes motivational and cognitive factors, such as attention (state and trait) or engagement. These factors are likely to be modulated by the factors of the third category that are related to the technology-acceptance, i.e., to the way users perceive the BCI system. This last category includes different states such as the level of anxiety, self-efficacy, mastery confidence, perceived difficulty or the sense of agency.

The challenge is thus to modulate these factors to optimize the user's state and trait and thus increase the probability of a good BCI performance and/or of an efficient learning. In order to modulate these factors—that can be either states (e.g., motivation) or malleable traits (e.g., SA), one can act on specific effectors: design artifacts or cognitive activities/training.

The effectors we will introduce hereafter are mainly based on theoretical hypotheses. Their impact on the users' states, traits and performance are yet to be quantified. Thus, although these links make sense from a theoretical point of view, they should still be considered with caution. We determined three types of links between the factors and effectors. "Direct influence on user state": these effectors are suggested to influence the user's state and, consequently, are likely to have a direct impact on performance. For instance, proposing a positively biased feedback—making users believe they are doing better than what they really are—has been suggested to improve (novice) users' sense of agency (i.e., the feeling of being in control, see Section 8.3.3.1 for more details) [62]. "Help for users with a specific profile": these effectors could help users who have a specific profile and consequently improve their performance. For instance, proposing an emotional support has been suggested to benefit highly tensed/anxious users [74] (see Section 8.3.3.2 for more details). "Improved abilities": this link connects effectors of type cognitive activities/exercises to abilities (malleable traits) that could be improved thanks to these activities. For instance, attentional neurofeedback has been suggested to improve attentional abilities [75]. For more details, see [71].

This model has been built based on the literature related to MI-based BCIs (and mainly to motor-imagery based BCIs). It would be interesting to investigate the relevance of this model for other BCI paradigms, such as BCIs based on steady-state visual evoked potentials or BCIs based on P300. It is noteworthy that, for instance, motivation has already been shown to modulate P300 amplitude and performance [58]. The effect of mastery confidence (which is included in the "technology-acceptance factors" in our model) on P300-based BCI performance has also been investigated [76]. The results of this study were not conclusive, which led the authors to hypothesize that either this variable had no effect on performance or that they may not have succeeded to manipulate participants' mastery confidence. Further investigation is now required. Besides, the same authors proposed a model of BCI performance [77]. This model gathers physiological, anatomical and psychological factors. Once again, it is interesting to see that, while organized differently, similar factors were included in the model. To summarize, it would be relevant to further investigate the factors influencing performance in different BCI paradigms and then investigate to which extent some of these factors are common to all paradigms (i.e., task-unspecific), while determining which factors are specific to the paradigm/task. Then, the next step would be to propose a full and adaptive model of BCI performance.

Now, from a signal-processing and machine-learning point of view, many challenges are remaining. We should aim at determining some physiological or neurophysiological correlates of the factors included in this model in order to be able to estimate, in real time, the state of the BCI user. Therefore, the signal-processing community should design tools to recognize these neural correlates in real time, from noisy signals. Besides, the model itself requires machine-learning expertise to be implemented, as detailed in Section 8.2.3.2. Then, one of the main challenges will be to determine, for each user, based on the recorded signals and performance, when the training procedure should be adapted in order to optimize the performance and learning process. Machine-learning techniques could be used in order to determine, based on a pool of previous data (e.g., using case-based reasoning) and on theoretical knowledge (e.g., using rule-based reasoning), when to make the training procedure evolve. In the field of ITS, where the object is to adapt the training protocol dynamically to the state (e.g., level of skills) of the learner, a popular approach is to use multiarm bandit algorithms [78]. Such an approach could be adapted for BCI training. The evolution of the training procedure could be either continuous or divided into different steps, in which case it would be necessary to determine relevant thresholds on users' states values, from which the training procedures should evolve, e.g., to become more complex, to change the context and propose a variation of the training tasks, and to go back to a previous step that may have not been assimilated correctly.

8.2.3.2 A computational model for BCI adaptation

As discussed in previous sections, it is necessary to identify the psychological factors, user skills and traits which will determine a successful BCI performance. Co-adaptive BCIs, i.e., dynamically adaptive systems which adjust to signal variabilities during a BCI task, and in such way adapt to the user, while the user adapts to the machine via learning—showed tremendous improvement in the system performance ([79] for MI; [80] for P300). However, these techniques dwell mostly within the signal variabilities, by only adjusting to them, without acknowledging and possibly influencing the causes of such variabilities—human factors. These factors, once acknowledged, should be structured in a conceptual framework as in [8] in order to be properly influenced or to be adapted upon. In this framework for adaptive BCI methods, the human psychological factors are grouped by their degree of stability or changeability in time, e.g., skills could take multiple sessions (months) to change, while attention drops operate within short time periods. All these changes might have certain EEG signatures, thus considering the time necessary for these factors to change, the machine could be notified to adapt accordingly and could predict and prevent negative behavior. To influence user behavior, the framework contains a BCI task model, arranged within the same time scales as the user's factors. Consequently, if the user does not reach a certain minimal threshold of performance for one BCI task, the system would switch to another, e.g., if kinesthetic imagination of hand movements is worse, then tongue than it would switch to tongue MI. Additionally, if the user shows MI illiteracy, after a session, then the system would switch to other paradigms and so on. Hence, the task model represents the possible BCI tasks managed by the exploration/exploitation ratio to adapt to the users and optimally influence them, within the corresponding time scales. Once identified and modeled theoretically, we need to search for computational models generic enough which could encompass such complex and unstable behavior, and enable us to design adaptive BCIs, whose signal processing, training tasks and feedback are dynamically adapted to these factors.

Several behavioral sciences and neuroscience theories strive to explain the brain's cognitive abilities based on statistical principles. They assume that the nervous system maintains internal probabilistic models that are updated by neural processing of sensory information using Bayesian probability methods. Kenneth Craik suggested in 1943 that the mind constructs "small-scale models" of reality—later named Mental Models [81]—that it uses to anticipate events. Using a similar principle, active inference, a generic framework based on Bayesian inference, models any adaptive system, as the brain, in a perception/action context [82]. Active inference describes the world to be in a true state which can never be completely revealed as the only information the adaptive system has are observations obtained through sensory input. The true state of the world is in fact hidden to the observers, and as such is set in their internal, generative model of the world, as hidden states. The true state of the world is inferred through sensory input or observations and is updated in a generative model, i.e., an internal representation of the world containing empirical priors and prior beliefs. The empirical priors are the hidden states and the possible actions to be made when an observation occurs. The event anticipation and action selection are defined with the free energy minimization principle or minimization of the surprise, and utility function, i.e., a measure of preferences over some set of outcomes. In other words, a set of possible actions which were previously generated in the internal model as empirical priors are favored in order to get a desired outcome. For instance, if a stranger—A—asks a person—B—to borrow him a phone in the street, the outcome of this event or the decision of B would depend on his model of the world or empirical priors that relate to such an event. B's decision will also depend on his prior beliefs, for instance a religious man would have principles such that one should always help those in need. B can never reveal the absolute truth about A's intentions. So, if B's experience, i.e., empirical priors were negative, and no prior beliefs or higher values govern his actions, he will be likely to refuse. However, if it was positive, B will be likely to accept to help A. Additionally, B's reaction time will depend on a specific prior which encodes the exploration/exploitation ratio. Hence, B anticipates an outcome and acts in such a way and in a certain time to reach that event imagined in the future. He **inferred** the true state—the stranger's intentions—with his empirical priors and he **acted** to achieve a desired outcome or comply to prior beliefs. The promotion of certain outcomes is encoded in the utility function and is set as prior beliefs. The free energy minimization principle relies on minimizing the Kullback–Leibler divergence or the relative entropy between two probability distributions—the current state and the desired state. It can be thought of as a prediction error that reports the difference between what can be attained from the current state and the goals encoded by prior beliefs. So, by favoring a certain action, one can reduce the prediction error, and in this way, the action becomes the cause of future sensory input. This computational framework enables us to model the causes of sensory input in order to better anticipate and favor certain outcomes, which is indeed what we are looking for in BCI systems.

A P300-speller is a communication BCI device which relies on a neurophysiological phenomenon—called the oddball effect—that triggers a peak in the EEG signal, around 300 ms after a rare and unexpected event—a P300. That is why, this type of BCI is also called a reactive BCI, as the machine elicits and detects ERPs, i.e., the brain's reaction to stimuli. In the case of P300-speller, a set of letters are randomly flashed and the users need to focus their visual attention on the letter they wish to spell. Once the target letter is flashed (as an unexpected and rare event) the brain reacts enabling the machine to detect the ERP and spell the desired letter.

Bayesian inference has been successfully used for instance in designing adaptive P300-spellers [83]. In this example, the outcome of a probabilistic classifier (two multivariate-Gaussian mixture) is updated online. In such a way, the machine spells a letter once it attains a certain confidence level, i.e., the decision speed or reaction time depends on the reliability of accumulated evidence. This permits the machine to stop at an optimal moment, while maximizing both speed and accuracy. However, as we mentioned earlier, this example is user-dependent and adaptive but does not go further by considering the cause of such EEG variability in order to reduce or anticipate it. To achieve this, we could endow the machine with a certain intelligence, with active inference [84]. As we explained, active inference is used to model cognitive behavior and decision-making processes. However, in our case, we wish to equip the machine with such generative models, in order to achieve a full symbiotic user-machine coadaptation. The true states, in this case, belong to the user characteristics and intentions and are in fact hidden to the machine. Concretely, the hidden states are the letters or words the user intends to spell with the BCI. In the beginning, all the letters have equal probability to be spelled, but the more the machine flashes letters, the more it accumulates empirical priors and becomes confident about the target letter. In such way, the user intentions are represented as empirical priors (hidden states) which the machine has to update through the accumulation of observations—the classifier output. Furthermore, the machine will act (flash) in such a way to achieve the desired outcome—to reveal the target letter in minimal time. Hence, by using these principles, we could not only achieve optimal stopping [83] but also optimal flashing [84], i.e., flashing such group of letters to maximize the P300 effect. The flashing would be in an intelligent order yet appear to the user to be in a random order, so that the oddball effect stays uncompromised.

The criteria of optimization, i.e., whether one would favor subjective user's experience over the system performance, depends on the purpose of the BCI system [8]. For example, for entertainment or rehabilitation purposes, it is important to motivate the user to keep playing or keep giving effort. To achieve this can be possible by using positively biased feedback. On the other hand, for controlling a wheelchair using a BCI, then the system's accuracy is of essential importance. Active inference could provide such adaptive power, setting the BCI goals within an intelligent artificial agent which would encode the utility function and would manipulate the exploration/exploitation factor, see Figure 8.1.

The remaining challenges comprise using active inference to adapt tasks of other BCI paradigms such as motor imagery. The ultimate goal would be to use active inference to create a fully adaptive and user customizable BCI. In this case, the hidden states which the machine needs to infer and learn would be more than trial-wise user intentions, but also user's states, skills and traits (measured with passive BCI for instance) and provided to the machine as additional (neuro)physiological observations. The convenience about active inference is that it is applicable to any adaptive system. So, we can use any information as input (higher level user observations) and tune the parameters (priors) to each user, in order to provide them with optimal tasks. The optimal tasks would be governed by the BCI purpose (control, communication, neuro-rehabilitation, etc.), paradigm (P300, MI, SSEP) and exercise (MI of hands, feet, or counting the number of flashes, etc.).

Regarding signal processing, the adaptive processes which arise such as adapting spatial or temporal filters should not only adjust to the signal variabilities but be also guided by the context and purpose of the BCI. This way, the signal-processing techniques could extend their adaptive power and be more applicable and flexible across contexts and users. Furthermore, the signal-processing pipeline would need to expand and include other possible (neuro)physiological measurements in order to measure high-level user factors. The machine-learning techniques will have to accommodate for more dimensions, not only the features extracted from EEG but also the variable states of the user should be taken into account. Active inference

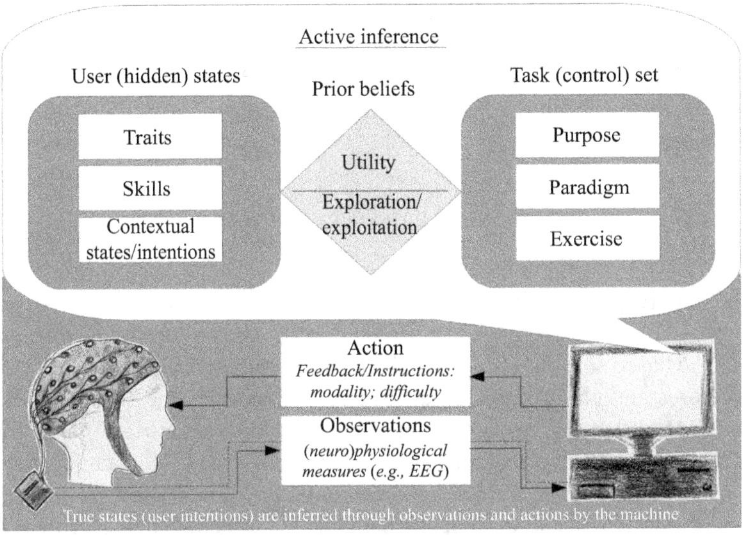

Figure 8.1 *A concept of how active inference could be used to implement a fully adaptive BCI. The machine "observes" one or several (neuro)physiological measurements which serve to infer the user's immediate intentions or states, skills and traits in longer periods of time. Depending on the purpose of the BCI, its paradigm and exercise, and considering the information the machine learned about the user, it will provide optimal action (feedback or instructions in different modalities or difficulty). An intelligent agent will encode the priors (utility and exploration/exploitation ratio) that are regulated for each user and specific context, favoring the optimal paradigm, exercise and action within specific timescales of adaptation*

would fit this landscape and add such a layer through an internal model of the various causes of signal variability and by its single cost function—free energy.

8.3 Improving BCI user training

Machine-learning and signal-processing tools can also be used to deepen our understanding of BCI user learning as well as to improve this learning. Notably, such tools can be used to design features and classifiers that are not only good to discriminate the different BCI commands but also good to ensure that the user can understand and learn from the feedback resulting for this classifier/features. This feedback can also be further improved by using signal-processing tools to preprocess it, in order to design an optimal display for this feedback, maximizing learning. Finally, rather than designing adaptive BCI algorithms solely to increase BCI command decoding accuracy, it seems also promising to adapt BCI algorithms in a way and at a rate that favor user learning. Altogether, current signal-processing and machine-learning algorithms should not be designed solely for the machine, but also with the user in mind, to ensure that the resulting feedback and training enable the user to learn efficiently. We detail these aspects below.

8.3.1 Designing features and classifiers that the user can understand and learn from

So far the features, e.g., the power in some frequency bands and channels, and classifiers, e.g., LDA or support vector machine (SVM), used to design EEG-based BCI are optimized based solely on the basis of their discriminative power [85–87]. In other words, features and classifiers are built solely to maximize the separation between the classes/MI tasks used to control the BCI, e.g., left versus right-hand imagined movement. Thus, a purely machine-oriented criteria—namely, data separation—is used to optimize features and classifiers, without any consideration for whether such features and classifiers lead to a feedback that (1) is understandable by the user and (2) can enable the user to learn to self-regulate those features. In the algorithms used so far, while the features are by design as separable as possible, there is no guarantee that they can become more separable with training. Actually, it is theoretically possible that some features with an initially lower discriminative power can be easier to learn to self-regulate. As such, while on the short-term selecting features that are initially as discriminant as possible makes sense, on the longer term, if the user can learn EEG self-regulation successfully, then it may make more sense to select features that will lead to a possibly even better discrimination after user learning. Similarly, while the classifier output, e.g., the distance between the input feature vector and the LDA/SVM discriminant hyperplane [88], is typically used as feedback to the user, it is also unknown whether such feedback signal variations can be understood or make sense for the user. Maybe a different feedback signal, possibly less discriminant, would be easier to understand and learn to control by the user.

Interestingly enough, there are very relevant research results from neuroscience, psychology and human–computer interaction that suggest that there are some constraints and principles that need to be respected so as to favor user learning of self-regulation or to enable users to understand as best as possible some visualization and feedback. In particular, it was shown with motor related invasive BCIs on monkeys, that using features that lie in the natural subspace of their motor-related activity, i.e., in their natural motor repertoire, leads to much more efficient learning of BCI control than using features that lie outside this subspace/natural repertoire [89,90]. This suggests that not all features have the same user-learning potential, and thus that features should be designed with such considerations in mind. Similarly, regarding feedback and visualization, humans perceive with more or less ease variations of a visual stimuli, depending on the spatial and temporal characteristics of these variations, e.g., how fast the stimuli changes and what the amplitude of this change is, see, e.g., [91] for an overview. For instance, it is recommended to provide visualizations that are consistent over time, i.e., whose meaning should be interpreted in the same way from one trial to the next and that vary smoothly over time [91]. This as well suggests that the feedback should ideally be designed while taking such principles into consideration. There are also many other human-learning principles that are in general not respected by current BCI designs, see notably [5] as well as Section 8.3.3. There is thus a lot of room for improvement.

The learning and feedback principles mentioned above could be used as constraints into the objective functions of machine-learning and signal-processing algorithms used in BCI. For instance, to respect human perception principles [91], we could add these perception properties as regularization terms in regularized machine-learning algorithms such as regularized spatial filters [92] or classifiers [85,93]. Similarly, regularization terms could be added to ensure that the features/classifier lie in the natural motor repertoire of the user, to promote efficient learning with motor-related BCIs. This could be achieved for instance, by transferring data between users [94,95], to promote features that were shown to lead to efficient learning in other users. In other words, rather than designing features/classifiers using objective functions that reflect only discrimination, such objective functions should consider both discrimination and human-learning/perception principles. This would ensure the design of both discriminative and learnable/understandable features.

It could also be interesting to explore the extraction and possibly simultaneous use of two types of features: features that will be used for visualization and feedback only (and thus that may not be optimal from a classification point of view), and features that will be used by the machine to recognize the EEG patterns produced by the user (but not used as user training feedback). To ensure that such features are related, and thus that learning to modulate them is also relevant to send mental commands, they could be optimized and extracted jointly, e.g., using multitask learning [96].

8.3.2 Identifying when to update classifiers to enhance learning

It is already well accepted that in order to obtain better performances, adaptive BCI systems should be used [8,9,97]. Due to the inherent variability of EEG signals, as well as to the change in users' states, e.g., fatigue or attention, it was indeed shown that adaptive classifiers and features generally gave higher classification accuracy than fixed ones [8,9]. Typically, this adaptation consists in reestimating parameters of the classifiers/features during online BCI use, in order to keep track of the changing features distribution. However, again, such adaptation is typically performed only from a machine perspective to maximize data discriminability, without considering the user in the loop. The user is using the classifier output as feedback to learn and to use the BCI. If the classifier is continuously adapted, this means the feedback is changing continuously, which can be very confusing for the user, or even prevent them from learning properly. Indeed, both the user and the machine need to adapt to each other—the so-called coadaptation in BCI [98]. A very recent and interesting work proposed a simple computational model to represent this interplay between the user learning and the machine learning, and how this coadaptation takes place [99]. While such work is only a simulation, it has nonetheless suggested that an adaptation speed that is either too fast or too slow prevents this coadaptation from converging, and leads to a decreased learning and performance.

Therefore, when to adapt, e.g., how often, and how to adapt, e.g., how much, should be made with the user in mind. Ideally, the adaptation should be performed at a rate and strength that suits each specific user, to ensure that it does not confuse users but rather that it helps them to learn. To do so seems to stress once more the need for a model of the user (discussed in Section 8.2). Such model would infer from the data, among other, how much change the user can deal with, to adapt the classifier accordingly. In this model, being able to measure the users' BCI skills (see also Section 8.2.2) would also help in that regard. It would indeed enable to know when the classifier should be updated because the user had improved and thus their EEG patterns have changed. It would also be interesting to quantify which variations in the EEG feature distribution would require an adaptation that may be confusing to the user—e.g., those changing the EEG source used—and those that should not be—e.g., just tracking change in feature amplitude. This would enable to perform only adaptation that is as harmless as possible for the user. A point that would need to be explored is whether classifiers and features should only be adapted when the user actually changes strategy, e.g., when the user has learned a better MI task. This indeed requires the classifier to be able to recognize such new or improved mental tasks, whether other adaptations may just add some feedback noise and would be confusing to the user.

8.3.3 Designing BCI feedbacks ensuring learning

Feedback is generally considered as an important facilitator of learning and skill acquisition [100,101] with a specific effect on the motivation to learn, see e.g., [102].

Black and Wiliam [103] proposed that to be effective, feedback must be directive (indicate what needs to be revised) and facilitative (provides suggestions to guide learners). In the same way, Kulhavy and Stock proposed that effective feedback must allow verification, i.e., specify if the answer is correct or incorrect and elaboration, i.e., provide relevant cues to guide the learner [104].

In addition to guidance, informative feedback had to be goal directed by providing learners with information about their progress toward the goal to be achieved. The feeling that the goal can be met is an important way to enhance the motivation and the engagement of the learners [105].

Feedback should also be specific to avoid being considered as useless or frustrating [106]. It needs to be clear, purposeful, meaningful [107] and to lead to a feeling of competence in order to increase motivation [108].

Another consideration that requires to be much more deeply explored is that feedback must be adapted to the characteristics of the learners. For example, [109] showed that elaborated feedback enhances performances of low-ability students, while verification condition enhance performances of high-ability students. In the BCI field, Kübler *et al.* showed that positive feedback (provided only for a correct response) was beneficial for new or inexperienced BCI users, but harmful for advanced BCI users [62].

As underlined in [5], classical BCI feedback satisfies few of such requirements. Feedback typically specifies if the answer is correct or not—i.e., the feedback is corrective—but does not aim at providing suggestions to guide the learner—i.e., it is not explanatory. Feedback is also usually not goal directed and does not provide details about how to improve the answer. Moreover, the feedback may often be unclear and meaningless, since it is based on a classifier build using calibration data recorded at the beginning of the session, during which the user does not master the MI task they must perform.

In [5], we discussed the limits of feedback used in BCI and proposed solutions, some of which having already yielded positive results [88,110]. A possibility would be to provide the user with richer and more informative feedback by using, for example, a global picture of his/her brain activity, e.g., a 2D or 3D topography of cortical activation obtained by inverse solutions. Another proposal is to collect a better information on the mental task achieved by the subject (e.g., by recording event-related desynchronization/synchronization activity) to evaluate users' progress and give them relevant insights about how to perform the mental task. Finally, it would be relevant to use more attractive feedback by using game-like, 3D or virtual reality, thus increasing user engagement and motivation [111,112].

In a recent study, [113] tested a continuous tactile feedback by comparing it to an equivalent visual feedback. Performance was higher with tactile feedback indicating that this modality can be a promising way to enhance BCI performances.

To conclude, the feedback used in BCI are simple and often poorly informative, which may explain some of the learning difficulties encountered by many users. Based on the literature identifying the parameters that maximize the effectiveness of feedback in general, BCI studies have already identified possible theoretical improvements. However, further investigations will be necessary to explore new research directions in order to make BCI accessible to a greater number of people. In particular, the machine-learning and signal-processing communities have the skills and tools necessary to design BCI feedback that are clearer, adaptive and adapted to the user, more informative and explanatory. In the following we provide more details on some of these aspects. In particular, we discuss the importance to design adaptive biased feedback, emotional and explanatory feedback and provide related research directions in which the machine-learning and signal-processing communities can contribute.

8.3.3.1 Designing adaptive biased feedback

As stated earlier in this chapter, it is essential to compute and understand the user's emotional, motivational and cognitive states in order to provide them with an appropriate, adapted and adaptive feedback that will favor the acquisition of skills especially during the primary training phases of the user [114]. Indeed, in the first stages, the fact that the technology and the interaction paradigm (through MI tasks) are both new for the users is likely to induce a pronounced computer anxiety associated with a low sense of agency. Yet, given the strong impact that the sense of agency (i.e., the feeling of being in control) has on performance—see Section 8.2.3.1—it seems important to increase it as far as possible. Providing the users with a sensory feedback informing them about the outcome of their "action" (MI task) seems to be necessary in order to trigger a certain sense of agency at the beginning of their training. This sense of agency will in turn unconsciously encourage users to persevere, increase their motivation and thus promote the acquisition of MI-BCI related skills, which is likely to lead to better performances [115–117]. This process could underlie the (experimentally proven) efficiency of positively biased feedback for MI-BCI user-training.

Positively biased feedback consists in leading users to believe that their performance was better than it actually was. The literature [62,118] reports that providing MI-BCI users with a biased (only positive) feedback is associated with improved performances while they are novices. However, that is no longer the case once they have progressed to the level of expert users. This result could be due to the fact that positive feedback provides users with an illusion of control which increases their motivation and will to succeed. As explained by [115], once users reach a higher level of performance, they also experience a high level of self-efficacy which leads them to consider failure no longer as a threat [119] but as a challenge. And facing these challenges leads to improvement. Another explanation is the fact that experts develop the ability to generate a precise predicted outcome that usually matches the actual outcome (when the feedback is not biased). This could explain why when the feedback is biased, and therefore the predicted and actual outcomes do not match, expert users attribute the discrepancy to external causes more easily. In other words, it can be hypothesized that experts might be disturbed by a biased feedback because they can perceive that it does not truly reflect their actions, thus decreasing their sense of being in control.

To summarize, it is noteworthy that the experience level of the user needs to be taken into account when designing the optimal feedback system, and more specifically the bias level. As discussed before, the user experience is nonetheless difficult to assess (see also Section 8.2.2). For instance, when using LDA to discriminate two classes, the LDA will typically always output a class, even if it is uncertain about it. This might lead to a class seemingly always recognized, even if the user does not do much. Hence, if both classes are equally biased, the user would most likely not gain motivation for the one always recognized—performing good, but could feel bored. Note that even if one class is always recognized (seemingly giving higher performances than the other class) that does not mean that the user is actually performing well when imagining such class, it can be due to the classifier being unbalanced and providing as output this class more often (e.g., due to a faulty electrode). On the other hand, if the biased feedback is applied to the class which is not well recognized the user would probably gain motivation. Thus, in [120] the task was adaptively biased, depending on the user performances in real time, e.g., positively for the class which was recognized less often and negatively for the one recognized more often, in order to keep the user engaged. This idea came from the Flow theory [121] which explains that the intrinsic motivation, full immersion in the task and concentration can be attained if the task is adapted to user skills. Following the requirements of Flow theory, in [120] the environment is designed to be engaging and entertaining, the goals clear with immediate visual and audio feedback, and task difficulty adapted to user performance in real time. It is shown that the users feel more in control, and more in flow when the task is adapted. Additionally, the offline performance and flow level correlated.

This suggests that adapting the task may create a virtuous loop, potentially increasing flow with performance.

The approach of providing an adapted and adaptive feedback, obtained by modulating the bias level, sounds very promising in order to maintain BCI users in a flow state, with a high sense of agency. Nonetheless, many challenges remain in order to optimize the efficiency of this approach. First, once more, it is necessary to be able to infer the state of the user, and especially their skill level, from their performance and physiological data. Second, we will have to determine the bias to be applied not only to the BCI output as a function of the evolution of the users' skills but also as a function of their profile. Indeed, the basic level of sense of agency is not the same for everybody. Also, as shown in our models [8,71], both the sense of agency and the flow are influenced by several factors: they do not depend only upon the performance. Thus, many parameters—related to users' states and traits—should be taken into account to know how to adapt the bias.

8.3.3.2 Designing adaptive emotional feedback

The functioning of the brain has often been compared to that of a computer, which is probably why the social and emotional components of learning have long been ignored. However, emotional and social contexts play an important role in learning [122,123]. The learner's affective state has an influence on problem-solving strategies [46] and motivational outcome [56]. Expert teachers can detect such emotional states and react accordingly to the latter to positively impact learning [124]. However, the majority of feedback used for BCI training users typically do not benefit from adaptive social and emotional feedback during BCI training. In [125], we added some social context to BCI training by creating a game where BCI users had to compete or collaborate against or with each other, which resulted in improved motivation and better BCI performances for some of the participants. Other studies tried to provide a nonadaptive emotional feedback, under the form of smileys indicating whether the mental command was successfully recognized [62,112]. No formal comparisons without such emotional feedback was performed though, making the efficiency of such feedback still unknown. ITS providing an emotional and motivational support can be considered as a substitute and have been used in distant learning protocols where such feedback components were also missing. Indeed, they have proven to be successful in improving learning, self-confidence and affective outcome [13]. We tested such method for BCI in [126], where we implemented a learning companion for BCI training purpose. The companion provided both an adapted emotional support and social presence. Its interventions were composed of spoken sentences and facial expressions adapted based on the performance and progress of the user. The results show that emotional support and social presence have a beneficial impact on users' experience. Indeed, users that trained with the learning companion felt that it was easier to learn and memorize than the group that only trained with the usual training protocol (i.e., with no emotional support or social presence). This learning companion did not lead to any significant increase in online classification performance so far, though, which suggests that it should be further improved. It could for example consider the user's profile which influences BCI performances [127] and monitor the user's emotional state and learning phase [47]. Indeed, both social and emotional feedback can have a positive, neutral or negative influence on learning depending on the task design, the type of feedback provided and the variables taken into account to provide the feedback [122,128]. In this context, machine learning could have a substantial impact for the future applications of ITS in BCI training applications. In particular, it seems promising to use machine learning to learn from the student's EEG, reactions and from its previous experience, what is the most appropriate emotional feedback it should provide to the user.

8.3.3.3 Designing explanatory feedback

As mentioned above, in many learning tasks—BCI included—the role of the feedback has been found to be essential in supporting learning and to make this learning efficient [107,129]. While feedbacks can be of several types, for BCI training, this feedback is almost always corrective only [5]. A corrective feedback is a feedback that tells the user whether the task they just performed is correct or incorrect. Indeed, in most BCIs, the feedback is typically a bar or a cursor indicating whether the mental task performed by the user was correctly recognized. Unfortunately, human-learning theories and instructional design principles all recommend to provide a feedback that is explanatory, i.e., which does not only indicate correctness, but also why it was correct or not. Indeed, across many learning tasks, explanatory feedback, which thus explains the reasons of the feedback, was shown to be superior to corrective one [107,129].

Consequently, it would be promising to try to design explanatory feedback for BCI. This is nonetheless a substantial challenge. Indeed, being able to provide explanatory feedback means being able to understand the cause of success or failure of a given mental command. So far, the BCI community has very little knowledge about these possible causes. Some works did identify some predictors of BCI performances [72,130–132]. However, most of these works identified predictors of performance variations across many trials and possibly many runs or sessions. Exceptions are [133] and [134], who showed respectively than cortical gamma activity in attentional networks as well as tension in forehead and neck muscles were correlated to single trial performances. In [134], we designed a first explanatory feedback for BCI, informing users of their forehead and neck muscle tension, identifying when it was too strong, to guide them to be relaxed. Unfortunately, this did not lead to significant increase in online BCI performance. Such work was however only a preliminary attempt that should thus be explored further, to identify new predictors of single trial performance and use them as feedback.

We denote features measuring causes of success or failure of a trial or group of trials as feedback features. We thus encourage the feedback community to design and explore new feedback features. This is another machine-learning and signal-processing problem, in which rather than classifying EEG as corresponding to a given mental command or another, we should classify them as predicting a successful or a failed trial. Thus, with different labels than before, machine learners can explore and design various tools to identify the most predictive feedback features. Such features

Table 8.1 Summary of signal-processing and machine-learning challenges to BCI user training and experience, and potential solutions to be explored

	Challenges	**Potential solutions**
Modeling the BCI user	Robust recognition of users' mental states from physiological signals	Exploring features, denoising and classification algorithms for each mental state
	Quantifying the many aspects of users' BCI skills	Riemannian geometry to go beyond classification accuracy
	Determining when to adapt the training procedure, based on the user's state, to optimize performance and learning	Case-based/rule-based reasoning algorithms; multiarm bandits to adapt automatically the training procedure
	Computationally modeling the users states and traits and adaptation tools	Exploiting active inference tools
Understanding and improving BCI user learning	Designing features and classifiers resulting in feedback favoring learning	Regularizers incorporating human-learning and perception principles
	Adapting classifiers with a way and timing favoring learning	Triggering adaptation based on a user's model
	Adapting the bias based on the user's level of skills to maintain their flow and agency	Triggering adaptation based on a model of the bias–skill relationship
	Adapting feedback to include emotional support and social presence	Build on the existing work of the ITS field
	Identifying/designing explanatory feedback features	Designing features to classify correct versus incorrect commands

could then be used as additional feedback during online BCI experiments, possibly supporting efficient BCI skills learning.

8.4 Conclusion

In this chapter, we tried to highlight to our readers that when designing BCIs, both the machine (EEG signal decoding) and the user (BCI skill learning and performance) should be taken into account. Actually, in order to really enable BCIs to reach their full potential, both aspects should be explored and improved. So far, the vast majority of the machine-learning community has worked on improving and robustifying the EEG signal decoding, without considering the human in the loop. Here, we hope we convinced our readers that considering the human user is necessary—notably to guide and boost BCI user training and performance—and that machine learning and signal-processing can bring useful and innovative solutions to do so. In particular, throughout the chapter we identified nine challenges that would need to be solved to enable users to use and to learn to use BCI efficiently and for each suggested potential machine-learning and signal-processing research directions to address them. These various challenges and solutions are summarized in Table 8.1.

We hope this summary of open research problems in BCI will inspire the machine-learning and signal-processing communities and will motivate their scientists to explore these less traveled but essential research directions. In the end, BCI research does need contributions from these communities to improve the user experience and learnability of BCI and enable them to become finally usable and useful in practice, outside laboratories.

Acknowledgments

This work was supported by the French National Research Agency with the REBEL project (grant ANR-15-CE23-0013-01), the European Research Council with the BrainConquest project (grant ERC-2016-STG-714567), the Inria Project-Lab BCI-LIFT and the EPFL/Inria International lab.

References

[1] Clerc M, Bougrain L, Lotte F. Brain–computer interfaces 2: technology and applications. London: ISTE-Wiley; 2016.

[2] Lotte F, Bougrain L, Clerc M. Electroencephalography (EEG)-based brain–computer interfaces. In: Wiley Encyclopedia on Electrical and Electronics Engineering. London: Wiley; 2015.

[3] Neuper C, Pfurtscheller G. Neurofeedback training for BCI control. In: Brain–Computer Interfaces. The Frontiers Collection. Springer-Verlag Berlin Heidelberg; 2010. p. 65–78.

[4] Jeunet C, Lotte F, N'Kaoua B. Human learning for brain–computer interfaces. In: Brain–Computer Interfaces 1: Foundations and Methods. London: Wiley Online Library; 2016. p. 233–250.

[5] Lotte F, Larrue F, Mühl C. Flaws in current human training protocols for spontaneous brain–computer interfaces: lessons learned from instructional design. Frontiers in Human Neuroscience. 2013;7(568).

[6] Lotte F, Jeunet C. Towards improved BCI based on human learning principles. In: 3rd International Brain–Computer Interfaces Winter Conference; 2015.

[7] Jeunet C, Jahanpour E, Lotte F. Why standard brain–computer interface (BCI) training protocols should be changed: an experimental study. Journal of Neural Engineering. 2016;13(3):036024.

[8] Mladenovic J, Mattout J, Lotte F. A generic framework for adaptive EEG-based BCI training and operation. In: Nam C, Nijholt A, Lotte F, editors. Handbook of Brain–Computer Interfaces. Boca Raton, FL, USA: Taylor & Francis; 2018.

[9] Shenoy P, Krauledat M, Blankertz B, Rao RPN, Müller KR. Towards adaptive classification for BCI. Journal of Neural Engineering. 2006;3(1):R13.

[10] Swan M. Sensor mania! the internet of things, wearable computing, objective metrics, and the quantified self 2.0. Journal of Sensor and Actuator Networks. 2012;1(3):217–253.

[11] Jovanov E, Milenkovic A, Otto C, De Groen PC. A wireless body area network of intelligent motion sensors for computer assisted physical rehabilitation. Journal of NeuroEngineering and Rehabilitation. 2005;2(1):6.

[12] Baca A, Kornfeind P. Rapid feedback systems for elite sports training. IEEE Pervasive Computing. 2006;5(4):70–76.

[13] Woolf BP, Arroyo I, Cooper D, Burleson W, Muldner K. Affective tutors: automatic detection of and response to student emotion. In: Advances in Intelligent Tutoring Systems. Berlin Heidelberg: Springer; 2010. p. 207–227.

[14] Fisk AD, Schneider W. Memory as a function of attention, level of processing, and automatization. Journal of Experimental Psychology: Learning, Memory, and Cognition. 1984;10(2):181.

[15] Keller JM. The ARCS model of motivational design. In: Motivational Design for Learning and Performance. New York, Dordrecht Heidelberg: Springer; 2010. p. 43–74.

[16] Wickens CD. Multiple resources and performance prediction. Theoretical Issues in Ergonomics Science. 2002;3(2):159–177.

[17] Posner MI, Boies SJ. Components of attention. Psychological Review. 1971;78(5):391.

[18] Cohen RA, Sparling-Cohen YA, O'Donnell BF. The neuropsychology of attention. New York: Springer; 1993.

[19] Shaw TH, Warm JS, Finomore V, *et al.* Effects of sensory modality on cerebral blood flow velocity during vigilance. Neuroscience Letters. 2009;461(3):207–211.

[20] Glaholt MG. Eye tracking in the cockpit: a review of the relationships between eye movements and the aviator's cognitive state. In: Scientific Report No. DRDC-RDDC-2014-R153. Defence Research and Development Canada (DRDC), Toronto, CA; 2014.

[21] Frey J, Mühl C, Lotte F, Hachet M. Review of the use of electroencephalography as an evaluation method for human–computer interaction. In: International Conference on Physiological Computing Systems; 2014.

[22] Klimesch W, Doppelmayr M, Russegger H, Pachinger T, Schwaiger J. Induced alpha band power changes in the human EEG and attention. Neuroscience Letters. 1998;244(2):73–76.

[23] Saavedra C, Bougrain L. Processing stages of visual stimuli and event-related potentials. In: The NeuroComp/KEOpS'12 Workshop; 2012.

[24] Grosse-Wentrup M, Schölkopf B, Hill J. Causal influence of gamma oscillations on the sensorimotor rhythm. NeuroImage. 2011;56(2):837–842.

[25] Grosse-Wentrup M, Schölkopf B. High gamma-power predicts performance in sensorimotor-rhythm brain–computer interfaces. Journal of Neural Engineering. 2012;9(4):046001.

[26] Baddeley AD, Hitch G. Working memory. Psychology of Learning and Motivation. 1974;8:47–89.

[27] Mayer RE. Multimedia learning (2nd ed.). Cambridge University Press, New York; 2009.

[28] Gerjets P, Walter C, Rosenstiel W, Bogdan M, Zander TO. Cognitive state monitoring and the design of adaptive instruction in digital environments: lessons learned from cognitive workload assessment using a passive brain–computer interface approach. Frontiers in Neuroscience. 2014;8.

[29] Harris Sr RL, Glover BJ, Spady Jr AA. Analytical techniques of pilot scanning behavior and their application. In: NASA Technical Report; 1986.

[30] Ahlstrom U, Friedman-Berg FJ. Using eye movement activity as a correlate of cognitive workload. International Journal of Industrial Ergonomics. 2006;36(7):623–636.

[31] de Greef T, Lafeber H, van Oostendorp H, Lindenberg J. Eye movement as indicators of mental workload to trigger adaptive automation. In: Foundations of Augmented Cognition Neuroergonomics and Operational Neuroscience; 2009. p. 219–228.

[32] Verwey WB, Veltman HA. Detecting short periods of elevated workload: a comparison of nine workload assessment techniques. Journal of Experimental Psychology: Applied. 1996;2(3):270.

[33] Wobrock D, Frey J, Graeff D, de la Rivière JB, Castet J, Lotte F. Continuous mental effort evaluation during 3D object manipulation tasks based on brain and physiological signals. In: Proceedings of INTERACT 15; 2015.

[34] Gevins A, Smith ME, Leong H, *et al.* Monitoring working memory load during computer-based tasks with EEG pattern recognition methods. Human Factors. 1998;40(1):79–91.

[35] Joseph RD. Contributions to perceptron theory. New York: Cornell University; 1961.

[36] Viglione S. Applications of pattern recognition technology. In: Mathematics in Science and Engineering, JM Mendel, KS Fu (Eds.). Elsevier; 1970;66:115–162.

[37] Blankertz B, Tangermann M, Vidaurre C, *et al.* The Berlin brain–computer interface: non-medical uses of BCI technology. Frontiers in Neuroscience. 2010;4.

[38] Grimes D, Tan DS, Hudson SE, Shenoy P, Rao RP. Feasibility and pragmatics of classifying working memory load with an electroencephalograph. In: Proceedings of the SIGCHI Conference on Human Factors in Computing Systems. ACM; 2008. p. 835–844.

[39] Mühl C, Jeunet C, Lotte F. EEG-based workload estimation across affective contexts. Frontiers in Neuroscience Section Neuroprosthetics. 2014;8:114.

[40] Ahn M, Cho H, Ahn S, Jun SC. High theta and low alpha powers may be indicative of BCI-illiteracy in motor imagery. PLoS One. 2013;8(11):e80886.

[41] Yamamoto S, Matsuoka S. Topographic EEG study of visual display terminal (VDT) performance with special reference to frontal midline theta waves. Brain Topography. 1990;2(4):257–267.

[42] Herff C, Heger D, Fortmann O, Hennrich J, Putze F, Schultz T. Mental workload during N-back task—quantified in the prefrontal cortex using fNIRS. Frontiers in Human Neuroscience. 2014;7:935.

[43] Peck EM, Afergan D, Yuksel BF, Lalooses F, Jacob RJ. Using fNIRS to measure mental workload in the real world. In: Advances in Physiological Computing. London: Springer; 2014. p. 117–139.

[44] Durantin G, Gagnon JF, Tremblay S, Dehais F. Using near infrared spectroscopy and heart rate variability to detect mental overload. Behavioural Brain Research. 2014;259:16–23.

[45] Philippot P, Schaefer A. Emotion and memory. In: Bonanno TJMGA, editor. Emotions and Social Behavior. Emotions: Current Issues and Future Directions. New York: Guilford Press; 2001. p. 82–122.

[46] Isen A. Positive affect and decision making, In: Lewis M, Haviland-Jones J, editors. Handbook of Emotions. New York, NY, US: Guilford Press; 2000. p. 417–435.

[47] Kort B, Reilly R, Picard RW. An affective model of interplay between emotions and learning: reengineering educational pedagogy-building a learning companion. In: Advanced Learning Technologies, 2001. Proceedings. IEEE International Conference on. IEEE; 2001. p. 43–46.

[48] Picard RW, Healey J. Affective wearables. In: Wearable Computers, 1997. Digest of Papers., First International Symposium on. IEEE; 1997. p. 90–97.

[49] Picard RW. Toward computers that recognize and respond to user emotion. IBM Systems Journal. 2000;39(3.4):705–719.

[50] Arroyo I, Cooper DG, Burleson W, Woolf BP, Muldner K, Christopherson R. Emotion sensors go to school. In: AIED. vol. 200; 2009. p. 17–24.

[51] Lin YP, Wang CH, Wu TL, Jeng SK, Chen JH. Multilayer perceptron for EEG signal classification during listening to emotional music. In: TENCON 2007—2007 IEEE Region 10 Conference. IEEE; 2007. p. 1–3.

[52] Murugappan M, Rizon M, Nagarajan R, Yaacob S, Hazry D, Zunaidi I. Time-frequency analysis of EEG signals for human emotion detection. In: 4th Kuala Lumpur International Conference on Biomedical Engineering 2008. Springer; 2008. p. 262–265.

[53] Murugappan M, Ramachandran N, Sazali Y. Classification of human emotion from EEG using discrete wavelet transform. Journal of Biomedical Science and Engineering. 2010;3(04):390.

[54] Heger D, Herff C, Putze F, Mutter R, Schultz T. Continuous affective states recognition using functional near infrared spectroscopy. Brain–Computer Interfaces. 2014;1(2):113–125.

[55] Harter S. A new self-report scale of intrinsic versus extrinsic orientation in the classroom: motivational and informational components. Developmental Psychology. 1981;17(3):300.

[56] Stipek DJ. Motivation to learn: from theory to practice. Needham Heights: ERIC; 1993.

[57] Knyazev GG. EEG delta oscillations as a correlate of basic homeostatic and motivational processes. Neuroscience & Biobehavioral Reviews. 2012;36(1):677–695.

[58] Kleih SC, Nijboer F, Halder S, Kübler A. Motivation modulates the P300 amplitude during brain–computer interface use. Clinical Neurophysiology. 2010;121(7):1023–1031.

[59] Miller NE. Some directions for clinical and experimental research on biofeedback. In: White L, Tursky B, editors. Clinical Biofeedback: Efficacy and Mechanisms. New York: Guilford Press[AMY]; 1982.
[60] Yates AJ. Biofeedback and the modification of behavior. Boston: Springer Science & Business Media; 2012.
[61] Kübler A, Kotchoubey B, Kaiser J, Wolpaw JR, Birbaumer N. Brain–computer communication: unlocking the locked in. Psychological Bulletin. 2001;127(3):358.
[62] Kübler A, Neumann N, Kaiser J, Kotchoubey B, Hinterberger T, Birbaumer NP. Brain–computer communication: self-regulation of slow cortical potentials for verbal communication. Archives of Physical Medicine and Rehabilitation. 2001;82(11):1533–1539.
[63] Hernandez LD, Rieger K, Koenig T. Low motivational incongruence predicts successful EEG resting-state neurofeedback performance in healthy adults. Neuroscience. 2018;378:146–154.
[64] Hammer EM, Halder S, Blankertz B, *et al.* Psychological predictors of SMR-BCI performance. Biological Psychology. 2012;89(1):80–86.
[65] Neumann N, Birbaumer N. Predictors of successful self control during brain–computer communication. Journal of Neurology, Neurosurgery & Psychiatry. 2003;74(8):1117–1121.
[66] Christensen JC, Estepp JR, Wilson GF, Russell CA. The effects of day-to-day variability of physiological data on operator functional state classification. NeuroImage. 2012;59(1):57–63.
[67] Thompson DE, Quitadamo LR, Mainardi L, *et al.* Performance measurement for brain–computer or brain–machine interfaces: a tutorial. Journal of Neural Engineering. 2014;11(3):035001.
[68] Hill NJ, Häuser AK, Schalk G. A general method for assessing brain–computer interface performance and its limitations. Journal of Neural Engineering. 2014;11(2):026018.
[69] Lotte F, Jeunet C. Online classification accuracy is a poor metric to study mental imagery-based BCI user learning: an experimental demonstration and new metrics. In: International Brain–Computer Interface Conference; 2017.
[70] Busemeyer JR, Diederich A. Cognitive modeling. USA: Sage; 2010.
[71] Jeunet C, N'Kaoua B, Lotte F. Towards a cognitive model of MI-BCI user training. In: International Graz BCI Conference; 2017.
[72] Jeunet C, N'Kaoua B, Lotte F. Advances in user-training for mental-imagery-based BCI control: psychological and cognitive factors and their neural correlates. Progress in Brain Research. 2016;228:3–35.
[73] Poltrock SE, Brown P. Individual differences in visual imagery and spatial ability. Intelligence. 1984;8(2):93–138.
[74] N'Kambou R, Mizoguchi R, Bourdeau J. Advances in intelligent tutoring systems. vol. 308. Berlin, Heidelberg: Springer Science & Business Media; 2010.
[75] Zander T, Battes B, Schölkopf B, Grosse-Wentrup M. Towards neurofeedback for improving visual attention. In: Proceedings of the Fifth International Brain–Computer Interface Meeting: Defining the Future. vol. 86; 2013.

[76] Kleih SC, Philippi M. Does mastery confidence influence P300 based brain–computer interface (BCI) performance? In: Systems, Man, and Cybernetics (SMC), 2016 IEEE International Conference on. IEEE; 2016. p. 001433–001436.

[77] Kleih SC, Kübler A. Psychological factors influencing brain–computer interface (BCI) performance. In: Systems, Man, and Cybernetics (SMC), 2015 IEEE International Conference on. IEEE; 2015. p. 3192–3196.

[78] Clement B, Roy D, Oudeyer PY, Lopes M. Multi-arm bandits for intelligent tutoring systems. Journal of Educational Data Mining. 2015;7(2):20–48.

[79] Schwarz A, Scherer R, Steyrl D, Faller J, Müller-Putz GR. A co-adaptive sensory motor rhythms brain–computer interface based on common spatial patterns and random forest. In: Engineering in Medicine and Biology Society (EMBC), 2015 37th Annual International Conference of the IEEE. IEEE; 2015. p. 1049–1052.

[80] Thomas E, Daucé E, Devlaminck D, *et al.* CoAdapt P300 speller: optimized flashing sequences and online learning. In: 6th International Brain–Computer Interface Conference; 2014.

[81] Johnson-Laird PN. Mental models: towards a cognitive science of language, inference, and consciousness. 6. Boston: Harvard University Press; 1983.

[82] Friston K, Schwartenbeck P, FitzGerald T, Moutoussis M, Behrens T, Dolan RJ. The anatomy of choice: active inference and agency. Frontiers in Human Neuroscience. 2013;7.

[83] Mattout J, Perrin M, Bertrand O, Maby E. Improving BCI performance through co-adaptation: applications to the P300-speller. Annals of Physical and Rehabilitation Medicine. 2015;58(1):23–28.

[84] Mladenović J, Joffily M, Frey J, Lotte F, Mattout J. Endowing the machine with active inference: a generic framework to implement adaptive BCI. In: NeuroAdaptive Technology Conference'17; 2017.

[85] Lotte F, Congedo M, Lécuyer A, Lamarche F, Arnaldi B. A review of classification algorithms for EEG-based brain–computer interfaces. Journal of Neural Engineering. 2007;4:R1–R13.

[86] Blankertz B, Tomioka R, Lemm S, Kawanabe M, Müller KR. Optimizing spatial filters for robust EEG single-trial analysis. IEEE Signal Processing Magazine. 2008;25(1):41–56.

[87] Lotte F. A tutorial on EEG signal-processing techniques for mental-state recognition in brain–computer interfaces. In: Guide to Brain–Computer Music Interfacing. London: Springer; 2014. p. 133–161.

[88] Pfurtscheller G, Neuper C. Motor Imagery and Direct Brain–Computer Communication. Proceedings of the IEEE. 2001;89(7):1123–1134.

[89] Sadtler PT, Quick KM, Golub MD, *et al.* Neural constraints on learning. Nature. 2014;512(7515):423–426.

[90] Hwang EJ, Bailey PM, Andersen RA. Volitional control of neural activity relies on the natural motor repertoire. Current Biology. 2013;23(5): 353–361.

[91] Ware C. Information visualization: perception for design. San Franciso: Elsevier; 2012.

[92] Lotte F, Guan CT. Regularizing common spatial patterns to improve BCI designs: unified theory and new algorithms. IEEE Transactions on Biomedical Engineering. 2011;58(2):355–362.

[93] Lotte F, Bougrain L, Cichocki A, *et al.* A review of classification algorithms for EEG-based brain–computer interfaces: a 10-year update. Journal of Neural Engineering. 2018;15(3):031005.

[94] Lotte F, Guan CT. Learning from other subjects helps reducing brain–computer interface calibration time. In: International Conference on Audio, Speech and Signal Processing (ICASSP'2010); 2010. p. 614–617.

[95] Lotte F. Signal processing approaches to minimize or suppress calibration time in oscillatory activity-based brain–computer interfaces. In: Proceedings of the IEEE; 2015.

[96] Caruana R. Multitask learning. In: Learning to Learn. Boston: Springer; 1998. p. 95–133.

[97] Millán JR, Mouriño J. Asynchronous BCI and local neural classifiers: an overview of the adaptive brain interface project. IEEE Transactions on Neural Systems and Rehabilitation Engineering. 2003. Special Issue on Brain–Computer Interface Technology. 2003;11(2):159–161.

[98] Vidaurre C, Sannelli C, Müller KR, Blankertz B. Machine-learning-based coadaptive calibration for brain–computer interfaces. Neural Computation. 2011;23(3):791–816.

[99] Müller JS, Vidaurre C, Schreuder M, Meinecke FC, von Bünau P, Müller KR. A mathematical model for the two-learners problem. Journal of Neural Engineering. 2017;14(3):036005.

[100] Azevedo R, Bernard RM. A meta-analysis of the effects of feedback in computer-based instruction. Journal of Educational Computing Research. 1995;13(2):111–127.

[101] Bangert-Drowns RL, Kulik CLC, Kulik JA, Morgan M. The instructional effect of feedback in test-like events. Review of Educational Research. 1991;61(2):213–238.

[102] Narciss S, Huth K. How to design informative tutoring feedback for multimedia learning. In: Instructional Design for Multimedia Learning; Münster. 2004. p. 181–195.

[103] Black P, Wiliam D. Assessment and classroom learning. Assessment in Education: Principles, Policy & Practice. 1998;5(1):7–74.

[104] Kulhavy RW, Stock WA. Feedback in written instruction: the place of response certitude. Educational Psychology Review. 1989;1(4):279–308.

[105] Fisher SL, Ford JK. Differential effects of learner effort and goal orientation on two learning outcomes. Personnel Psychology. 1998;51(2):397–420.

[106] Williams SE. Teachers' written comments and students' responses: a socially constructed interaction. In: Proceedings of the Annual Meeting of the Conference on College Composition and Communication; 1997.

[107] Hattie J, Timperley H. The power of feedback. Review of Educational Research. 2007;77:81–112.

[108] Ryan R, Deci EL. Self-determination theory and the facilitation of intrinsic motivation, social development, and well-being. American Psychologist. 2000;55(1):68–78.
[109] Hanna GS. Effects of total and partial feedback in multiple-choice testing upon learning. The Journal of Educational Research. 1976;69(5): 202–205.
[110] Hwang HJ, Kwon K, Im CH. Neurofeedback-based motor imagery training for brain–computer interface (BCI). Journal of Neuroscience Methods. 2009;179(1):150–156.
[111] Lécuyer A, Lotte F, Reilly RB, Leeb R, Hirose M, Slater M. Brain–computer interfaces, virtual reality and videogames. IEEE Computer. 2008;41(10): 66–72.
[112] Leeb R, Lee F, Keinrath C, Scherer R, Bischof H, Pfurtscheller G. Brain–computer communication: motivation, aim and impact of exploring a virtual apartment. IEEE, Transactions on Neural Systems & Rehabilitation Engineering. 2007;15(4):473–482.
[113] Jeunet C, Vi C, Spelmezan D, N'Kaoua B, Lotte F, Subramanian S. Continuous tactile feedback for motor-imagery based brain–computer interaction in a multitasking context. In: Proc. Interact 2015; 2015.
[114] McFarland DJ, McCane LM, Wolpaw JR. EEG-based communication and control: short-term role of feedback. IEEE Transactions on Rehabilitation Engineering. 1998;6(1):7–11.
[115] Achim N, Al Kassim A. Computer usage: the impact of computer anxiety and computer self-efficacy. Procedia-Social and Behavioral Sciences. 2015;172:701–708.
[116] Saadé RG, Kira D. Computer anxiety in e-learning: the effect of computer self-efficacy. Journal of Information Technology Education. 2009;8:177–191.
[117] Simsek A. The relationship between computer anxiety and computer self-efficacy. Online Submission. 2011;2(3):177–187.
[118] Barbero Á, Grosse-Wentrup M. Biased feedback in brain–computer interfaces. Journal of Neuroengineering and Rehabilitation. 2010;7(1):34.
[119] Kleih S, Kaufmann T, Hammer E, *et al.* Motivation and SMR-BCI: fear of failure affects BCI performance. In: Proceedings of the Fifth International Brain–Computer Interface Meeting 2013. Verlag der Technischen Universität Graz Austria; 2013.
[120] Mladenović J, Frey J, Bonnet-Save M, Mattout J, Lotte F. The impact of flow in an EEG-based brain–computer interface. In: International BCI Conference; 2017.
[121] Csikszentmihalyi M. Toward a psychology of optimal experience. In: Flow and the Foundations of Positive Psychology. Dordrecht: Springer; 2014. p. 209–226.
[122] Johnson DW, Johnson RT. An educational psychology success story: social interdependence theory and cooperative learning. Educational Researcher. 2009;38(5):365–379.

[123] Salancik GR, Pfeffer J. A social information processing approach to job attitudes and task design. Administrative Science Quarterly. 1978;23(2):224–253.

[124] Goleman D. Emotional intelligence. Brockman Inc, New York; 1995.

[125] Bonnet L, Lotte F, Lécuyer A. Two brains, one game: design and evaluation of a multiuser BCI video game based on motor imagery. IEEE Transactions on Computational Intelligence and AI in Games. 2013;5(2):185–198.

[126] Pillette L, Jeunet C, Mansencal B, N'Kambou R, N'Kaoua B, Lotte F. PEANUT: personalised emotional agent for neurotechnology user-training. In: 7th International BCI Conference; 2017.

[127] Jeunet C, N'kaoua B, Subramanian S, Hachet M, Lotte F. Predicting mental imagery-based BCI performance from personality, cognitive profile and neurophysiological patterns. PLoS One. 2015;10(12):e0143962.

[128] Kennedy J, Baxter P, Belpaeme T. The robot who tried too hard: social behaviour of a robot tutor can negatively affect child learning. In: Proceedings of the Tenth Annual ACM/IEEE International Conference on Human-Robot Interaction. ACM; 2015. p. 67–74.

[129] Shute VJ. Focus on formative feedback. Review of Educational Research. 2008;78:153–189.

[130] Ahn M, Jun SC. Performance variation in motor imagery brain–computer interface: a brief review. Journal of Neuroscience Methods. 2015;243: 103–110.

[131] Grosse-Wentrup M. What are the causes of performance variation in brain–computer interfacing? International Journal of Bioelectromagnetism. 2011;13(3):115–116.

[132] Blankertz B, Sannelli C, Halder S, *et al.* Neurophysiological predictor of SMR-based BCI performance. NeuroImage. 2010;51(4):1303–1309.

[133] Grosse-Wentrup M, Schölkopf B, Hill J. Causal influence of gamma oscillations on the sensorimotor rhythm. NeuroImage. 2010;56(2):837–842.

[134] Schumacher J, Jeunet C, Lotte F. Towards explanatory feedback for user training in brain–computer interfaces. In: IEEE International Conference on Systems Man and Cybernetics (IEEE SMC). 2015, p. 3169–3174.

Chapter 9
Feedforward artificial neural networks for event-related potential detection

Hubert Cecotti[1]

Abstract

The detection of brain responses at the single-trial level in the electroencephalogram (EEG) such as event-related potentials (ERPs) is a difficult problem that requires different processing steps to extract relevant discriminant features as the input signal is noisy and the brain responses can be different over time. In brain–computer interface, single-trial detection is primarily applied to distinguish the presence of large ERP components such as the P300. Because the characteristics of the P300 can depend on the parameters of the oddball paradigm, the type of stimuli, and as it can vary across subjects and over time during experiments, a reliable classifier must take into account this variability for the detection of the P300. While most of the signal and classification techniques for the detection of brain responses are based on linear algebra, different pattern recognition techniques such as convolutional neural network (CNN), as a type of deep learning technique, have shown some interest as they are able to process the signal after limited preprocessing. In this chapter, we propose to investigate the performance of different feedforward neural networks in relation of their architecture and in relation to how they are evaluated: a single system for each subject or a system for all the subjects. More particularly, we want to address the change of performance that can be observed between specifying a neural network to a subject, or by considering a neural network for a group of subjects, taking advantage of a larger number of trials from different subjects. The results support the conclusion that a CNN trained on different subjects can lead to an AUC above 0.9 by using an appropriate architecture using spatial filtering and shift invariant layers.

9.1 Introduction

In machine learning, the methods based on deep learning have gained a great success in classification problems, optimization control, and time series analysis. In particular, deep learning methods for pattern recognition have won a large number of

[1]Department of Computer Science, College of Science and Mathematics, Fresno State, USA

competitions [1]. These approaches typically use prior information about the problem for the creation of their architecture, such as the number of main dimensions in the input signal, and the type of variations that can occur across trials. Convolutional neural networks (i.e., conv nets or CNN) have been initially proposed and evaluated for handwritten character recognition during the 1990s, extending the hierarchical approaches of Fukushima with learning [2]. They provided state-of-the-art results, and they have been popular in the research community of document analysis and recognition [3–5]. Because the architecture of a conv net must be set in relation to a particular problem and specific connections between units in the network, its implementation must be specific as each hidden layer includes more characteristics, i.e., the connections to local subsets of inputs from the previous layer, than a fully connected (FC) hidden layer in a multilayer perceptron (MLP). Thanks to graphical processing units [6,7] and recent open-source libraries that allow to only focus on the choice of the architecture, and not the implementation of the conv net itself, more architectures and deeper architectures can be tested rapidly on various databases [8].

While conv nets have been successfully used in computer vision, they have not been fully exploited in electroencephalogram (EEG) signal processing, for applications such as brain–computer interfaces (BCIs). First, the low number of available trials to train a model and the low signal-to-noise ratio do not allow to easily capture the manifold where lie the data. Most of the signal processing and classification techniques rely on linear algebra, e.g., linear classifiers, for the classification of event-related potentials (ERPs). Some studies using conv nets have been proposed for the detection of ERPs in the EEG signal, e.g., the P300 [9], with applications to the P300 speller and rapid serial visual presentation (RSVP) tasks for target detection [10]. These architectures typically include a first convolutional layer that acts as a spatial filtering layer, then the next layers decrease the number of features through filtering in the time domain and/or subsampling functions. Finally, the last hidden layers may include an FC hidden layer. It is worth noting that as linear discriminant analysis does provide relevant results for ERP detection, it is not necessary to use an FC hidden layer at the last stages of the architecture, i.e., such as an MLP after the extracted features through a succession of convolution and pooling functions. Special architectures have also been proposed for the detection of steady-state visual evoked potentials, where spatial filtering was achieved through a convolutional layer and then the Fourier transform was inserted between two hidden layers to transfer the power spectrum of the signal to higher levels in the network [11]. In all these cases, the architecture used sigmoid functions as activation units as opposed to the most recent rectified linear unit (ReLU) function [12], which provides a reduced likelihood of vanishing gradient.

In this paper, we propose to compare different architectures of conv nets by using state-of-the-art functions that are readily available online with current libraries and toolboxes (e.g., with MATLAB®). This paper focuses therefore on the choice of the architecture, and it does not aim at providing new convolution or pooling functions. The performance is evaluated on a database of healthy participants who had to search a target image during a rapid serial presentation task [13,14]. We propose to investigate the performance of CNNs in relation of their architecture, and in relation to

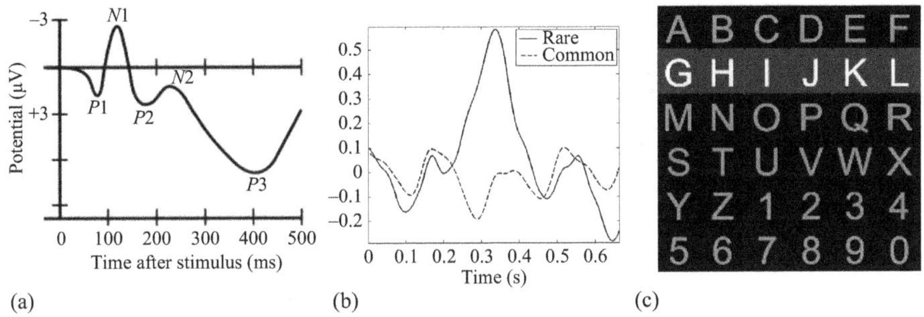

Figure 9.1 (a) Example of event-related potential components, (b) example of a P300 ERP component on the sensor location Pz, (c) the P300 speller based on the oddball paradigm

how they are evaluated: a single system for each subject or for all the subjects. More particularly, we want to address the change of performance that can be observed between specifying a neural network to a subject, or by considering a neural network for a group of subjects, taking advantage of a larger number of trials from different subjects. The extent to which it is better to have a subject-specific system or not is important in the field of BCI because the user experience must be enhanced for any user, including patients who would need to use a BCI daily. The remainder of this paper is organized as follows. ERPs are presented in Section 9.2. Feedforward neural networks such as MLP and conv nets are described in Section 9.3. The dataset and the models that were implemented are detailed in Section 9.4. The results are given in Section 9.6. Finally, the impact of the results is discussed in Section 9.7.

9.2 Event-related potentials

An ERP is the measured brain response evoked by specific sensory, cognitive, or motor event [15]. It corresponds to any stereotyped electrophysiological response to a stimulus. The brain response to a single stimulus is usually not directly visible in the EEG of a single trial because EEG measurements reflect thousands of simultaneous post-synaptic neural activations [16]. One of the most widely studied ERP components is the P300 ERP component, which is a large positive wave that peaks around 300 ms after the stimulus onset [17] (see Figure 9.1). The P300 can be used to index cognitive processing related to stimulus categorization, and as an input signal in BCI systems [18,19]. The P300 has been traditionally used with a two-stimulus oddball task where an infrequent "oddball" target stimulus is presented among a series of frequent nontarget distractor stimuli, and only the infrequent target stimulus requires a response (behavioral or cognitive) by the observer [20]. The latency of the P300 is correlated with the reaction time to a target stimulus, suggesting the presence of a temporal variability across trials that contain a P300 [21,22]. It is better to not assume that an averaged ERP waveform represents the single-trial waveform prototype as it

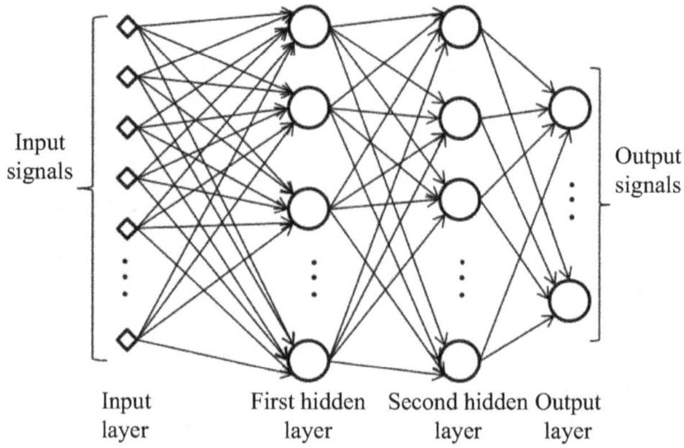

Figure 9.2 Example of a multilayer perceptron with two hidden layers

can be the accumulation of different time shifted ERPs. The difficulty of the task is directly related to the reaction times and the P300 peak latencies. Other ERPs components such as the N200 (a negative wave that peaks 150–350 ms post-stimulus) have been used with single-trial detection as a single component, or with the P300. Multivariate approaches have been used since the early days of ERP analysis (e.g., the quantitative analysis of averaged evoked potentials [23,24]). The difference between two types of ERPs should be ideally achieved at the single-trial level in order to obtain a high information transfer rate between the user and the machine. The problem of ERP detection can be formulated as a binary classification problem with two classes: rare and frequent nontargets. This problem has inputs with a high number of features (e.g., 8,192 with 64 sensors and a 1-s window at 128 Hz) and unbalanced class distribution due to the oddball paradigm.

9.3 Feedforward neural networks

Feedforward neural networks are a class of artificial neural networks in supervised learning that allows to solve regression and pattern recognition problems. They are able to learn from examples and to generalize the acquired knowledge to process new unseen examples. The MLP (Figure 9.2), as a class of feedforward neural networks, receives as an input a single vector and transforms this input vector through a series of hidden layers. Each hidden layer (l) contains a set of neurons, and each neuron is FC to all the neurons in the previous layer (l-1). The last layer, the output layer, is typically FC and the value of each neuron from this layer corresponds to the score of the corresponding class. Each layer can be decomposed into feature maps that represent multiple logical entities from a single hidden layer. These feature maps can be implemented through tensors where orders of the components can be related to feature maps. There are as many neurons as classes in the output layer, unless an additional neuron can be added representing the rejected examples, which is a class on its own. Each neuron is described by its state $s = f(\sigma)$, a set of weights equal to the number of inputs, and an optional bias depending on the type connections between layers.

9.3.1 Activation functions

The activation function of a neuron defines the output of that neuron given an input or set of inputs.

- The logistic function with an output range of (0, 1):

$$f(\sigma) = \frac{1}{1+e^{-\sigma}} \qquad f'(\sigma) = f(\sigma)(1 - f(\sigma)) \tag{9.1}$$

- The hyperbolic tangent with an output range of $(-1, 1)$:

$$f(\sigma) = \tanh(\sigma) \qquad f'(\sigma) = 1 - f^2(\sigma) \tag{9.2}$$

- ReLU with an output range of

$$f(\sigma) = \begin{cases} 0 & \text{for } \sigma < 0 \\ \sigma & \text{for } \sigma \geq 0 \end{cases} \qquad f'(\sigma) = \begin{cases} 0 & \text{for } \sigma < 0 \\ 1 & \text{for } \sigma \geq 0 \end{cases} \tag{9.3}$$

where σ corresponds to the weighted sum of the inputs of the neuron. The state of a neuron i in the layer l of the feature map m, $\sigma_{i,l,m}$, before the activation function can be defined as

$$\sigma_{i,l,m} = w_0 + \sum_{j=1}^{N_{input}} x_{j,l-1,m_k} w_j \tag{9.4}$$

where N_{input} is the number of inputs of the considered neuron, and $x_{j,l-1,m_k}$, $1 \leq j \leq N_{input}$ represent the set of inputs of the neuron in the previous layer, coming from possibly different feature maps m_k, $1 \leq m_k \leq N_{l-1}$, N_{l-1} is the number of feature maps of the layer $l-1$. The set of weights w_j, with $1 \leq j \leq N_{input}$, and w_0 is the bias. In the network, the weights are initialized randomly. For instance, they can be initialized with a Gaussian distribution with a mean of 0 and a standard deviation of 0.01. The set of parameters Θ that contains the set of weights and biases of network is updated through supervised learning by using the backpropagation algorithm, which is typically achieved by using stochastic gradient descent. The weights and biases of the neurons are updated to minimize the error function by taking small steps in the direction of the negative gradient of the loss function:

$$\Theta_{it+1} = \Theta_{it} - \alpha \nabla E(\Theta_{it}) \tag{9.5}$$

where it corresponds to the iteration number, $\alpha \in \mathbb{R}$ is the learning rate, and $E(\Theta_{it})$ is the loss function. $\nabla E(\Theta_{it})$ is evaluated using the whole training set, and the standard gradient descent algorithm considers the entire data set at once whereas the stochastic gradient descent algorithm evaluates the gradient and updates the parameters by using a subset of the training set. This subset is called a mini-batch. Each evaluation of the gradient using the mini-batch is an iteration, taking one step toward minimizing the loss function. A full pass over the entire training set using mini-batches is defined as an epoch. For training a network, the available dataset with a groundtruth is split into two parts: the first part (training dataset) is dedicated to the weight updates with the backpropagation, the second part (validation dataset) is used to evaluate the generalization performance of the network and determine when to stop learning, i.e., when the ability of the network to generalize drops. It is worth noting that only the training and test dataset are provided, and the person setting the parameters has to split the training database into a part for training only and one part for the validation.

9.3.2 Error evaluation

The typical functions to minimize are the mean square error (MSE) and the mean cross entropy error (MCEE):

$$MSE = \frac{1}{N_{train} \cdot N_{class}} \sum_{j=1}^{N_{train}} \sum_{i=1}^{N_{class}} (out(i)_j - tar(i)_j)^2 \tag{9.6}$$

$$MCEE = -\frac{1}{N_{train} \cdot N_{class}} \sum_{j=1}^{N_{train}} \sum_{i=1}^{N_{class}} log(out(i)_j) \cdot tar(i)_j \tag{9.7}$$

where $out(i)_j$ represents the value of the output neuron i for the example j, and $tar(i)$ represents the value of the target output for the neuron i that is provided by the ground truth of the training dataset. If the example j belongs to the class i, then $tar(i) = 1$ otherwise $tar(i) = 0$. N_{train} is the number of examples that are evaluated and N_{class} represents the total number of class in the problem. While the values for *out* and *tar* can be swapped for MSE, as the values of *tar* are typically 0 or 1, it is better to not put them in the log for MCEE to avoid problems, i.e., log(0). In addition to MSE or MCEE that can be evaluated on a validation database as a stopping condition for training, the area under the ROC (receiver operating characteristic) curve (AUC) can be estimated at each iteration and used as a stopping condition [25].

9.3.3 Architectures

When considering the detection of ERPs with trials of 1 s at a sampling rate of 128 Hz and 64 sensors, a single FC neuron in the first hidden layer of an MLP would have $128 \times 64 = 8{,}192$ different weights. This number can be significantly higher for high density EEG (e.g., 256 sensors) and a high sampling rate (1 kHz). In such a case, with a large number of hidden neurons and/or hidden layers, the number of parameters becomes very high and it can lead to overfitting. These numbers have to be set in relation to the typical ERP experiment or calibration session, with 20 min and a stimulus onset interval of 200 ms, we obtain only 6,000 trials. If we add the assumption of a low target probability (e.g., 10%), a class can be represented by only 600 examples. As there are a large number of features, and it is possible to take into account the separations between the dimensions in time and space, it is possible to adapt the architecture and specify a subset of inputs for the each neuron.

We consider a mask h of size $n_1 \times n_2$ to be applied on a 2D signal. n is an odd number to center the mask with the same number of elements on both sides. The

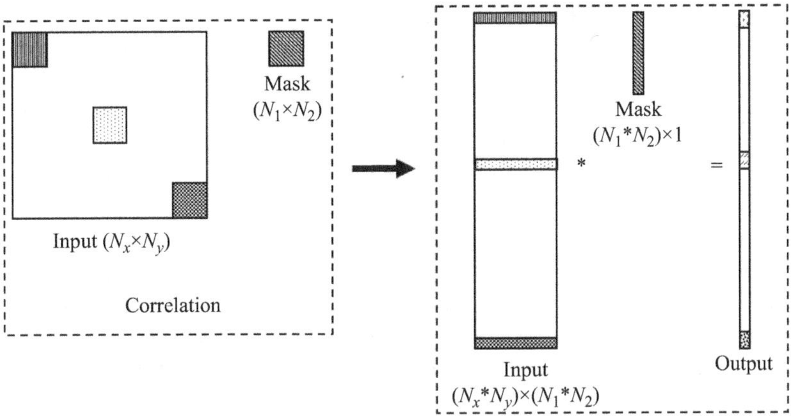

Figure 9.3 Unrolling the correlation operation in a 2D convolutional layer to produce a matrix–matrix product version

application of this mask on a point (i,j) of the input f can be defined as a correlation g_{corr}, or a convolution g_{conv}:

$$g_{corr}(i,j) = \sum_{k=-n_1/2}^{n_1/2} \sum_{l=-n_2/2}^{n_2/2} h(k,l)f(i+k,j+l) \tag{9.8}$$

$$g_{conv}(i,j) = \sum_{k=-n_1/2}^{n_1/2} \sum_{l=-n_2/2}^{n_2/2} h(k,l)f(i-k,j-l) \tag{9.9}$$

In the subsequent sections, the term convolution is employed as the kernels of the filters are learnt and the orientation of the mask has no importance. However, this difference has an implication when they are displayed or given for the initialization of a network as the user must know if they should be flipped or not. When a convolutional layer is included in a network, one aspect of the training phase is to learn the weights of the mask. This key feature is one of the main reason why conv nets are efficient tools for pattern recognition as the feature extraction stage, through the hidden layers, is both data driven and task driven, and the whole training process of the different stages is performed as a whole. While a natural approach would be to apply the mask on each point as presented in the previous equations, a better approach that considers matrix multiplication should be used, as depicted in Figure 9.3. This approach with 2D inputs can by generalized to tensors.

Conv nets take into account the underlying dimensions of the inputs. If the inputs are represented as tensors, the order of the tensor can be used for the creation of the architecture. LeNet was one of the very first CNNs which helped propel the field of Deep Learning. This pioneering work by Yann LeCun was named LeNet5 after many previous successful iterations since the 1980s. LeNet architectures were tailored for handwritten character recognition tasks such as reading zip codes for automatic document analysis. Despite some major contributions with CNNs at the beginning of the 2000s [5], this type of system was not favored due to the architecture engineering requirement, and it was therefore not seen as a pure learning system that does not need any knowledge from the domain. In addition, the rise of more powerful CPUs and the availability of GPUs shifted the implementation paradigm from a connectionist spirit, i.e., an interconnected network of simple units, toward a more computational approach with parallel computing [6]. In 2012, AlexNet, which was a deeper and much wider version of the LeNet architectures, won the difficult ImageNet competition with outstanding results. AlexNet scaled the insights of LeNet models into a larger and deeper model that could be used to learn to classify more complex objects. The success of AlexNet significantly enhanced the interests for CNNs in the computer vision community.

The main recent contributions related to conv nets can be resumed as follows:

- The ReLUs.
- The dropout technique to randomly ignore single neurons during training to avoid overfitting.
- Max pooling to avoid the averaging effects of average pooling.
- GPUs to reduce training time, and therefore to test more architectures.
- The availability of free open-source libraries in various programming languages that allow to focus on the architecture without worrying about the implementation of a large number of components in the system.

Other architectures can be coupled with conv nets, such as recurrent neural networks with long short-term memory [26,27] and generative adversarial networks [28].

A conv net is a sequence of layers that are stacked and transform their inputs. The transformation should be differentiable to allow optimization through gradient descent. Three types of layers are usually considered to create a conv net:

- Convolutional layer: for feature extraction through the learnt kernels, to enhance discriminant features.
- Pooling layer: for feature reduction and local invariance.
- FC layer: no prior information is given between two layers.

Therefore, a conv net starts with an input layer, a succession of convolutional, pooling, and FC layers. If the input is an image, there are two dimensions (width and height) or three (e.g., the addition of color channels). With the EEG or MEG signals, it is common to consider two dimensions (time and space). Thanks to the knowledge of the main dimensions of the input signal, it is possible to take advantage of it and constrain the architecture to process the signal dimension per dimension or to take into account the relationships within a dimension. The number of possible transformations in each layer is defined as the number of feature maps or the depth of the layer. For instance, if we consider 10 spatial filters for the first convolutional layer, then the number of maps is 10 and the depth of this layer is 10.

The following architecture describes an example of conv net for ERP detection:

- The input [64×128] holds the values of the different time points from the different sensors after the stimulus onset; here, the trial has 64 sensors and 1 s length with a sampling rate of 128 Hz.
- The conv layer computes the output of neurons that are connected to subsets of neurons in the previous layer, in this case the input layer. Each neuron computes a dot product between their weights and a subset of neurons they are connected to. The structure of the local connections is usually similar across the different neurons (e.g., a matrix with a constant size to have a constant number of inputs across the neurons). If we consider 10 maps and a filter of size [64×2], we obtain a set of neurons of size [1×64×10].
- An activation function is applied on all the neurons of the first convolutional layer (e.g., ReLU function). The set of neurons is unchanged, only the state of each neuron has changed.
- The pooling layer performs a downsampling operation along the time dimension. [1×32×10] with a downsampling of factor 2.
- The FC layer computes the class scores, resulting in set of neurons of size [1×1×2] or [2], corresponding to target and nontargets.

In this architecture, the conv net transforms the EEG signal layer by layer from the values in μV to the scores of the target and nontargets. It is worth mentioning that some layers include a set of parameters. Both the conv and FC layers perform transformations that are based on the weights and biases of the neurons, which are trained by gradient descent. The ReLU and pooling layers correspond both to a fixed function; hence, no training is required for these layers. Therefore, a conv net architecture can be defined as a sequence of layers that transform the EEG signal into another signal through a succession of convolution, activation, and pooling functions.

The convolutional layer parameters correspond to linear filters that are learnt. A filter is usually small in relation to the size of the input size. For images, it is common to consider filters of size 3×3 or 5×5. For EEG signals, it depends on the sampling rate and what is meant to be extracted. Because there is no notion of neighborhood in the spatial domain, as all the sensors are placed in a single dimension, the size of the filter in the spatial dimension can be set as the number of sensors. However, the spatial dimension can be decomposed into two dimensions to capture the 2D aspect of the scalp. In this case, the input signal can be represented as a 3D structure [height × width × timepoints], where the information in the height and width dimensions represent the coordinates of the sensors. If the input is [64×128] (64 sensors, 128 time points) and we consider a spatial filter of size [64×1], then during the forward propagation, the filter is convolved across the different time points of the input signal, computing a dot product between the values of all the sensors at a particular time point and the weights of the filter. As the filter is slid over the time dimension of the input signal, it produces a one-dimensional activation map that gives the output of this spatial filter: a signal of size [1×128]. The same way, if we consider a temporal filter of size [1×4] applied on the input signal, then we can obtain a two-dimensional activation map of size [64×32] with a stride equal to the size of the filter, i.e., without overlapping.

With spatial filtering, the output $y(i)$ of the ith neuron in the feature map m of the layer l can be defined as

$$y(i) = \sum_{j=1}^{N_{m,l}} x(i,j) \cdot w_j \tag{9.10}$$

where $x(i,j) \in \mathbb{R}^{N_t \times N_s}$ is the EEG trial, where N_t is the number of sampling points, and N_s is the number of sensors. w_j with $1 \leq j \leq N_{m,l}$ represent the weights of the filter, $N_{m,l}$ is the number of neurons.

With convolutional layers, there are key parameters for arranging the connections between a layer and its next layer after the convolution. These parameters will determine how many neurons are present in the next layer. More complex structures can be created by mixing feature maps such as in the layer C3 of LeNet5 [4]. In the next sentences, the input/output layers represent the input/output of the convolutional layer and not the input/output of the whole network.

- Depth (number of feature maps): the depth corresponds to the number of filters to be created, i.e., the number of transformations of the input.
- Stride: this parameter represents how the filter is going to be moved across the input. If the stride is 1, the filter will visit all the neurons of the input. If the stride is 2, the filter will skip one neuron at a time while moving across the input. With a stride of 2, it is possible to reduce the size of the features in the output layer.
- Zero-padding: due to the size of the filter, it can be necessary to add extra values around the input in order to obtain the same number of neurons in the output as in the input layer.

Depth, stride, and zero-padding must be chosen to always keep an integer number of neurons in each layer. The parameters must be chosen to better capture the content of the signal and to set the number of features judiciously through the layers until the final output layer. The stride parameter can be therefore used in convolutional layers for avoiding pooling layers [5].

The size of a dimension d of the output of a convolutional layer can be determined by

$$O_d = 1 + \frac{(I_d - R_d + 2P_d)}{S_d} \tag{9.11}$$

where I_d is the number of neurons, R_d is the size of the receptive field, S_d is the applied stride, and P_d the amount of zero padding, all in the dimension d of the input. For example, with an input of [64×128] and a receptive field of size [64×5], with padding [0×2], stride [1×2], we obtain an output of size [1×65]. The dimension of the different outputs has to be properly adjusted to maintain integer values through the different layers. For inputs with a large number of features, a pooling layer can be introduced between conv layers in order to reduce the number of features while introducing some local invariance. The goal of the pooling layer is to reduce the size of the representation in order to limit the amount of parameters and hence the computation in the conv net. The pooling layer is performed independently on every feature map of the input and resizes it spatially, using the Max operation. It is worth noting that for the sole purpose of reducing the number of features through the network, it is possible to set the stride parameter in the convolutional layer to 2 or more if necessary.

9.4 Methods

We consider inputs of size $N_t \times N_c$ where N_t and N_c represent the number of time points and the number of channels, respectively. The architecture of a conv net contains several hidden layers, including at least one convolutional layer. In this type of network, the weights of the layer are shared across the different inputs. The weight sharing model reduces the number of parameters to learn in the network, making a conv net faster to train. The convolution on a 2D signal such as an image behaves like the application of a linear filter, follows by an activation function (e.g., sigmoid or ReLU function). The architecture of a conv net depends on the problem and the type of variations across trials. With images, it is easy to get a good feeling about the geometric deformations that can be applied on the signal and keep the same label. For the choice of the architecture with images, it is also possible to get inspired by the visual system in the human brain. However, for the classification of brain responses, it is more difficult to determine what type of features must be extracted. For the detection of ERPs, the experimental protocol has a central role as it will determine what will be the ERP components and their characteristics that can vary between two conditions (i.e., target vs. nontarget). For the classification of ERPs for BCI that include the P300, a large ERP component, it is important to determine the type of changes that can occur over time. A first assumption is the stationarity of the spatial location of the brain responses of interest. We assume that there exists a finite subset of spatial distribution where we can find discriminant information for the task. This first stage can be achieved by spatial filtering, where the input signals acquired from different channels is then projected to one or several "virtual" channels. Spatial filtering can be achieved through a convolutional layer of size $[1 \times N_c]$. It is worth noting that the spatial filtering stage can include more layers, e.g., by grouping sensors that are close, taking into account the spatial location of each sensor. In such a case, weight sharing may not be the ideal choice as each neighborhood may rely on a specific function. The next processing step can deal with the reduction of the number of time points. This stage is necessary if the signal has a high sampling rate and the signal has not been downsampled. In recent BCI competitions related to ERPs detection, the signal was often bandpassed to frequencies in the alpha band, suggesting the number of time points could be limited if the main ERP component is the P300. Other signal processing steps that can be included through the architecture of the conv net include the extraction of shift invariant features.

9.5 Experimental protocol

In this study, we consider the data from an RSVP task. In BCI, various RSVP tasks have been proposed to increase information throughput for image analysis by sorting and triaging images [13,14,29–32], and face recognition tasks [33,34]. The EEG database was previously used in [35,36]. The experimental protocol is described thereafter [Figure 9.4(a)]. Participants were seated 75 cm from a Dell P2210 monitor. They viewed a series of simulated images from a desert metropolitan environment using an RSVP paradigm. Images (960×600 pixels, 96 dpi, subtending 36.3×22.5) were presented using E-prime software on a Dell Precision T7400 PC. Images were presented with a stimulus onset asynchrony of 0.5 s, with no inter-stimulus interval. Images contained either a scene without any people (nontarget) or a scene with a person holding a gun (target). 110 target images and 1,346 nontarget images were presented to each participant. Scenes in which a target appeared were also presented without the person in the nontarget condition. All stimuli appeared within 6.5 degrees of center of the monitor. The purpose of the task was to discriminate target images from nontarget images. In the considered data, 16 participants responded to targets by silently counting the number of targets. Electrophysiological recordings were digitally sampled at 1,024 Hz from 64 scalp electrodes arranged in a 10-10 montage using a BioSemi Active Two system (Amsterdam, The Netherlands). Impedances were kept below 25 kΩ. External leads were placed on the outer canthus of both eyes, and above and below the right orbital fossa to record the electrooculogram signal. The signal was then bandpassed between 0.1 and 21.33 Hz using a fourth-order Butterworth filter. Finally, the signal was downsampled to 64 Hz. For each stimulus, we selected the signal corresponding to 800 ms after the stimulus onset, i.e., 51 time points. Hence, each example in the database is a matrix of size 51×64, which is about the same size of the images in computer vision problems.

9.5.1 Conv nets

We propose 12 neural networks architectures in order to highlight the effect of some particular hidden layers. In all the models, the activation unit is an ReLU function ($f(x) = \max(0, x)$), the number of neurons in the output layer is set to 2, i.e., the number of classes, and the outputs are normalized with a softmax function. The first architecture NN_1 has no hidden layers: the inputs are directly connected to the output layer (6,530 parameters in the model). In MLP_1, MLP_2, MLP_3, and MLP_4, there is a single hidden layer with 10, 50, 100, and 200 neurons, respectively. It corresponds to 32,672 parameters for MLP_1 and up to 653,402 parameters for MLP_4. For CNN_1, CNN_2, CNN_3, and CNN_4, we use only a single convolutional layer with a spatial filtering purpose ([1×64]) with 4, 8, 16, and 32 feature maps, respectively. For CNN_1, the number of parameters is 666 while it reaches 5,314 for CNN_4. With CNN_5, we consider a first convolutional layer with 8 maps, the convolution window is set to [1×64], it corresponds to a spatial filtering function. In the second convolutional layer, the layer has only a single dimension (in time). We use 16 maps with a convolution window set to [48×1]; hence, this layer is almost a classification

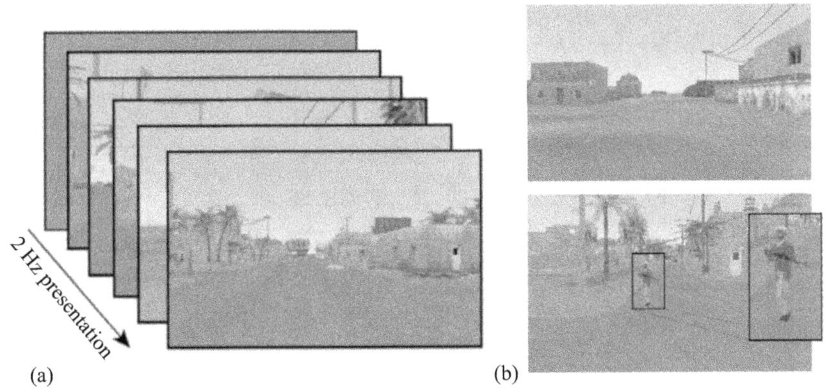

Figure 9.4 *(a) Rapid serial visual presentation task. (b) Representative examples of stimuli on target (bottom) and nontarget trials (top). The inset showing a target is for illustration purposes only, it did not appear in actual stimuli*

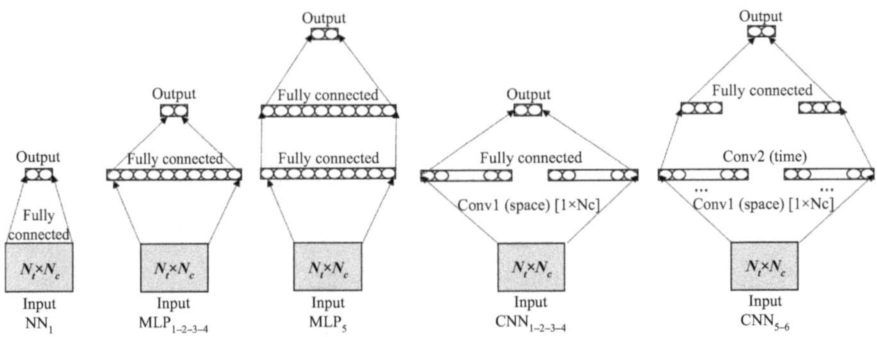

Figure 9.5 *The different architectures*

layer as it considers most of the inputs. Thanks to this convolution, we obtain a set of outputs that corresponds to different shifts of a part of the input features. In CNN_6, the second convolutional layer is set to $[3 \times 1]$ and is equivalent to some temporal filtering. CNN_5 and CNN_6 have, respectively, 1,378 and 2,098 parameters. Finally, this layer is FC to the output layer. In MLP_5, we consider two hidden layers of 50 neurons, corresponding to 165,902 parameters. The different architectures are depicted in Figure 9.5.

The selection of the best model is based on the maximization of the AUC by using a validation dataset. Because the number of target trials is significantly inferior to the number of nontarget trials, the target trials have been replicated in the training dataset for each trial. Before running the tests, the size of the mini-batch was set based different evaluations (mini-batch size: 5, 10, 50, 100, 200, 250, 300) with an MLP.

Table 9.1 Single-trial performance (AUC × 10e3) for different classifiers (each model is trained for each subject)

Subject	NN_1	MLP_1	MLP_2	MLP_3	MLP_4
1	889 ± 20	858 ± 20	856 ± 18	840 ± 32	849 ± 18
2	877 ± 45	799 ± 110	798 ± 66	826 ± 60	820 ± 63
3	967 ± 18	969 ± 14	967 ± 6	967 ± 8	963 ± 14
4	924 ± 28	868 ± 63	879 ± 58	877 ± 76	887 ± 64
5	873 ± 55	874 ± 75	882 ± 63	861 ± 94	876 ± 76
6	869 ± 37	842 ± 44	844 ± 42	849 ± 50	848 ± 54
7	946 ± 15	934 ± 16	942 ± 10	931 ± 20	920 ± 9
8	716 ± 132	791 ± 100	834 ± 94	805 ± 62	732 ± 75
9	948 ± 22	956 ± 23	956 ± 31	958 ± 25	961 ± 22
10	778 ± 53	692 ± 95	703 ± 73	725 ± 98	727 ± 94
11	897 ± 49	892 ± 76	871 ± 58	898 ± 66	881 ± 92
12	955 ± 21	949 ± 23	948 ± 30	946 ± 27	943 ± 26
13	964 ± 18	971 ± 19	971 ± 20	970 ± 19	969 ± 27
14	951 ± 23	942 ± 46	949 ± 41	944 ± 48	943 ± 47
15	883 ± 45	854 ± 73	851 ± 73	850 ± 74	844 ± 88
16	908 ± 45	890 ± 46	884 ± 39	892 ± 49	877 ± 46
Mean	897 ± 39	880 ± 53	883 ± 45	884 ± 51	878 ± 51
SD	68 ± 28	76 ± 32	72 ± 25	68 ± 27	75 ± 29

The mini-batch size that maximized the AUC was 250 (AUC = 0.895, the worst size was the smallest one: 5 with AUC = 0.823).

9.5.2 Performance evaluation

For the performance evaluation, we consider two conditions. In the first condition, a model is trained for each participant. We consider a 5-fold cross validation procedure, where 1 fold is used for the test, 1 fold is used for the validation, and the 3 remaining folds are used for training the model. In the second condition, a model is trained for all the participants. We consider a 16-fold crossvalidation procedure where 1 fold, i.e., 1 subject is dedicated to the test, 1 fold is used for the validation, and the other blocks are used to train the model. With the last condition, the trials from different subjects are not mixed; hence, the classifier is never trained on data from a subject that is used for the test or the validation. The system was implemented with Matlab2017a using an Intel i7-6700K, 32Gb, and an NVidia GTX1080 graphic card.

9.6 Results

The results for the models that were trained for each subject are presented in Tables 9.1–9.3. For each subject and architecture, the mean and standard deviation

Table 9.2 Single-trial performance (AUC × 10e3) for different CNNs (each model is trained for each subject)

Subject	**CNN$_1$**	**CNN$_2$**	**CNN$_3$**	**CNN$_4$**
1	857 ± 37	858 ± 36	847 ± 20	829 ± 65
2	812 ± 44	839 ± 43	824 ± 41	844 ± 53
3	969 ± 21	970 ± 13	971 ± 12	952 ± 40
4	879 ± 26	882 ± 40	878 ± 34	875 ± 39
5	876 ± 68	883 ± 65	874 ± 70	872 ± 62
6	872 ± 48	862 ± 50	870 ± 45	830 ± 83
7	926 ± 33	939 ± 27	935 ± 28	936 ± 24
8	820 ± 48	814 ± 60	824 ± 64	817 ± 37
9	954 ± 35	947 ± 37	956 ± 37	954 ± 44
10	801 ± 61	806 ± 67	779 ± 62	791 ± 55
11	881 ± 64	883 ± 78	883 ± 68	892 ± 67
12	953 ± 23	956 ± 22	959 ± 24	958 ± 26
13	969 ± 18	974 ± 18	973 ± 18	974 ± 20
14	956 ± 26	954 ± 39	959 ± 31	962 ± 29
15	868 ± 70	877 ± 55	880 ± 70	860 ± 81
16	895 ± 45	890 ± 54	888 ± 48	893 ± 50
Mean	893 ± 42	896 ± 44	894 ± 42	890 ± 48
SD	5617	55 ± 19	59 ± 20	59 ± 19

Table 9.3 Single-trial performance (AUC×10e3) for different models (each model is trained for each subject)

Subject	**CNN$_5$**	**CNN$_6$**	**MLP$_5$**
1	845 ± 22	834 ± 59	852 ± 26
2	825 ± 46	773 ± 103	832 ± 70
3	970 ± 16	965 ± 24	976 ± 10
4	842 ± 32	816 ± 30	881 ± 60
5	886 ± 54	869 ± 40	882 ± 75
6	861 ± 57	856 ± 14	828 ± 74
7	943 ± 22	920 ± 15	935 ± 14
8	840 ± 78	824 ± 75	803 ± 49
9	953 ± 37	950 ± 42	961 ± 29
10	799 ± 45	802 ± 16	694 ± 71
11	892 ± 65	895 ± 74	873 ± 66
12	957 ± 29	950 ± 25	949 ± 25
13	969 ± 24	968 ± 26	971 ± 25
14	951 ± 36	887 ± 47	932 ± 52
15	872 ± 90	849 ± 92	862 ± 83
16	885 ± 31	820 ± 89	892 ± 49
Mean	893 ± 43	874 ± 48	883 ± 49
SD	57 ± 21	62 ± 30	74 ± 24

Table 9.4 Single-trial performance (AUC×10e3) for different CNN architectures (each CNN model is trained with 14 subjects)

Subject	NN_1	MLP_1	MLP_2	MLP_3	MLP_4	MLP_5
1	903	904	910	912	894	893
2	784	828	805	813	821	812
3	954	971	969	967	963	975
4	904	909	914	908	910	912
5	915	895	894	916	913	902
6	849	852	861	859	865	860
7	914	924	923	920	910	924
8	703	717	705	758	720	687
9	948	937	948	953	964	944
10	712	719	720	722	684	701
11	899	895	910	889	895	912
12	961	951	967	969	974	969
13	947	947	945	935	966	951
14	918	931	911	908	915	930
15	821	844	873	863	846	869
16	925	942	925	944	930	927
Mean	879	885	886	890	885	885
SD	82	77	79	72	84	85

across the 5 folds are given. In the first table, the best performance is achieved with the most simple method, with a performance of 0.897 ± 0.039. The second table depicts the trend that can be observed by changing the number of spatial filters. Overall, the number of spatial filters has not a significant impact for this evaluation.

The results for the models that are trained with several subjects are presented in Tables 9.4 and 9.5. Each row i corresponds to the result for when the test data is from the subject i and the training data are from the other subjects. The best performance is obtained with CNN_4 with an average AUC of 0.902. Contrary to the first evaluation where the results were relatively similar between the different methods, there are more differences with the evaluation considering all the subjects.

9.7 Discussion

Artificial neural networks have been proven efficient in a large number of pattern recognition tasks, in particular in computer vision by reaching human performance in some datasets. However, to transfer this efficiency into problems related to the detection of brain responses, it is important to evaluate the extent to which the training inputs are likely to present an underlying structure that can be discovered and exploited for optimizing the classification accuracy. Contrary to images, the gain that can be obtained with conv nets can be limited, as it depends on how the dataset is evaluated.

Table 9.5 Single-trial performance (AUC×10e3) for different CNN architectures (each CNN model is trained with 14 subjects)

Subject	CNN_1	CNN_2	CNN_3	CNN_4	CNN_5	CNN_6
1	906	919	917	922	881	897
2	792	810	808	827	812	817
3	974	970	971	969	974	971
4	931	918	907	915	908	907
5	914	913	923	920	931	898
6	836	850	860	866	852	862
7	906	942	947	943	937	943
8	739	769	739	783	782	769
9	967	965	960	963	959	950
10	742	744	716	719	754	786
11	901	887	903	901	926	906
12	964	976	981	977	984	980
13	927	952	936	946	952	938
14	918	942	953	937	938	922
15	869	868	888	903	888	905
16	931	925	940	940	940	929
Mean	888	897	897	902	901	899
SD	74	71	79	71	69	62

In the presented experiment, the advantage of conv nets was only justified when the model had to capture the differences that exist across several subjects, showing that there exists a variability between users that require an appropriate architecture. With a model that is specified for a single participant, the obtained results indicate that simple models could be sufficient. These results are counter intuitive based on previous studies that indicated a significant enhancement thanks to the conv net approaches. This effect can be due to the high quality of the recordings and the low variability of the signal across the trials.

Methods based on linear algebra such as xDAWN [37] for spatial filtering and linear discriminant analysis and its variant for classification have offered in different BCI applications, such as the P300 speller, state-of-the-art performance. With the recent availability of tools that allow biomedical engineers to focus only on the neural network architecture, conv nets provide now a reliable alternative to linear algebra based techniques for neural engineering and neuroimaging processing. In the present paper, we have focused on the evaluation of deep artificial neural networks, more particularly CNNs. We have shown that it is possible to achieve a better performance by training a model with a large number of subjects, instead of learning a model for each individual, by taking advantage of a higher number of training samples. These results are significant because a key problem in BCI, and in many human–computer interaction applications, is the need of a calibration session to model the characteristics of the user. A calibration session before each use of the system can significantly decrease the users comfort and impact the potential of a commercial application. Thanks to a model that is able to capture a large variability across trials, it is then possible to model a multisubject classifier. While the type of system may require a long time for the estimation of the model, i.e., to train the classifier, the performance can be robust enough for the creation of a generic classifier. Different preprocessing steps can be included within the architecture of a conv net, such a spatial filtering, and the extraction of shift invariant features.

9.8 Conclusion

Artificial neural networks and in particular CNNs have demonstrated their high performance in different pattern recognition and artificial intelligence tasks. Complex and deep architectures have the possibility to reach human performance in computer vision tasks. For the detection of brain responses and for the use of CNN in neuroimaging, a key challenge is to understand the type of variability that can be observed across trials. While previous studies during recent years have focused on the decrease of the calibration session duration in BCI, new opportunities arise with CNNs as it is possible to create models for ERP detection that can be used by several users for a performance as good as models that are specifically tailored for an individual. Further work will be dedicated on the optimization of the architecture and the addition of artificial trials to extend the size of the training database. Finally, while very deep architectures could provide better results than what was presented in this paper, the choice of the architecture should be ideally justified and its performance could help to better understand the variations that occur across trials during an experimental task.

References

[1] Cireşan D, Meier U, Gambardella LM, *et al.* Convolutional Neural Network Committees For Handwritten Character Classification. In: Proc. of the 11th Int. Conf. on Document Analysis and Recognition (ICDAR); 2011. p. 1135–1139.

[2] Fukushima K, Miyake S, Ito T. Neocognitron: a neural network model for a mechanism of visual pattern recognition. IEEE Transactions on Systems, Man, and Cybernetics. 1983;SMC-13:826–834.

[3] LeCun Y, Boser B, Denker J, *et al.* Handwritten Digit Recognition with a Back-Propagation Network. In: Advances in Neural Information Processing Systems (NIPS). vol. 2; 1990. p. 396–404.

[4] LeCun Y, Bottou L, Bengio Y, *et al.* Gradient-Based Learning Applied to Document Recognition. Proceedings of the IEEE. 1998 Nov;86(11): 2278–2324.

[5] Simard P, Steinkraus D, Platt JC. Best Practices for Convolutional Neural Networks Applied to Visual Document Analysis. In: Proc. of the 7th Int. Conf. Document Analysis and Recognition (ICDAR); 2003. p. 958–962.

[6] Chellapilla K, Puri S, Simard PY. High Performance Convolutional Neural Networks for Document Processing. In: Proc. 10th Int. Workshop Frontiers in Handwriting Recognition; 2006.

[7] Steinkraus D, Buck I, Simard PY. GPUs for Machine Learning Algorithms. In: Proc. ICDAR. vol. 2; 2005. p. 1115–1120.

[8] Manor R, Geva AB. Convolutional neural network for multi-category rapid serial visual presentation BCI. Frontiers in Computational Neuroscience. 2015;9(146).

[9] Cecotti H, Sato-Reinhold J, Sy JL, *et al.* Impact of Target Probability on Single-Trial EEG Target Detection in a Difficult Rapid Serial Visual Presentation Paradigm Task. In: 33nd Int IEEE Conf. of the Eng. in Medicine and Biology Soc.; 2011.

[10] Cecotti H, Eckstein MP, Giesbrecht B. Single-trial classification of event-related potentials in rapid serial visual presentation tasks using supervised spatial filtering. IEEE Transactions on Neural Networks and Learning Systems. 2014 Nov;25(11):2030–2042.

[11] Cecotti H. A time-frequency convolutional neural network for the offline classification of steady-state visual evoked potential responses. Pattern Recognition Letters. 2011;32(8):1145–1153.

[12] Jarrett K, Kavukcuoglu K, Ranzato M, *et al.* What is the Best Multi-Stage Architecture for Object Recognition? In: Proc. of the 12th Int. Conf. on Computer Vision (ICCV'09); 2009. p. 2146–2153.

[13] Gerson A, Parra L, Sajda P. Cortically-coupled computer vision for rapid image search. IEEE Transactions on Neural Systems and Rehabilitation Engineering. 2006;14(2):174–179.

[14] Pohlmeyer EA, Wang J, Jangraw DC, *et al.* Closing the loop in cortically-coupled computer vision: a brain-computer interface for searching image databases. Journal of Neural Engineering. 2011;8:036025.

[15] Luck SJ. An Introduction to the Event-Related Potential Technique. Cambridge, MA: The MIT Press; 2005.

[16] Childers DG, Perry NW, Fischler IA, *et al.* Event-related potentials: a critical review of methods for single-trial detection. Critical Reviews in Biomedical Engineering. 1987;14(3):185–200.

[17] Sutton S, Braren M, Zubin J, *et al.* Evoked-potential correlates of stimulus uncertainty. Science. 1965;26(150):1187–1188.

[18] Donchin E, Spencer KM, Wijesinghe R. The mental prosthesis: assessing the speed of a P300-based brain-computer interface. IEEE Transactions on Rehabilitation Engineering. 2000;8(2):174–179.

[19] Donnerer M, Steed A. Using a P300 brain-computer interface in an immersive virtual environment. Presence: Teleoperators and Virtual Environments. 2010;19(1):12–24.

[20] Polich J. Updating P300: an integrative theory of P3a and P3b. Clinical Neurophysiology. 2007;118:2128–2148.

[21] Dien J, Spencer KM, Donchin E. Parsing the late positive complex: mental chronometry and the ERP components that inhabit the neighborhood of the P300. Psychophysiology. 2004;41(5):665–678.

[22] Folstein JR, Van Petten C. After the P3: late executive processes in stimulus categorization. Psychophysiology. 2011;48(6):825–841.

[23] Donchin E. A multivariate approach to the analysis of average evoked potentials. IEEE Transactions on Biomedical Engineering. 1966;13(3):131–139.

[24] Donchin E. Discriminant analysis in average evoked response studies: the study of single trial data. Electroencephalography and Clinical Neurophysiology. 1969;27:311–314.

[25] Fawcett T. An introduction to ROC analysis. Pattern Recognition Letters. 2006;27:861–874.

[26] Hochreiter S, Schmidhuber J. Long short-term memory. Neural Computation. 1997;9(8):1735–1780.

[27] Sainath TN, Vinyals O, Senior A, *et al.* Convolutional, Long Short-Term Memory, Fully Connected Deep Neural Networks. In: Proc. IEEE Int. Conf. on Acoustics, Speech and Signal Processing (ICASSP); 2015. p. 4580–4584.

[28] Goodfellow IJ, Pouget-Abadie J, Mirza M, *et al.* Generative Adversarial Networks. arxivorg. 2014.

[29] Bigdely-Shamlo N, Vankov A, Ramirez RR, *et al.* Brain activity-based image classification from rapid serial visual presentation. IEEE Transactions on Neural Systems and Rehabilitation Engineering. 2008;16(5):432–441.

[30] Parra LC, Christoforou C, Gerson AD, *et al.* Spatio-temporal linear decoding of brain state: application to performance augmentation in high-throughput tasks. IEEE Signal Processing Magazine. 2008;25(1):95–115.

[31] Pohlmeyer EA, Jangraw DC, Wang J, *et al.* Combining Computer and Human Vision Into a BCI: Can the Whole be Greater Than the Sum of Its Parts? In: Proc. of the 32nd Int. IEEE EMBC Conf. Buenos Aires, Argentina; 2010. p. 138–41.

[32] Yu K, Ai-Nashash H, Thakor N, *et al.* The analytic bilinear discrimination of single-trial EEG signals in rapid image triage. PLoS One. 2014;9(6):e100097.

[33] Cai B, Xiao S, Jiang L, *et al.* A Rapid Face Recognition BCI System Using Single-Trial ERP. In: Proc. of the 6th Int. Neural Eng. IEEE/EMBS Conf.; 2013. p. 89–92.

[34] Touryan J, Gibson L, Horne HJ, *et al.* Real-time measurement of face recognition in rapid serial visual representation. Frontiers in Psychology. 2011;2(42):1–8.

[35] Marathe AR, Ries AJ, Lawhern VJ, *et al.* The effect of target and non-target similarity on neural classification performance: a boost from confidence. Frontiers in Neuroscience. 2015;9:1–11.

[36] Cecotti H, Marathe A, Ries AJ. Optimization of single-trial detection of event-related potentials through artificial trials. IEEE Transactions on Biomedical Engineering. 2015;62(9):2170–2176.

[37] Rivet B, Souloumiac A. Optimal linear spatial filters for event-related potentials based on a spatio-temporal model: asymptotical performance analysis. Signal Processing. 2013;93(2):387–398.

Chapter 10

Signal models for brain interfaces based on evoked response potential in EEG

Yeganeh M. Marghi[1], Paula Gonzalez-Navarro[1], Fernando Quivira[1], James McLean[1], Bruna Girvent[1], Mohammad Moghadamfalahi[1], Murat Akcakaya[2], and Deniz Erdogmus[1]

Abstract

Electroencephalography (EEG)-based brain–computer interfaces (BCIs) are developed to provide access channels for alternative communication and control systems to people with severe speech and physical impairments. Designs that exploit evoked response potentials (ERPs) in EEG constitute the majority of research efforts dedicated to noninvasive BCIs. Visual, auditory, and tactile stimulation paradigms are used to actively probe the user's brain to collect EEG evidence towards inferring intent in the context of the particular application. As assistive technology devices, however, existing EEG-based BCIs lack sufficient speed and accuracy to safely and reliably restore function at acceptable levels. This is mainly because the recorded EEG signals are not only noisy with a low signal-to-noise ratio, but are also nonstationary, due to physiological or environmental artifacts, sensor failure, and user fatigue. In this chapter, we address how reliable intent inference engines with reasonable speed and accuracy can be developed using parametric modeling. Examples of real-world data in the framework of the ERP-based BCI paradigm are provided to exemplify our detection and classification methods.

10.1 ERP-based BCIs

Electroencephalography (EEG)-based brain–computer interface (BCI) systems restore the users' abilities in communication and control by providing a direct

[1]Electrical and Computer Engineering Department, Northeastern University, USA
[2]Electrical and Computer Engineering Department, University of Pittsburgh, USA

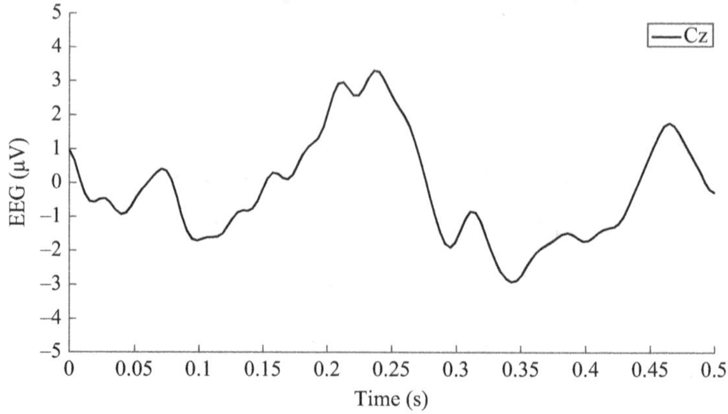

Figure 10.1 Example of a P300 waveform from an actual data recorded from electrode Fz and averaged across 100 trials in a rapid serial visual presentation (RSVP) task

electronic interface and measurement of the brain activities. These systems work based on detection of evoked response potentials (ERPs), i.e., a signature change in EEG pattern in response to specific events or stimuli [1]. ERPs are time-locked brain responses to sensory, motor, or cognitive events that can represent neural activities related to brain processes. Depending on the timing and polarity of their peaks, ERPs are divided into different types. Typically, for human brain signals, fast responses (early waves) correspond to sensory or perceptual processes that are function of the physical parameters of the stimulus. On the other hand, latent ERPs correspond to the cognitive processes [1,2]. One of the extensively studied ERPs is the P300 wave, which is associated with a positive deflection in EEG signals in response to a target, often an unexpected and less frequent stimulus, e.g., in an oddball paradigm. Depending on the attention of the subject and task difficulty, the amplitude and latency of this waveform could change (Figure 10.1). Since its peak has a positive polarity and it typically happens 300 ms after the stimulus onset, it is called P300 [3]. The P300 waveform can be elicited using different forms of stimulus, most commonly visual and more recently, auditory [4,5] and tactile [6,7].

In ERP-based BCIs, stimuli are treated as impulse inputs to a dynamic system that is assumed to be at rest. EEG that is time-locked to stimulus onset is used to estimate the intent. The user intent is usually classified on a per trial basis where each stimulus is considered to be a trial. In most designs, a sequence of trials with overlapping temporal windows of EEG are used as a measurement conditioned to the label of the stimulus whose onset corresponds to the time-zero of the window. The P300 waveform typically occurs within 0–1,000 ms after stimuli onsets. In BCI design, the labeled EEG is collected in supervised calibration sessions, and the supervised data is then used to model the EEG and to develop intent inference engines.

10.1.1 Multidimensional EEG classification

In ERP-based BCI systems, EEG signals are often recorded from multiple EEG channels. Multidimensionality in BCI trials is mainly associated with two things: (i) multichannel EEG measurement and (ii) latencies with respect to stimulus onset (temporal sampling of delayed brain responses). Raw data being naturally high dimensional, there is a need for dimension reduction techniques that are appropriate for the purpose, e.g., principal component analysis (PCA) for band-energy-aware adaptive finite impulse response (FIR) filterbanks or signal-model-aware goal-oriented projections.

Better estimation of conditional EEG feature distributions not only improves user intent estimation performance but also helps to make better simulation-based design choices. Artifact- and interference-free brain signals in EEG can be well represented by nonstationary Gaussian processes over space and time, hence linear filtering and dimension reduction operations do not disturb Gaussianity of feature vectors after such processing [8]. Feature vectors extracted from the supervised EEG data can be used to estimate mean and covariance of corresponding conditional Gaussian distributions. Alternatively, mean and autocorrelation sequences in time can be estimated and used to form structured model parameters. Due to the high dimensionality of the feature vectors or parameters to be estimated, many labeled data samples are needed for accurate and reliable estimates; this necessitates long calibration data collection phases—a problem that can be mitigated through modeling of signal statistics at the population level. Longer acquisition sessions might also not always be possible, due to practicality or user fatigue considerations. Consequently, in the presence of only small amounts of calibration data from short calibration sessions, sample covariance estimates for high-dimensional feature vectors can be rank-deficient (or equivalently, temporal autocorrelations may be underestimated) [9,10]; therefore the resulting intent classifiers lack generalization accuracy. Linear discriminant analysis (LDA), step-wise LDA, quadratic discriminant analysis (QDA), and regularized (quadratic) discriminant analysis (RDA) are popularly used in the literature to build classifiers that offer decision boundaries between discrete user intent labels in the given context (e.g., in a binary selection paradigm, target event vs. nontarget event) [7,11–16]. Although these methods have delivered promising results, they may be oversmoothing the decision boundary in an attempt to tackle the aforementioned rank-deficiency issue, since they typically optimize parameters based on a small number of samples. In Section 10.2.2, an EEG model is introduced in which the covariance matrix is modeled as a Kronecker product of temporal and spatial covariances [17]. The results will illustrate the model fit/complexity trade-off given small number of calibration samples.

10.1.2 Nonstationarities in EEG signals

Even though multichannel ERP-based BCIs are promising systems to provide communications between humans and machines, because of nonstationarities they are not sufficiently reliable to perform at acceptable speed and accuracy levels. Nonstationarities in the EEG reduce the effectiveness of these systems in real-world applications. These nonstationarities change the underlying distribution of the EEG data; therefore, an intent inference engine designed based on a training dataset may not always work with the predicted accuracy or speed. If unaccounted for, these changes in statistical characteristics may drastically hinder the system performance. In recent years, there is an increased interest in developing adaptive methods for the design of EEG-based BCI systems to enable these systems handle nonstationarities in the EEG. Different adaptation methods have been proposed and their performance have been demonstrated either using offline datasets or through experimental studies that are performed in controlled laboratory settings. Such BCI systems employ the following techniques for adaptation: (i) semisupervised and unsupervised learning utilizing unlabeled EEG data [18–23], (ii) designing coadaptation sessions to improve BCI usability [24–27], (iii) various learning methods for subject-to-subject and session-to-session knowledge transfer [28–31], and (iv) robust and unsupervised feature extraction to handle nonstationarities in the EEG [32–36]. Specifically, online adaptation of variations of linear and QDA models under both single Gaussian distribution assumption (such as adaptive LDA, adaptive pool mean, adaptive pool covariance) [37–43] and finite Gaussian mixture models (GMMs) [42,44,45], adaptive support vector machine-based methods [46–49], Kalman filter-based approaches [50–52], and recursive Bayesian techniques [53,54] constitute the state-of-the-art for adaptive EEG-based BCIs.

Even though promising, existing adaptive EEG-based BCIs suffer from the following main shortcomings: (i) adaptation of the systems is performed without direct screening of EEG nonstationarity introducing potentially additional computational burden when adaptation is not required; (ii) the performance of the most adaptation techniques were demonstrated through offline analysis, but real-time application of such techniques may not be feasible; (iii) systems that rely on coadaptation still require cue-guided data collection sessions before the free usage of the system and such an approach may not guarantee the robustness of the system to potential EEG nonstationarities; (iv) online adaptation of the intent inference engines mostly rely on the estimated labels for the unlabeled data, but the estimation accuracy could drastically decrease as EEG starts to show nonstationarities; (v) the adaptive intent inference engines (classifiers) generally assume that the separation surface structures among different classes of EEG data remain the same throughout the test phase which may not be true as the nonstationarities in EEG could change the statistical characteristics of EEG such that the data could follow totally different distributions than the initially assumed and could render the separation structures of classifiers invalid; and (vi) even if various learning techniques are utilized to improve subject-to-subject and session-to-session knowledge transfer, it is not possible to train classifiers that will be robust to all possible nonstationary behaviors in EEG.

10.1.3 Noise in the class labels

Another source resulting in loss of performance for ERP-based BCIs is noisy class labels for calibration data. Standard supervised classifiers are trained using the calibration data under the assumption that the corresponding class labels are absolutely correctly known. In many BCI settings, this assumption breaks down since the human in the loop may behave contrary to instructions intentionally or unintentionally (e.g., attention drifts may occur during calibration). Consequently, the model parameter estimation process, in general, should not take provided labels as absolutely correct but rather treat them as another piece of noisy information. If ignored, the level of error in labels may lead to significantly detrimental outcomes in BCI performance. In Section 10.2.4, we will illustrate how a noisy label approach can be followed for calibration in the context of tactile BCIs—a setting where this issue is most prevalent.

10.2 ERP-based inference

10.2.1 ERP detection

In order to capture the P300 while omitting the possible motor EEG [13] and to improve the signal-to-noise ratio, we need to preprocess EEG signals for the user intent detection. Our preprocessing module consists of:

1. An FIR linear-phase bandpass filter passing [1.5, 42] Hz with zero DC gain and a notch filter at 60 Hz
2. A time window over [0, 500) ms after each stimulus onset for each trial
3. A downsampler
4. A projection to a lower dimensional space through PCA

After passing the EEG signal trough the preprocessing module, finally data from all the channels were concatenated to form a feature vector for each trial.

In ERP-based BCIs, each trial signal can be modeled as a distribution conditioned on the presence of ERP

$$p(\mathbf{y}|s) = N(\mathbf{y}; \mu_s, \Sigma_s)$$

with class-dependent mean and covariance. This is in accordance to standard EEG analysis that models the signals with a Gaussian distribution. EEG evidences from EEG and language model are fused using a Naïve Bayes assumption to make a joint decision using MAP inference. Assume D as the dictionary of all candidates for user intent and $s \in D$. We define s^* as the random variable which represents the actual user intent and $\hat{s}$ is the system estimation. Then, the decision rule is as follows:

$$\begin{aligned}\hat{s} &= \arg\max_{s\in\mathscr{D}} P\left(s|\mathbf{y}; C\right) \\ &= \arg\max_{s\in\mathscr{D}} P\left(\mathbf{y}|s\right) P\left(s|C\right)\end{aligned} \tag{10.1}$$

where $\mathbf{y}$ represents the recorded EEG evidence which was obtained during the presentation of a sequence of symbols and C demonstrates the evidence provided by the language model. Assuming that conditioned on s and the language model evidences are independent from each other, maximization of the posterior distribution as illustrated in the first line of (10.1) is equivalent to the maximization of the multiplication of conditional EEG evidence distribution $P\left(\mathbf{y}|s\right)$ and context prior $P\left(s|C\right)$. Since a symbol presented to a user is either a target or a nontarget, the intent inference can be formulated as a binary decision problem. Accordingly, $P(\mathbf{y}|s)$ is learned for target and nontarget symbol classes using the calibration data.

Mathematical Notations: In the rest of this chapter, we assume that $c \in \{1, \ldots, M\}$ denote the class labels for M BCI tasks. We denote the calibration dataset collected for each class as $\mathbf{Y}^c = \{\mathbf{y}_n, r_n\}$ for $n = 1, \ldots, N$, with $\mathbf{y}_n$ as the EEG measurements (evidence), N as the number of measurements in the calibration data, and $r_n = c$ as the class indicator.

10.2.2 Linear model and covariance matrix structures

The multichannel EEG can be modeled as a linear forward model from unknown brain sources. It can be shown that this modeling approach is equivalent to using an ARMA model over the multichannel EEG under certain assumptions. Linear forward model describes the EEG recorded from N_{ch} channels as a linear combination of signals originated from N_s brain sources. Under certain assumptions, the brain source signals are modeled as: (i) a stationary signal, (ii) an autoregressive process of order one AR(1), and (iii) an autoregressive process of order zero AR(0) to impose more structure on the temporal covariance and hence reduce the number of parameters even further.

To define the multichannel EEG signal as a linear combination of unknown brain sources, assume

$$\mathbf{b}^i[n] = \left[b_1^i[n]\ b_j^i[n]\ \cdots\ b_{N_b}^i[n]\right]^T \in \mathbb{R}^{N_b}$$

as the vector of brain source signals, where N_b is the unknown number of brain sources and $b_j^i[n]$ is the generated signal from source j at time sample n for the ith trial. Then, assume the multivariate measurement at the time instant n, $\mathbf{v}^i[n]$, is defined as

$$\mathbf{v}^i[n] = \left[v_1^i[n]\ v_2^i[n]\ \cdots\ v_{N_{ch}}^i[n]\right]^T \in \mathbb{R}^{N_{ch}}$$

where $v_{ch}^i[n]$ is the nth time sample recorded at channel ch for trial i. Then, the linear model can be defined as

$$\mathbf{v}^i[n] = \mathbf{H}\mathbf{b}^i[n] \tag{10.2}$$

where

$$\mathbf{H} = \begin{bmatrix} h_{1,1} & \cdots & h_{1,N_b} \\ h_{2,1} & \cdots & h_{2,N_b} \\ \vdots & & \vdots \\ h_{N_{ch},1} & \cdots & h_{N_{ch},N_b} \end{bmatrix} \in \mathbb{R}^{N_{ch} \times N_b} \tag{10.3}$$

is the forward combination matrix. It is assumed that the number of brain sources is more than or equal to the number of channels, $N_{ch} \leq N_b$, which means that the matrix $\mathbf{H}$ is a full row rank matrix.

Then, the feature vector of brain sources, $\mathbf{x}^i \in \mathbb{R}^{N_b \cdot N_t}$ is defined as

$$\mathbf{x}^i = \left[\mathbf{b}^i[1]^T\ \mathbf{b}^i[j]^T\ \cdots\ \mathbf{b}^i[N_t]^T\right]^T \tag{10.4}$$

Similarly, we can define $\mathbf{y}^i \in \mathbb{R}^{N_{ch} \cdot N_t}$ as the EEG feature vector as

$$\mathbf{y}^i = \left[\, \mathbf{v}^i[1]^T \; \mathbf{v}^i[2]^T \; \cdots \; \mathbf{v}^i[N_t]^T \,\right]^T \tag{10.5}$$

where N_t is the number of time samples in $[0, w)$ ms time window.

Assuming that the matrix $\mathbf{H}$ is constant through time, the linear model of the EEG feature vector, $\mathbf{y}^i$, in (10.5) becomes

$$\mathbf{y}^i = \mathbf{H}_f \mathbf{x}^i \tag{10.6}$$

where $\mathbf{H}_f \in \mathbb{R}^{(N_t \cdot N_{ch}) \times (N_t \cdot N_b)}$ is a time invariant block diagonal matrix and the $N_{ch} \times N_b$ blocks are equal to the matrix $\mathbf{H}$.

$$\mathbf{H}_f = \begin{bmatrix} \mathbf{H} & 0 & \cdots & 0 \\ 0 & \mathbf{H} & 0 & 0 \\ \vdots & 0 & \ddots & 0 \\ 0 & 0 & 0 & \mathbf{H} \end{bmatrix} \tag{10.7}$$

We also assumed that $\mathbf{b}^1[n], \; \ldots, \; \mathbf{b}^i[n]$ are independent, identically distributed (i.i.d.) samples that come from a Gaussian distribution $\mathcal{N}_{N_b}(\boldsymbol{\mu}_\mathbf{b}[n], \boldsymbol{\Sigma}_\mathbf{b}[n,n])$, where

$$\begin{aligned} \boldsymbol{\mu}_\mathbf{b}[n] &= E[\mathbf{b}[n]] \\ \boldsymbol{\Sigma}_\mathbf{b}[n,n] &= Cov[\mathbf{b}^i[n], \mathbf{b}^i[n]] \end{aligned} \tag{10.8}$$

According to (10.2), $\mathbf{v}^i[n]$ is expressed as $\mathbf{v}^i[n] \sim \mathcal{N}_{N_b}(\boldsymbol{\mu}_\mathbf{v}[n], \boldsymbol{\Sigma}_\mathbf{v}[n,n])$ for which we have

$$\begin{aligned} \boldsymbol{\mu}_\mathbf{v}[n] &= E[\mathbf{v}[n]] \\ \boldsymbol{\Sigma}_\mathbf{v}[n,n] &= Cov[\mathbf{v}^i[n], \mathbf{v}^i[n]] = \mathbf{H}\boldsymbol{\Sigma}_\mathbf{b}[n,n]\mathbf{H}^T \end{aligned} \tag{10.9}$$

Therefore, one should consider that $\mathbf{x}^i$ and $\mathbf{y}^i$ are also generated from $\mathcal{N}(\boldsymbol{\mu}_\mathbf{x}, \boldsymbol{\Sigma}_\mathbf{x})$ and $\mathcal{N}(\boldsymbol{\mu}_\mathbf{y}, \boldsymbol{\Sigma}_\mathbf{y})$, respectively. To estimate $\boldsymbol{\Sigma}_\mathbf{x}$ and $\boldsymbol{\Sigma}_\mathbf{y}$, several spatial/temporal structures have been applied on the brain sources [17].

10.2.2.1 Imposing a spatial characteristics on the brain sources

It is first assumed that the unknown brain sources are statistically independent from each other. This implies that $\boldsymbol{\Sigma}_\mathbf{b}[m,n] \;\forall\, 0 \le m, n \le N_t$ is a diagonal matrix.

$$\boldsymbol{\Sigma}_\mathbf{b}[m,n] = \begin{bmatrix} c_b[m,n]_{1,1} & \cdots & 0 \\ \vdots & \ddots & \vdots \\ 0 & \cdots & c_b[m,n]_{N_b,N_b} \end{bmatrix} \tag{10.10}$$

where $c_b[m,n]_{j,j}$ is a scalar value and represents the correlation between the mth and nth time samples of brain source jth. By assuming that all brain sources have the same covariance structure, $c_b[m,n]$, we have

$$\boldsymbol{\Sigma}_\mathbf{b}[m,n] = c_b[m,n] \cdot \mathbb{I}_{N_b} \tag{10.11}$$

where $\mathbb{I}_{N_b}$ is an identity matrix of size N_b. According to (10.6) and (10.9), the covariance of $\mathbf{y}^i$ is defined in (10.12), which represents that under independence assumption, the class conditional covariance matrices can be factorized as a Kronecker product [17].

$$\boldsymbol{\Sigma}_{\mathbf{y}} = \mathbf{H}_f \boldsymbol{\Sigma}_{\mathbf{x}} \mathbf{H}_f^T = \begin{bmatrix} \mathbf{H}\boldsymbol{\Sigma}_{\mathbf{b}}[1,1]\mathbf{H}^T & \cdots & \mathbf{H}\boldsymbol{\Sigma}_{\mathbf{b}}[1,N_t]\mathbf{H}^T \\ \mathbf{H}\boldsymbol{\Sigma}_{\mathbf{b}}[2,1]\mathbf{H}^T & \cdots & \mathbf{H}\boldsymbol{\Sigma}_{\mathbf{b}}[2,N_t]\mathbf{H}^T \\ \vdots & & \vdots \\ \mathbf{H}\boldsymbol{\Sigma}_{\mathbf{b}}[N_t,1]\mathbf{H}^T & \cdots & \mathbf{H}\boldsymbol{\Sigma}_{\mathbf{b}}[N_t,N_t]\mathbf{H}^T \end{bmatrix}$$

$$= \begin{bmatrix} \mathbf{H}c_b[1,1]\mathbb{I}_{N_b}\mathbf{H}^T & \cdots & \mathbf{H}c_s[1,N_t]\mathbb{I}_{N_b}\mathbf{H}^T \\ \mathbf{H}c_b[2,1]\mathbb{I}_{N_b}\mathbf{H}^T & \cdots & \mathbf{H}c_b[2,N_t]\mathbb{I}_{N_b}\mathbf{H}^T \\ \vdots & & \vdots \\ \mathbf{H}c_b[N_t,1]\mathbb{I}_{N_b}\mathbf{H}^T & \cdots & \mathbf{H}c_b[N_t,N_t]\mathbb{I}_{N_b}\mathbf{H}^T \end{bmatrix} \tag{10.12}$$

$$= \begin{bmatrix} c_b[1,1] & \cdots & c_b[1,N_t] \\ c_b[2,1] & \cdots & c_b[2,N_t] \\ \vdots & & \vdots \\ c_b[N_t,1] & \cdots & c_b[N_t,N_t] \end{bmatrix} \otimes \mathbf{H}\mathbf{H}^T$$

The matrix $\mathbf{H}\mathbf{H}^T \in \mathbb{R}^{N_{ch}\times N_{ch}}$ is defined as the spatial covariance matrix, $\boldsymbol{\Sigma}_{ch}$. To guarantee the invertibility of the spatial matrix, it is required to assume $N_{ch} \leq N_b$.

The temporal covariance matrix, $\boldsymbol{\Sigma}_t$, is defined as follows:

$$\boldsymbol{\Sigma}_t = \begin{bmatrix} c_b[1,1] & \cdots & c_b[1,N_t] \\ c_b[2,1] & \cdots & c_b[2,N_t] \\ \vdots & & \vdots \\ c_b[N_t,1] & \cdots & c_b[N_t,N_t] \end{bmatrix} \in \mathbb{R}^{N_t\times N_t} \tag{10.13}$$

According to (10.12) and (10.13), the full covariance matrix of the EEG feature vectors can be expressed as a Kronecker product of a spatial and temporal covariances matrices as

$$\boldsymbol{\Sigma}_{\mathbf{y}} = \boldsymbol{\Sigma}_t \otimes \boldsymbol{\Sigma}_{ch} \tag{10.14}$$

Under this covariance structure, the number of parameters that need to estimated is

$$N_{p1} = \frac{N_{ch}(N_{ch}+1)}{2} + \frac{N_t(N_t+1)}{2} \tag{10.15}$$

10.2.2.2 Imposing a temporal characteristics on the brain sources

Brain sources as stationary processes

By assuming brain sources as stationary processes in time, we have $\boldsymbol{\Sigma}_\mathbf{b}[m,n] = \boldsymbol{\Sigma}_\mathbf{b}[m-n]$ and $c_b[m,n] = c_b[m-n]$. Then, the temporal covariance matrix becomes a Toeplitz matrix

$$\boldsymbol{\Sigma}_t = \begin{bmatrix} c_b[0] & \cdots & c_b[N_t-1] \\ \vdots & \ddots & \vdots \\ c_b[N_t-1] & \cdots & c_b[0] \end{bmatrix} \tag{10.16}$$

and accordingly the full covariance matrix is characterized by

$$N_{p2} = \frac{N_{ch}(N_{ch}+1)}{2} + N_t \tag{10.17}$$

number of parameters [17].

Brain sources as an autoregressive (AR) model

The pth order AR model of the multivariate signal $\mathbf{b}^i[n]$ can be expressed as

$$\mathbf{b}^i[n] = \sum_{k=1}^{p} A_k \mathbf{b}^i[n-k] + e^i[n] \tag{10.18}$$

where A_k represents the $N_b \times N_b$ weight matrix at time lag k, and $e^i[n]$ is an i.i.d. zero mean Gaussian additive noise that is independent from $\mathbf{b}^i[n]$ [17]. It also can be assumed that not only the brain sources are statistically independent but also the AR model among all the sources is invariant. Then, (10.18) can be written as

$$\mathbf{b}^i[n] = \sum_{k=1}^{p} a_k \cdot \mathbf{b}^i[n-k] + e^i[n] \tag{10.19}$$

where a_k is a scalar weight for the signal at time lag k. By considering $p = 1$ and assuming $\mathbf{b}[0] \sim \mathcal{N}_{N_b}(\boldsymbol{\mu}_\mathbf{b}[0], \boldsymbol{\Sigma}_\mathbf{b}[0,0])$, we have

$$\mathbf{b}^i_n = a_1 \cdot \mathbf{b}^i[n-1] + e^i[n] \tag{10.20}$$

and for $m \leq n$, the mean and the covariance can be written as

$$\mathbf{b}^i[n] = (a_1)^n \cdot \mathbf{b}[0] \Rightarrow E[\mathbf{b}^i[n]] = (a_1)^n \cdot \boldsymbol{\mu}_\mathbf{b}[0], \tag{10.21}$$

$$\begin{aligned} \boldsymbol{\Sigma}_\mathbf{b}[m,n] &= E\{\mathbf{b}^i[n]\mathbf{b}^i[m]^T\} - E\{\mathbf{b}^i[n]\}E\{\mathbf{b}^i[m]\}^T \\ &= a_1^{|(n-m)|} \cdot (E\{\mathbf{b}^i[m]\mathbf{b}^i[m]^T\} - \boldsymbol{\mu}_\mathbf{b}[m]\boldsymbol{\mu}_\mathbf{b}[m]^T) \\ &= a_1^{|(n-m)|} \cdot \boldsymbol{\Sigma}_\mathbf{b}[m,m] \\ &= a_1^{|(n-m)|} \cdot \boldsymbol{\Sigma}_\mathbf{b}[0,0] \end{aligned} \tag{10.22}$$

Since the $\boldsymbol{\Sigma}_\mathbf{b}$ is a symmetric matrix, by substituting (10.22) in (10.4), the covariance matrix of the brain source signals $\boldsymbol{\Sigma}_\mathbf{x}$ becomes

$$\boldsymbol{\Sigma}_\mathbf{x} = \begin{bmatrix} \boldsymbol{\Sigma}_\mathbf{b}[1,1] & \cdots & \boldsymbol{\Sigma}_\mathbf{b}[1,N_t] \\ \boldsymbol{\Sigma}_\mathbf{b}[2,1] & & \boldsymbol{\Sigma}_\mathbf{b}[2,N_t] \\ \vdots & & \vdots \\ \boldsymbol{\Sigma}_\mathbf{b}[N_t,1] & \cdots & \boldsymbol{\Sigma}_\mathbf{b}[N_t,N_t] \end{bmatrix} = \begin{bmatrix} 1 & a_1^{(1)} & \cdots & a_1^{(N_t-1)} \\ a_1^{(1)} & 1 & \cdots & a_1^{(N_t-2)} \\ \vdots & \vdots & \ddots & \vdots \\ a_1^{(N_t-1)} & a_1^{(N_t-2)} & \cdots & 1 \end{bmatrix} \otimes \boldsymbol{\Sigma}_\mathbf{b}[0,0] \tag{10.23}$$

Considering the assumptions introduced in Section 10.2.2.1, we have $\boldsymbol{\Sigma}_\mathbf{b}[m,n] = c_b[m,n] \cdot \mathbb{I}_{N_b}$, then from (10.22), we have

$$\boldsymbol{\Sigma}_\mathbf{b}[0,0] = c_b[0] \cdot \mathbb{I}_{N_b} \tag{10.24}$$

$$\boldsymbol{\Sigma}_\mathbf{b}[m,n] = a_1^{(|n-m|)} \cdot \boldsymbol{\Sigma}_\mathbf{b}[0] = a_1^{(|n-m|)} \cdot c_b[0] \cdot \mathbb{I}_{N_b} \tag{10.25}$$

Using the linear forward model and the estimated covariance in (10.25), the full covariance matrix of $\boldsymbol{\Sigma}_\mathbf{y}$ can be written as

$$\begin{aligned} \boldsymbol{\Sigma}_\mathbf{y} &= \mathbf{H}_f \boldsymbol{\Sigma}_\mathbf{x} \mathbf{H}_f^T \\ &= \begin{bmatrix} \mathbf{H}c_b[0]\mathbb{I}_{N_b}\mathbf{H}^T & \cdots & \mathbf{H}a_1^{(N_t-1)}c_b[0]\mathbb{I}_{N_b}\mathbf{H}^T \\ \mathbf{H}a_1^{(1)}c_b[0]\mathbb{I}_{N_b}\mathbf{H}^T & \cdots & \mathbf{H}a_1^{(N_t-2)}c_b[0]\mathbb{I}_{N_b}\mathbf{H}^T \\ \vdots & & \vdots \\ \mathbf{H}a_1^{(N_t-1)}c_b[0]\mathbb{I}_{N_b}\mathbf{H}^T & \cdots & \mathbf{H}c_b[0]\mathbb{I}_{N_b}\mathbf{H}^T \end{bmatrix} \\ &= c_b[0] \cdot \begin{bmatrix} 1 & a_1^{(1)} & \cdots & a_1^{(N_t-1)} \\ a_1^{(1)} & 1 & \cdots & a_1^{(N_t-2)} \\ \vdots & \vdots & \ddots & \vdots \\ a_1^{(N_t-1)} & a_1^{(N_t-2)} & \cdots & 1 \end{bmatrix} \otimes \mathbf{H}\mathbf{H}^T \end{aligned} \tag{10.26}$$

where $\mathbf{HH}^T$ is the spatial covariance matrix. The temporal covariance matrix is defined in the following equation:

$$\boldsymbol{\Sigma}_t = c_b[0] \cdot \begin{bmatrix} 1 & a_1^{(1)} & \cdots & a_1^{(N_t-1)} \\ a_1^{(1)} & 1 & \cdots & a_1^{(N_t-2)} \\ \vdots & \vdots & \ddots & \vdots \\ a_1^{(N_t-1)} & a_1^{(N_t-2)} & \cdots & 1 \end{bmatrix} \tag{10.27}$$

$c_b[0]$ is a constant value which is absorbed in the spatial covariance. N_{p_3} represents the number of spatial and temporal parameters for this covariance estimation. According to (10.28) the number of temporal parameters is equal to one.

$$N_{p_3} = \frac{N_{ch}(N_{ch}+1)}{2} + 1 \tag{10.28}$$

Under this assumption, if we consider $p = 0$ in (10.19), then $\boldsymbol{\Sigma}_\mathbf{b}[m, n] = 0$ and the temporal covariance $\boldsymbol{\Sigma}_t$ is

$$\boldsymbol{\Sigma}_t = I_{N_t} \tag{10.29}$$

This structure requires $N_{p_4} = N_{p_3} - 1$ number of parameters to characterize the full covariance [17].

One can observe that by imposing more restricted structures on the covariance matrix of the feature vectors, the number of parameters needed to estimate more decreases, which does not require a lot of samples for a reliable estimation. If we assume N_{p_0} is the number of parameters of the nonstructured (NS) full covariance, then

$$N_{p_0} = \frac{N_{ch}N_t(N_{ch}N_t+1)}{2} > N_{p_1} > N_{p_2} > N_{p_3} > N_{p_4} \tag{10.30}$$

10.2.3 Nonstationarities detection

In order to detect the nonstationarities in the EEG measurements during BCI operation, we assume that a classifier is built by employing the user-specific calibration data $\mathbf{Y}_0^c$ for $c = 1, \ldots, M$, and this classifier is used by that specific user to perform various BCI tasks. Additionally, in order to screen nonstationarities during the performance of BCI tasks, we use the calibration data to learn the class conditional EEG evidence distributions.

After modeling the initial distributions, the task is to develop learning algorithms that can effectively model the nonstationarities caused by the changes in the observed EEG. An important step in that endeavor is to detect whether or not there is any change in the statistical properties of the learned class conditional distributions. Assuming the change point to be deterministic but unknown, we calculate it by minimizing the supremum of the average detection delay conditioned on the observed EEG evidence with a constraint on the mean time between false alarms.

Mathematically, this means that using the batches of EEG evidence observed up until the measurement time n, $\mathbf{Y}_0^c, \ldots, \mathbf{Y}_n^c$, we estimate the time when the change in the statistical characteristics of the data from class c is occurring by employing the following constrained optimization problem [55]

$$\begin{aligned} &\operatorname{argmin}_{\tau}\ \sup_{n\geq 1}\ \operatorname{ess\ sup}\ E_n\left[(\tau-n)^{+}|\mathbf{Y}_0^c,\ \ldots,\ \mathbf{Y}_{n-1}^c\right] \\ &\qquad\text{such that}\qquad E_{\infty}[\tau]\geq\alpha \end{aligned} \tag{10.31}$$

where τ is the stopping time, when it takes a value k (it means that there is a change point at or prior to time k); sup stands for supremum; ess sup is the essential supremum of a set of random variables that we define below in more detail; $x^{+}=\max(0,x)$ for any variable x; $E_n[\cdot]$ is the expectation taken with respect to a distribution p_n^c (such that under p_n^c, $\{\mathbf{Y}_0^c,\ \ldots,\ \mathbf{Y}_{n-1}^c\}$ are i.i.d. with a fixed marginal distribution for $c\in\{1,2,\ldots,M\}$); $E_{\infty}[\tau]$ represents the mean time between false alarms assuming that change never happens in the data stream; and α is a predefined threshold. Essential supremum of a set of random variables $\mathbf{X}$ is a random variable Z with the following properties: (i) $P(Z\geq X)=1,\ \forall X\in\mathbf{X}$; and (ii) $\{P(Y\geq X)=1,\ \forall X\in\mathbf{X}\}\rightarrow P(Y\geq Z)=1,\forall X\in\mathbf{X}$ [55]. In light of the definitions above, the solution to the constraint optimization problem in (10.31) minimizes the supremum of the average delay conditioned on the worst case realization of $\{\mathbf{Y}_0^c,\ \ldots,\ \mathbf{Y}_{n-1}^c\}$ overall p_n, $n\geq 1$ [55]. When the p_n^c is known before and after the change in statistical characteristics, the optimum solution to (10.31) is found using cumulative sum (CUSUM) method [56,57]. Note that when there is no prior knowledge about the family of the distribution of the EEG evidence in $\mathbf{Y}_0^c$, we can employ an extended version of the CUSUM method [56]. In this approach, we divide the set $\mathbf{Y}_0^c=\{y_t|r_t=c\}$ for $t=1,\ \ldots,\ N$ into subsets with cardinality K and compute

$$z_k=\frac{1}{K}\sum_{\nu=(k-1)K+1}^{kK} y_{\nu},\qquad k=1,\ \ldots,\ N/K \tag{10.32}$$

to form $\mathscr{Z}_0^c=\{z_k|r_k=c\}$ for $k=1,\ \ldots,\ N/K$. For a sufficiently large K, z_k can be approximated as a Gaussian distributed random variable with an unknown parameter set θ_0^c, denoted as $p_{\theta_0^c}(\cdot)$. If there is a change in the statistical properties of the EEG evidence, the change is from a Gaussian random variable with parameter set θ_0^c to another Gaussian random variable with parameter set θ_1^c, denoted as $p_{\theta_1^c}(\cdot)$. This type of change fits in the framework of CUSUM. Therefore, for extended CUSUM, we can utilize the following method to estimate the change point

$$\tau=\inf\left\{n\geq 1:\left(g(\mathscr{Z}_0^c,\ \ldots,\ \mathscr{Z}_n^c)=R_n-\min_{1\leq k\leq n}R_k\right)\geq b\right\} \tag{10.33}$$

where $R_k=\sum_{t=1}^{k}\ln(p_{\theta_1^c}(\mathscr{Z}_t^c)/p_{\theta_0^c}(\mathscr{Z}_t^c))$; and $\mathscr{Z}_t^c$ is calculated from $\mathscr{Y}_t^c$ using the method in (10.32). To estimate the change point using (10.33), we estimate θ_0^c using the training data $\mathscr{Z}_0^c$ and estimate θ_1^c using the confidence interval extremes of $\hat{\theta}_0^c$ estimator. We also compute $b=\max_{1\leq k\leq N/K} g(\mathscr{Z}_0^c)$ recalling that $\mathscr{Z}_0^c=z_k|r_k=c$ for $k=1,\ \ldots,\ N/K$ [56]. In order to calculate the sequence of R_k values defined in (10.33), θ_c^1 values; that is, the bound distributions need to be calculated assuming some significance level α, test sensitivity γ, and using the sample mean M_0^c, and the sample covariance C_0^c. For each class c, the two bound means and covariances are as follows [56]:

$$M_1^c = M_0^c \pm \gamma\sqrt{\frac{n}{N}}Z_{\alpha/2}\sqrt{\text{diag}(C_0^c)} \tag{10.34}$$

$$C_{1,\min}^c = C_0^c - \gamma\frac{(N/n)-1}{\chi^2_{\alpha/2;N/n-1}}C_0^c \tag{10.35}$$

$$C_{1,\max}^c = C_0^c - \gamma\frac{(N/n)-1}{\chi^2_{1+\alpha/2;N/n-1}}C_0^c \tag{10.36}$$

For each class c, this yields to four distributions due to the two bound means and two bound covariances. The log likelihood CUSUM needs to be computed for each of these four distributions. If the null hypothesis that there is no nonstationarity in the data is rejected in favor of any of these four distributions for any class c, then a change is assumed to have occurred.

10.2.4 Decoupling the class label from ERP detection

As discussed in Section 10.2.1, each trial signal can be modeled as a distribution conditioned on the presence of ERP with class-dependent mean and covariance (10.1). This is in accordance to standard EEG analysis that models the signals with a Gaussian distribution. By allowing the ERP to be a random state related but not equivalent to whether a target was presented to the user, we proposed a model to decouple the user distraction from the user intention, character or stimulus confusion, among others.

We applied a GMM-based classification method to model the uncertainty in the class labels. To perform inference, we compute the posterior probability given EEG evidence $\mathbf{y}$. In (10.1), the conditional probability $p(s|c)$ can be learned from calibration data and corresponds to the probability of observing an ERP given a target or nontarget stimulus. This conditional distribution has two degrees of freedom and is parameterized as

$$\begin{aligned} p(s|c=0) &= (1-\gamma_0)(1-s)+\gamma_0 s \\ p(s|c=1) &= (1-\gamma_1)(1-s)+\gamma_1 s \end{aligned} \tag{10.37}$$

Learning the set of parameters $\{\mu_s, \Sigma_s, \gamma_c\}$ for $c \in \{0,1\}$ and $s \in \{0,1\}$ does not have explicit closed form solution. Expectation maximization is used to evaluate the complete likelihood and update the parameters, iteratively. For a dataset comprised of EEG signals and label pairs $\{\mathbf{y}_n, c_n\}_{n=1}^N$ for all calibration trials $n \in 1,\ldots,N$, the Q-function for the E-step is defined as

$$\begin{aligned} Q(s|\mathbf{y}_n, c_n) &= p(s|\mathbf{y}_n, c_n, \Theta^k) \\ &\propto p(\mathbf{y}_n|s,\Theta^k)p(s|c_n,\Theta^k) \\ &= N(\mathbf{y}_n; \mu_s^k, \Sigma_s^k)((1-\gamma_{c_n}^k)(1-s)+\gamma_{c_n}^k s) \end{aligned} \tag{10.38}$$

where Θ^k corresponds to the parameter set at the kth iteration. For the M-step, the update equation for each parameter is defined as follows:

$$\begin{aligned}
\mu_s^k &= \tfrac{1}{\eta_s}\textstyle\sum_n p(s|\mathbf{y}_n, c_n)x_n \\
\Sigma_s^k &= \tfrac{1}{\eta_s}\textstyle\sum_n p(s|\mathbf{y}_n, c_n)\,(\mathbf{y}_n - \mu_s^k)^T(\mathbf{y}_n - \mu_s^k) \\
\gamma_0^k &= \tfrac{1}{\eta_s}\textstyle\sum_n p(s|\mathbf{y}_n, c_n)(1-s)c_n \\
\gamma_1^k &= \tfrac{1}{\eta_s}\textstyle\sum_n p(s|\mathbf{y}_n, c_n)(1-c_n)s \\
\eta_s &= \textstyle\sum_n p(s|\mathbf{y}_n, c_n)
\end{aligned} \tag{10.39}$$

with $p(s|\mathbf{y}_n, c_n)$ obtained by normalizing the equation in (10.38).

Although this model better captures the stochastic relationship between targetness and ERP generation, estimating Σ_s for high-dimensional EEG from only a limited number of samples resulted in poor classification performance. We applied L1 regularization on the covariance estimate by replacing the equation above with the following update:

$$\begin{aligned}
S_s^k &= \tfrac{1}{\eta_s}\textstyle\sum_n p(s|\mathbf{y}_n, c_n)\,(\mathbf{y}_n - \mu_s^k)^T(\mathbf{y}_n - \mu_s^k) \\
C_s^k &= \arg\min_C - \log|C| + \mathrm{tr}(CS_s^k) + \lambda||C||_1 \\
\Sigma_s^k &= (C_s^k)^{-1}
\end{aligned} \tag{10.40}$$

where C corresponds to the regularized precision matrix, while S corresponds to the weighted sample covariance. The L1 norm of a matrix is defined as

$$||C||_1 = \sum_{i,j} |C_{i,j}|$$

which can be solved iteratively via graphical lasso [58].

10.3 Experimental results and discussions

Two ERP-based BCI paradigms have been developed as testbeds to assess the performance of the BCI systems under discussed conditions in Section 10.2.

10.3.1 ERP-based BCI typing system

One of the better known ERP-based BCI typing interface testbeds is RSVP Keyboard™ which stands for Rapid serial visual presentation. The RSVP Keyboard is a language-model-assisted typing BCI that features three different presentation paradigms: rapid serial visual presentation paradigm and two different matrix presentation paradigms (row column and single symbol flash).

In RSVP Keyboard, a set of pseudorandomly ordered stimuli are presented on prefixed location of the screen in a rapid serial manner. In RSVP, each stimulus is a trial and a set of trials which have been presented with no time gap in between, is called a sequence. Every sequence contains only a single target stimulus. Time series analysis can be used to identify which stimulus the attention (target) is placed on. In order to estimate the posterior probability in (10.1) in this paradigm, the context prior $P(s|C)$ is provided by an n-gram language model, which estimates the conditional probability of every letter in the alphabet based on $n-1$ previously typed letters in a Markov model framework.

The RSVP Keyboard inference model is explained in details in [13].

The RSVP Keyboard has two important system operation modes that we utilize in this manuscript:

1. *Calibration mode*: During calibration, the users are asked to attend to predefined target symbols within randomly ordered sequences to record labeled EEG data. The data acquired in this mode are then used in the estimation of class conditional EEG evidence distributions.
2. *Copy phrase task mode*: In this task, the users are given a set of predefined phrases. Each phrase includes a missing word and the users are asked to complete these words. This task is designed to assess the system and/or user performance in terms of speed and accuracy in the presence of a language model.

10.3.1.1 BCI performance under spatial–temporal structures

The area under the receiver operating characteristics (ROC) curve, i.e., AUC, was calculated for a calibration dataset using a 10-fold cross validation assuming signal models with different covariance structures in Section 10.2.2. AUC can be used as an indicator for the classifier performances. Figure 10.2 shows the AUC results

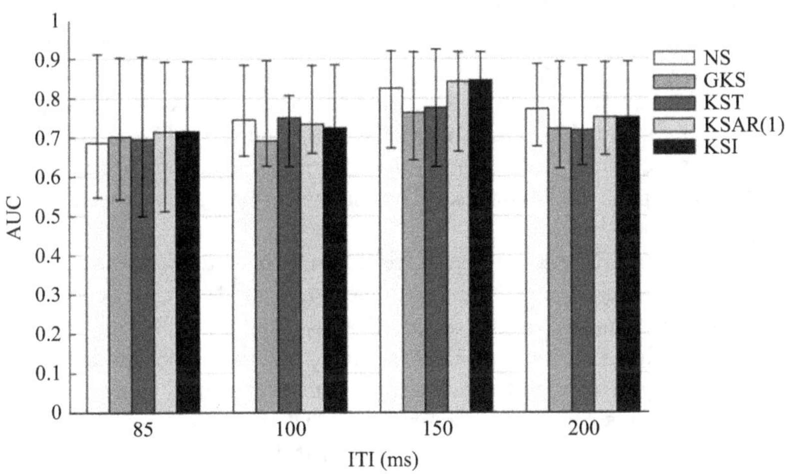

Figure 10.2 Bar graph of the median AUC in RSVP Keyboard™ with rapid serial visual presentation paradigm. The error bars indicate the maximum and minimum values for 12 users calculated by use of different signal models and ITI combinations when the classifiers are trained with all training data. EEG data were collected from 16 channels: Fp1, Fp2, F3, F4, Fz, Fc1, Fc2, Cz, P1, P2, C1, C2, Cp3, Cp4, P5, and P6

for four introduced Kronecker product covariance structures (KS) with full spatial covariances and different temporal covariances in Section 10.2.2: (i) full (GKS), (ii) Toeplitz (KST), (iii) identity (KSI) temporal, and (iv) Kronecker-product structure with AR(1) temporal covariance [KSAR(1)]. Moreover, we compute the class conditional NS full covariance matrices for every dataset to be compared with four structured covariance models. In Figure 10.2, the classification performance is illustrated for 12 healthy users for four different inter trial interval (ITI) values, which is defined as the time interval between consecutive trials (ITI $\in$ {200; 150; 100; 85} ms). Since the number of supervised observations in every dataset is proportionally lower than the number of parameters of every covariance structure, RDA has been applied to regularize the estimated class conditional covariance matrices. As can be seen, for RSVP paradigm, although the assumed structures does not demonstrate significant performance enhancement, the models with KSAR(1) and KSI covariance structures performed equal to or better than NS estimator with lower model order complexity, and they require shorter BCI calibration sessions (less number of samples) to achieve the same performance as a model that uses no covariance structure.

10.3.1.2 BCI performance under nonstationarities in EEG signals

In order to analyze the performance of the nonstationarity detection method, an experiment has been performed with the RSVP Keyboard [13]. Using the data collected in these experiments, we utilize the nonstationarity detection method described in Section 10.2.3 to analyze the effect of nonstationarities on the system performance. Specifically, we use the data collected in calibration session to train the detection method formulated in (10.32) and (10.33). Using the supervised calibration data, we learn the sample mean and covariance for target and nontarget symbol classes, M_0^c and C_0^c for $c = 1$ and $c = 2$ representing the nontarget and target classes, respectively.

After calibration, the participants were asked to complete a copy phrase task with eight sentences for four ITI values. During the copy phrase task, the system marks a phrase as unsuccessful if more than four wrong letter selections occur in row, and the next phrase is presented to the user. The user performance is reported in terms of typing speed and accuracy.

The nonstationarity detection results are presented in Figures 10.3 and 10.4, where the total number of sequences presented to the users, and task completion accuracy as a function of the number of nonstationarity detection, are illustrated respectively. In these scatter plots, each point represents one session of copy phrase task. The points on these figures are coded with different marker shapes based on the users' typing accuracy levels which are represented by AUC values. These AUC values are calculated based on the crossvalidation of the classifier's performance on the training datasets. Δ, $+$, and $\circ$ markers are used to represent AUC values greater than 85 per cent, AUC values between 75 per cent and 85 per cent, and AUC values below 75 per cent, respectively. The results in these figures show that for high AUC values nonstationarity is usually not detected, but as the AUC values decrease, there are more nonstationarity detections. As the number of nonstationarity detections increases, the typing duration increases and task completion accuracy decreases resulting in a poorer BCI performance. These results suggest that online adaptation of BCIs

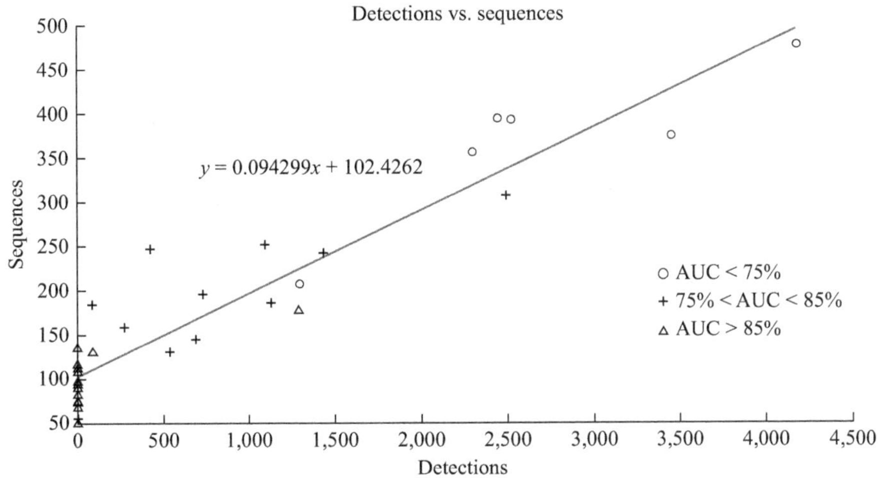

Figure 10.3 Number of sequences presented to the users as a function of the number of detected nonstationarities. EEG data were collected from 16 channels: Fp1, Fp2, F3, F4, Fz, Fc1, Fc2, Cz, P1, P2, C1, C2, Cp3, Cp4, P5, and P6

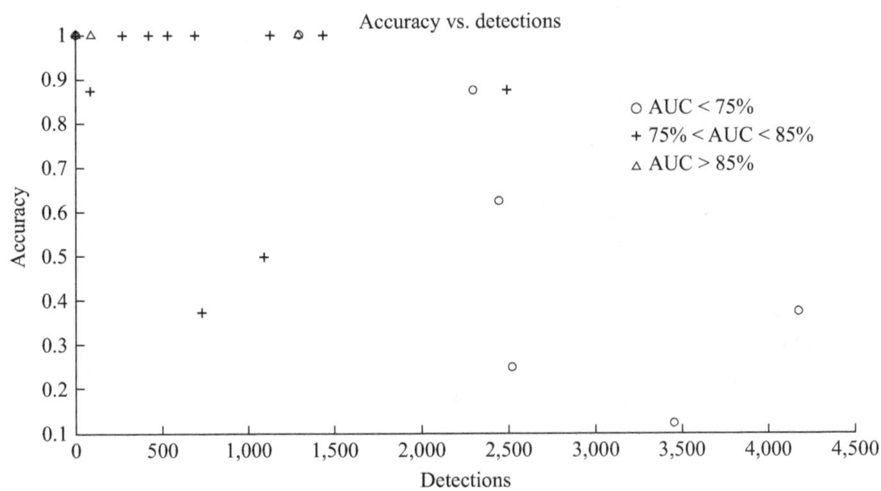

Figure 10.4 Task completion accuracy as a function of the number of detected nonstationarities

is user dependent and adaptation may not always be necessary. Participants with poor BCI task accuracy could benefit more from the online adaptations. Moreover, online adaptations may not need to be continuous but could be performed when more nonstationarities are detected in the EEG.

10.3.2 ERP-based BCI with tactile stimuli

Recently, using tactile stimuli in BCIs have captured popularity due to the advantages of using tactile sense rather than visual or auditory senses, which is suitable for people with impaired visual or auditory systems. Tactile P300 BCIs encounter significant obstacles in online classification accuracy [6,7] and decision speed compared to the visual based interfaces [59].

The BCI paradigm consequently controls how the user traverses the task. A number of different paradigms have been investigated in the search for a combination that yields high offline performance. Many studies have examined the connection between performance and tactor placement [60,61]. Many studies have also examined the impact of discrimination tasks on user performance and shown evidence that user performance was significantly worse for dual-task paradigms when compared with single-task paradigms [61,62]. Furthermore, many studies have validated that dual-task paradigms yield high training accuracies, but consistently fail during online testing [6,7,63].

We designed a tactile stimulation framework such that system actions are associated to each tactor and the system delivers vibrations in quick succession. The ERP is evoked when the user concentrates on the target tactor corresponding to their decision. Figure 10.5 illustrates the stimulus presentation and important design parameters for a typical tactile BCI task. Epochs are designed to be relatively short with breaks in between each one to mitigate mental fatigue. Two distinct BCI tasks were used to evaluate how particular paradigm changes would affect subject ERPs and performance during discrimination tasks. All subjects were asked to count the number of rare stimuli in each epoch and then report that number at the end of the epoch. If the user lost track and counted incorrectly, the data from that epoch was discarded.

- *Task (I) – Four tactor simultaneous oddball*
  This discrimination task replicates popular paradigms that use "distractors" to create multiple rare tactors. The target response is the ERP evoked by a stimulus delivered by the oddball tactor that the user is attentive to, while the nontarget response is the presumably flat ERP following an ignored stimulus delivered by the other oddball tactor. At the start of each epoch, the user is instructed to count the pulses delivered to one hand and to ignore all other stimuli. User intent is determined by classifying the set of ERPs corresponding to each oddball tactor as either the target or nontarget class. The tactor that the user counts is alternated every two epochs to generalize the training algorithm. The two distractors deliver stimuli in a random order but at equal rates so that neither are rare with respect to the other.
- *Task (II) – Three tactor sequential oddball*
  This task is designed to optimize the consistency of the user's response to target and nontarget stimulus across trials by completely separating the objectives of responding and ignoring rare stimuli. For a given epoch, the user either counts the pulses delivered to the right hand (target) *or* performs a mental math exercise and ignores the stimuli delivered to the left hand (nontarget). Only three tactors are utilized during a given epoch (*R,D1,D2* or *L,D1,D2*). In the context of a *yes* or

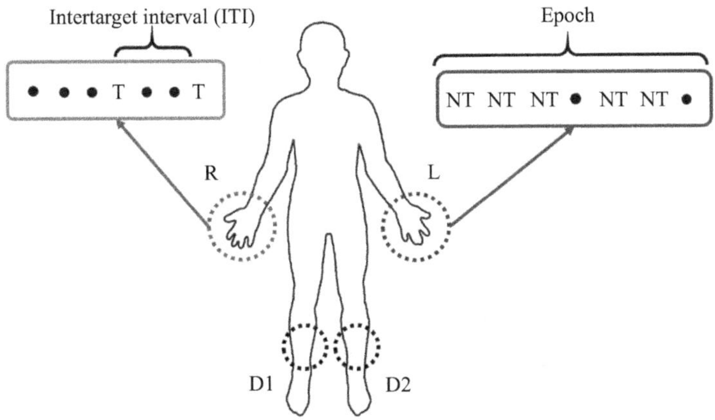

Figure 10.5 Representation of typical tactile BCI sequence and important design parameters. The main tactors (R and L) and distractors (D1 and D2) are secured at locations indicated by the dashed circles. Each square represents either a target (T) or nontarget (NT) trial. The trial itself begins at the stimulus onset and lasts through the duration of the ERP, ending at the onset of the next stimulus. A sequence of subsequent trials over which user intent is constant is called the epoch. EEG data were collected from seven channels, i.e., Fz, FC1, FC2, Cz, CP1, CP2, and Pz. ITI is uniformly random between 8 and 14 trials (4.8 and 8.4 s). The duration of each vibrational pulse is 0.1 s. The pause between the end of one stimulus and the start of the next, i.e., interstimulus interval (ISI), is uniformly random between 0.1 and 0.5 s to maximize unpredictability. For discrimination tasks where there are multiple rare tactors, the user receives 240 rare stimulations (120 target, 120 nontarget) over a variable number of epochs. The number of targets and/or nontargets presented during each epoch is uniformly random between 4 and 7

no question, the *yes* response is encoded in action of counting and being attentive, while the *no* response is encoded in the action of ignoring the stimulus. This is different from Task (I) where the action of counting and being attentive to either the right or left hand *and* the action of ignoring the other hand encodes the user's response. Because each response is trained separately, there is no need to alternate the target/nontarget tactor assignment.

10.3.2.1 Classification performance under noisy labels

To quantify the classification performance for each BCI task, two different classification tools were used: GMM and RDA. The final AUC of the subject for a given task was calculated (Table 10.1) of the subject's corresponding ROC curve. Classification performance varied widely across task and classification method for each subject.

Table 10.1 AUC for all subjects in Tasks (I) and (II) using RDA or the noisy-label GMM analysis for classification

	Task (I)		Task (II)	
Subject	**RDA**	**GMM**	**RDA**	**GMM**
1	59.74	58.65	78.53	90.87
2	72.23	91.37	88.96	98.50
3	55.99	76.30	75.57	88.50
4	67.30	64.71	78.36	98.42

Table 10.2 Estimations of the label noise separated by class and task for all subjects.

	Task (I)		Task (II)	
Subject	**Target**	**Nontarget**	**Target**	**Nontarget**
1	–	–	10.45	30.77
2	38.24	24.53	23.53	11.32
3	–	–	33.90	62.26
4	44.54	34.40	44.44	23.81

Values refer to the percentage of mislabeled trials. Dashes indicate when classification accuracy was too low to make a reasonable estimate of the label noise

Almost every subject showed improvement in Task (II) over Task (I). GMM classification, in comparison with the RDA classifier, resulted in improved or equivalent performance in the ERPs detection for all subject. Table 10.1 shows the AUC results for fours subjects. Label noise was assessed by processing the results of Tasks (I) and (II) using the RDA classification pipeline and comparing the class-specific, average ERPs of correctly and incorrectly classified trials. To obtain a rough estimate of label noise, the percentage of trials classified incorrectly using the RDA pipeline was calculated for each class, task, and subjects. Table 10.2 shows the classification results for Task (I) and Task (II) using RDA classifier. This is a rough estimate because it implies the assumption that all incorrectly classified trials are the result of the label noise which we know from examination of individual trials is not the case. This assumption is close to true for classification accuracies of at least 65 per cent, but it breaks down for performances below that level which is why there are no estimates in Task (I) for Subjects 1 and 3. Despite this, the results demonstrate that label noise was reduced in Task (II) compared to Task (I) for all subjects.

The results of Tasks (I) and (II) suggest that, as hypothesized, label noise plays a significant role in tactile BCI performance. The results of the proposed method,

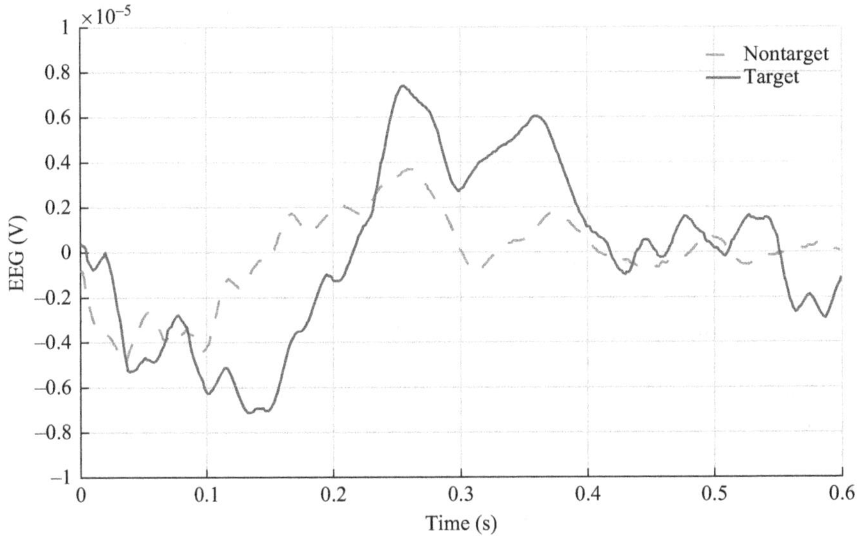

Figure 10.6 Mean ERPs in Task (II) for one of the subjects, at channel Fz. The solid line, target-related brain response has a large P300 component that is distinctly different from the dash line, nontarget-related brain response

which incorporates label noise into both the BCI paradigm and inference models, all subjects increased performance over traditional methods. Furthermore, a direct correlation between improved performance and a global decrease in label noise was observed. These trends are reflected the average ERPs from Tasks (I) and (II), which are shown for a single subject in Figure 10.6.

According to the results, the key component in the success of the Task (II) paradigm was that when the user experienced a *rare* stimulus, they already knew beforehand whether they should be attentive or not. In some ways, this feels counter-intuitive, since P300 relies on effectively *surprising* the user. What this study found, however, was that randomness in stimulus timing was perfectly sufficient for most users to elicit a distinguishable P3 and that randomness in target/nontarget appearance actually decreases the probability that the user elicited the EEG response associated with their intent. Moreover, the presented results suggest that a label noise model must be incorporated into the paradigm for any tactile stimulation based BCI system to achieve relevant online classification performance.

10.4 Summary

ERP-based BCIs that rely on visual, auditory, and tactile stimuli are frequently used in exploring possible solutions to the problem of devising alternative communication and control pathways for individuals with disabilities. For best performance outcomes with ERP-based BCIs, it is important to accurately detect the user state including intent from EEG/ERP features. In this chapter, our goal was to demonstrate how by taking advantage of signal modeling and machine-learning techniques, one can improve the models for the user intent inference. We addressed three main issues that impact the speed, accuracy, and reliability of ERP-based BCIs: (i) high dimensionality of EEG feature vectors; (ii) nonstationarities in EEG signals; (iii) noise and uncertainty in calibration sample labels.

To tackle the issue of high dimensionality, we considered several spatiotemporal structures for brain activity sources and resulting EEG feature vector covariance matrices. Imposing stronger assumptions on the covariance matrix structure resulted in lower degrees of freedom in covariance models, but with model complexity required number of calibration samples (therefore calibration phase durations) decrease, without significantly compromising BCI performance for short-term use.

To address the nonstationarity problem, an adaptive nonstationarity screening method was demonstrated. This method uses a change detection approach based on the extended CUSUM test to determine if the incoming data is undergoing a distribution drift, before indiscriminately adapting to the EEG distribution changes. This not only eliminates the burden of unnecessary computational complexity associated with model parameter updates that may not be necessary but also improves classification performance in the event there is detectable nonstationarity effects in EEG statistics. The tactile stimulation paradigm was specifically selected to illustrate the impact of noisy labels and how a simple hierarchical generative model can be incorporated into the design process to calibrate models using data with label uncertainties, specifically resulting in GMM class conditionals for available label classes in calibration data. Illustrations on data from RSVP Keyboard experiments and their auditory and tactile stimulus variations are provided, to assess the performance of these methods.

References

[1] Sur S, Sinha V. Event-related potential: an overview. Industrial Psychiatry Journal. 2009;18(1):70.

[2] Woodman GF. A brief introduction to the use of event-related potentials in studies of perception and attention. Attention, Perception, & Psychophysics. 2010;72(8):2031–2046.

[3] Picton TW. The P300 wave of the human event-related potential. Journal of Clinical Neurophysiology. 1992;9(4):456–479.

[4] Akcakaya M, Peters B, Moghadamfalahi M, *et al.* Noninvasive brain–computer interfaces for augmentative and alternative communication. IEEE Reviews in Biomedical Engineering. 2014;7:31–49.

[5] Hill, NJ, Lal TN, Bierig K, Birbaumer N, Schölkopf B. An auditory paradigm for brain-computer interfaces. Advances in Neural Information Processing Systems. 2005;17:569–576.

[6] Kaufmann T, Holz EM, Kübler A. Comparison of tactile, auditory, and visual modality for brain–computer interface use: a case study with a patient in the locked-in state. Frontiers in Neuroscience. 2013;7:129.

[7] Lugo ZR, Rodriguez J, Lechner A, *et al.* A vibrotactile p300-based brain–computer interface for consciousness detection and communication. Clinical EEG and Neuroscience. 2014;45:14–21.

[8] Rasmussen CE, Williams CK. Gaussian processes for machine learning. vol. 1. MIT Press, Cambridge; 2006.

[9] Stoica P, Viberg M. Maximum likelihood parameter and rank estimation in reduced-rank multivariate linear regressions. IEEE Transactions on Signal Processing. 1996;44(12):3069–3078.

[10] Ledoit O, Wolf M. Improved estimation of the covariance matrix of stock returns with an application to portfolio selection. Journal of Empirical Finance. 2003;10(5):603–621.

[11] Ryan DB, Frye GE, Townsend G, *et al.* Predictive spelling with a P300-based brain–computer interface: increasing the rate of communication. International Journal of Human-Computer Interaction. 2010;27(1):69–84.

[12] Blankertz B, Losch F, Krauledat M, *et al.* The Berlin brain–computer interface: accurate performance from first-session in BCI-Naïve subjects. IEEE Transactions on Biomedical Engineering. 2008;55(10):2452–2462.

[13] Moghadamfalahi M, Gonzalez-Navarro P, Akcakaya M, *et al.* The effect of limiting trial count in context aware BCIs: a case study with language model assisted spelling. In: International Conference on Augmented Cognition. Springer; 2015. p. 281–292.

[14] Treder MS, Blankertz B. (C)overt attention and visual speller design in an ERP-based brain–computer interface. Behavioral and Brain Functions : BBF. 2010;6(1):28.

[15] Orhan U, Hild KE, Erdogmus D, *et al.* RSVP keyboard: an EEG based typing interface. In: International Conference on Acoustics, Speech and Signal Processing (ICASSP). IEEE; 2012. p. 645–648.

[16] Orhan U, Erdogmus D, Roark B, *et al.* Offline analysis of context contribution to ERP-based typing BCI performance. Journal of Neural Engineering. 2013;10(6):066003.

[17] Gonzalez-Navarro P, Moghadamfalahi M, Akçakaya M, *et al.* Spatio-temporal EEG models for brain interfaces. Signal Processing. 2017;131:333–343.

[18] Kindermans PJ, Verstraeten D, Schrauwen B. A Bayesian model for exploiting application constraints to enable unsupervised training of a P300-based BCI. PLoS One. 2012;7(4):e33758.

[19] Spüler M, Rosenstiel W, Bogdan M. Online adaptation of a c-VEP brain–computer interface (BCI) based on error-related potentials and unsupervised learning. PLoS One. 2012;7(12):e51077.

[20] Spüler M, Rosenstiel W, Bogdan M. Unsupervised online calibration of a c-VEP brain–computer interface (BCI). In: International Conference on Artificial Neural Networks. Springer; 2013. p. 224–231.

[21] Ferrante A, Gavriel C, Faisal A. Towards a brain-derived neurofeedback framework for unsupervised personalisation of brain–computer interfaces. In: 2015 7th International IEEE/EMBS Conference on Neural Engineering (NER). IEEE; 2015. p. 162–165.

[22] Shin Y, Lee S, Lee HN. Dictionary update based adaptive EEG classification for real time brain–computer interface applications. In: Information and Communication Technology Convergence (ICTC), 2015 International Conference on. IEEE; 2015. p. 566–569.

[23] Zink R, Hunyadi B, Van Huffel S, *et al.* Exploring CPD based unsupervised classification for auditory BCI with mobile EEG. In: 2015 7th International IEEE/EMBS Conference on Neural Engineering (NER). IEEE; 2015. p. 53–56.

[24] Vidaurre C, Sannelli C, Müller KR, *et al.* Machine-learning-based coadaptive calibration for brain–computer interfaces. Neural Computation. 2011;23(3):791–816.

[25] Faller J, Vidaurre C, Solis-Escalante T, *et al.* Autocalibration and recurrent adaptation: towards a plug and play online ERD-BCI. IEEE Transactions on Neural Systems and Rehabilitation Engineering. 2012;20(3):313–319.

[26] Faller J, Solis-Escalante T, Costa U, *et al.* Online co-adaptive brain–computer interfacing: preliminary results in individuals with spinal cord injury. In: Neural Engineering (NER), 2013 6th International IEEE/EMBS Conference on. IEEE; 2013. p. 977–980.

[27] Scherer R, Faller J, Opisso E, *et al.* Bring mental activity into action! An enhanced online co-adaptive brain–computer interface training protocol. In: 2015 37th Annual International Conference of the IEEE Engineering in Medicine and Biology Society (EMBC). IEEE; 2015. p. 2323–2326.

[28] Nicolas-Alonso LF, Corralejo R, Gomez-Pilar J, *et al.* Adaptive semi-supervised classification to reduce intersession non-stationarity in multiclass motor imagery-based brain–computer interfaces. Neurocomputing. 2015;159:186–196.

[29] Lu S, Guan C, Zhang H. Unsupervised brain computer interface based on intersubject information and online adaptation. IEEE Transactions on Neural Systems and Rehabilitation Engineering. 2009;17(2):135–145.

[30] Arvaneh M, Guan C, Ang KK, *et al.* EEG data space adaptation to reduce intersession nonstationarity in brain–computer interface. Neural Computation. 2013;25(8):2146–2171.

[31] Zink R, Hunyadi B, Van Huffel S, *et al.* Tensor-based classification of an auditory mobile BCI without a subject-specific calibration phase. Journal of Neural Engineering. 2016;13(2):026005.

[32] Rodríguez-Bermúdez G, García-Laencina PJ. Automatic and adaptive classification of electroencephalographic signals for brain computer interfaces. Journal of Medical Systems. 2012;36(1):51–63.

[33] Kimovski D, Ortega J, Ortiz A, *et al.* Parallel alternatives for evolutionary multi-objective optimization in unsupervised feature selection. Expert Systems with Applications. 2015;42(9):4239–4252.

[34] Li M, Cui Y, Hao D, *et al.* An adaptive feature extraction method in BCI-based rehabilitation. Journal of Intelligent & Fuzzy Systems. 2015;28(2):525–535.

[35] Myrden A, Chau T. Feature clustering for robust frequency-domain classification of EEG activity. Journal of Neuroscience Methods. 2016;262:77–84.

[36] Sannelli C, Vidaurre C, Müller KR, *et al.* Ensembles of adaptive spatial filters increase BCI performance: an online evaluation. Journal of Neural Engineering. 2016;13(4):046003.

[37] Llera A, Gómez V, Kappen HJ. Adaptive classification on brain–computer interfaces using reinforcement signals. Neural Computation. 2012;24(11): 2900–2923.

[38] Vidaurre C, Schlogl A, Cabeza R, *et al.* A fully on-line adaptive BCI. IEEE Transactions on Biomedical Engineering. 2006;53(6):1214–1219.

[39] Vidaurre C, Schlogl A, Cabeza R, *et al.* Study of on-line adaptive discriminant analysis for EEG-based brain computer interfaces. IEEE Transactions on Biomedical Engineering. 2007;54(3):550–556.

[40] Vidaurre C, Blankertz B. Towards a cure for BCI illiteracy. Brain Topography. 2010;23(2):194–198.

[41] Vidaurre C, Kawanabe M, von Bünau P, *et al.* Toward unsupervised adaptation of LDA for brain–computer interfaces. IEEE Transactions on Biomedical Engineering. 2011;58(3):587–597.

[42] Llera A, Gómez V, Kappen HJ. Adaptive multiclass classification for brain computer interfaces. Neural Computation. 2014;26(6):1108–1127.

[43] Perdikis S, Leeb R, d R Millán J. Context-aware adaptive spelling in motor imagery BCI. Journal of Neural Engineering. 2016;13(3):036018.

[44] Buttfield A, Ferrez PW, Millan JR. Towards a robust BCI: error potentials and online learning. IEEE Transactions on Neural Systems and Rehabilitation Engineering. 2006;14(2):164–168.

[45] Liu G, Huang G, Meng J, *et al.* Improved GMM with parameter initialization for unsupervised adaptation of brain–computer interface. International Journal for Numerical Methods in Biomedical Engineering. 2010;26(6):681–691.

[46] Gu Z, Yu Z, Shen Z, *et al.* An online semi-supervised brain–computer interface. IEEE Transactions on Biomedical Engineering. 2013;60(9):2614–2623.

[47] Liao X, Yao D, Li C. Transductive SVM for reducing the training effort in BCI. Journal of Neural Engineering. 2007;4(3):246.

[48] Li Y, Guan C, Li H, *et al.* A self-training semi-supervised SVM algorithm and its application in an EEG-based brain computer interface speller system. Pattern Recognition Letters. 2008;29(9):1285–1294.

[49] Rakotomamonjy A, Guigue V. BCI competition III: dataset II—ensemble of SVMs for BCI P300 speller. IEEE Transactions on Biomedical Engineering. 2008;55(3):1147–1154.

[50] Sykacek P, Roberts SJ, Stokes M. Adaptive BCI based on variational Bayesian Kalman filtering: an empirical evaluation. IEEE Transactions on Biomedical Engineering. 2004;51(5):719–727.

[51] Wang Y, Gao X, Hong B, *et al.* Brain–computer interfaces based on visual evoked potentials. IEEE Engineering in Medicine and Biology Magazine. 2008;27(5):64–71.

[52] Tsui CSL, Gan JQ, Roberts SJ. A self-paced brain–computer interface for controlling a robot simulator: an online event labelling paradigm and an extended

Kalman filter based algorithm for online training. Medical & Biological Engineering & Computing. 2009;47(3):257–265.

[53] Kindermans PJ, Verschore H, Schrauwen B. A unified probabilistic approach to improve spelling in an event-related potential-based brain computer interface. IEEE Transactions on Biomedical Engineering. 2013;60(10):2696–2705.

[54] Bryan MJ, Martin SA, Cheung W, *et al.* Probabilistic co-adaptive brain–computer interfacing. Journal of Neural Engineering. 2013;10(6):066008.

[55] Poor HV, Hadjiliadis O. Quickest detection. vol. 40. Cambridge University Press, Cambridge; 2009.

[56] Alippi C, Roveri M. Just-in-time adaptive classifiers—part I: detecting non-stationary changes. IEEE Transactions on Neural Networks. 2008;19(7): 1145–1153.

[57] Ritov Y. Decision theoretic optimality of the CUSUM procedure. The Annals of Statistics. 1990;18:1464–1469.

[58] Friedman J, Hastie T, Tibshirani R. The elements of statistical learning. vol. 1. Springer series in statistics New York; 2001.

[59] Kaufmann T, Herweg A, Kübler A. Toward brain–computer interface based wheelchair control utilizing tactually-evoked event-related potentials. Journal of NeuroEngineering and Rehabilitation. 2014;11(1):7.

[60] Brouwer AM, Van Erp JBF. A tactile P300 BCI and the optimal number of tactors: effects of target probability and discriminability. In: Proceedings of the 4th International Brain–Computer Interface Workshop and Training Course; 2008. p. 280–285.

[61] Thurlings ME, Van Erp JBF, Brouwer AM, *et al.* Controlling a tactile ERP-BCI in a dual task. IEEE Transactions on Computational Intelligence and AI in Games. 2013;5(2):129–140.

[62] Adler J, Giabbiconi CM, Müller MM. Shift of attention to the body location of distracters is mediated by perceptual load in sustained somatosensory attention. Biological Psychology. 2009;81(2):77–85.

[63] van der Waal M, Severens M, Geuze J, *et al.* Introducing the tactile speller: an ERP-based brain–computer interface for communication. Journal of Neural Engineering. 2012;9(4):045002.

Chapter 11

Spatial filtering techniques for improving individual template-based SSVEP detection

Masaki Nakanishi[1], Yijun Wang[2], and Tzyy-Ping Jung[1]

Abstract

In the past decade, the performance of brain–computer interfaces based on steady-state visual evoked potentials (SSVEPs) has been significantly improved due to advances in signal analysis algorithms. For example, efficient target-identification methods based on template matching, in which individual templates are obtained by averaging the training data across trials, have been proposed to improve the performance of SSVEP detection. In template-based methods, spatial filtering plays an important role in improving the performance by enhancing the signal-to-noise ratio of SSVEPs. In conventional studies, several spatial-filtering approaches have been introduced for electroencephalogram analysis. However, the optimal spatial-filtering approach for individual template-based SSVEP detection still remains unknown. This chapter reviews the spatial-filtering approaches for improving the template-based SSVEP detection and evaluates their performance through a direct comparison using a benchmark dataset of SSVEPs.

11.1 Introduction

Visual evoked potentials (VEPs) are time-locked electroencephalogram (EEG) responses to visual stimuli such as flashlights or sudden changes in image pattern [1]. They consist of positive and negative potential deflections in EEG signals observed in the occipital scalp area over or near the primary visual cortex. The VEPs elicited by transient stimuli are called transient VEPs. In the 1960s, Regan attempted an experiment to produce continuous VEPs using sinusoidally modulated repetitive visual stimulation [2]. The stimulus successfully produced a stable oscillation, which is

[1]Swartz Center for Computational Neuroscience, Institute for Neural Computation, University of California San Diego, United States
[2]State Key Laboratory on Integrated Optoelectronics, Institute of Semiconductors, Chinese Academy of Sciences, China

named steady-state VEPs (SSVEPs). An SSVEP is also known as a photic driving response characterized by a sinusoidal-like waveform at the stimulus frequency and its harmonics [2]. Because the frequency components in SSVEPs are nearly stationary in amplitude and phase over a long period, the stimulus frequency can be reliably identified by observing the elicited SSVEP in the frequency domain. Due to its robust characteristics, SSVEPs have been widely used in the field of visual neuroscience, ophthalmology, and neural engineering including brain–computer interfaces (BCIs) [1].

An SSVEP-based BCI has attracted increasing attention because of its advantages of high information transfer rate (ITR), ease-of-use, and little user training [3]. Researchers have been using an SSVEP-based BCI for developing a variety of applications such as a text speller [4–6], a phone-dialing system [7], and a game controller [8]. In an SSVEP-based BCI, users are instructed to gaze at one of the multiple visual stimuli modulated with different stimulus properties (e.g., frequencies and/or phases), resulting in SSVEPs that exhibit the same properties as the target stimulus (Figure 11.1). Therefore, the target stimulus, which the user is gazing at, can be identified by analyzing the recorded SSVEPs using target-identification algorithms. In this way, the system can indirectly translate users' intentions into commands for controlling external devices. Due to intensive studies from many laboratories, the performance of an SSVEP-based BCI has been dramatically improved in the past decades [5]. In general, the performance of an SSVEP-based BCI is attributed to the following three major factors [4]:

1. Stimulus presentation: Visual stimuli can be presented as flashing light-emitting diodes or flickers on a computer monitor. In recent studies, a monitor-based stimulator has been widely used in an SSVEP-based BCI because the stimulation parameters (e.g., amount, size, color, pattern, and position of stimuli) can be flexibly configured. However, stimulus frequencies that can be presented on a monitor are limited by its refresh rate [9]. For example, the frequencies by which the refresh rate is not dividable (e.g., an 11 Hz under a 60 Hz) cannot be presented on a monitor. To address this issue, two successful approaches were introduced including the frequency approximation approach proposed by Wang *et al.* [10] and the sampled-sinusoid stimulation approach proposed by Chen *et al.* [11]. Both approaches demonstrated that flickers with any frequencies (up to the half of the refresh rate) and initial phases can be precisely rendered on a computer display, leading to robust SSVEPs with consistent properties [9–11].
2. Multiple target coding: Increasing the number of BCI commands is of great importance to realize a complex communication system such as a text speller. The aforementioned progress in stimulus presentation made it possible to render a large number of flickers with different frequencies. However, the number of frequencies is still limited because a sufficient frequency interval between two neighboring stimulus frequencies should be kept to remain the discriminability of elicited SSVEPs [11]. Recently, various target-coding methods have succeeded in rendering a large number of visual stimuli with different properties using limited numbers of available frequencies [12]. For example, the frequency-shift

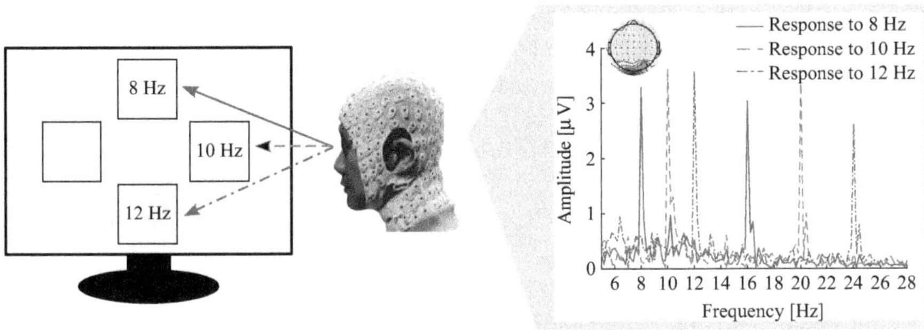

Figure 11.1 Diagram of an SSVEP-based BCI

keying (FSK) method, in which information is encoded using code words through sequential changes of discrete frequencies, was applied to present ten visual stimuli using only two stimulus frequencies [12]. Several other studies have further shown that frequency and phase information can be simultaneously embedded in SSVEPs and discriminated accurately [3,4,13]. With the hybrid frequency and phase coding approach, 32-command and 40-command BCI spellers were realized without compromising the discriminability of SSVEPs [4,5].

3. Target-identification algorithms: Proposing effective target-identification algorithms is one of the most important research topics to improve the performance of SSVEP-based BCIs. The most straightforward approach is to detect the peak amplitude of the spectrum of the EEG signals using power spectrum density analysis (PSDA) such as discrete Fourier transform [14,15]. In recent literature, temporal features of SSVEPs have been employed to develop more efficient methods than features in the frequency domain [16]. For instance, an algorithm based on template matching, which uses individual templates obtained by averaging training data across trials, achieved significantly better performance than the PSDA-based method [4,16].

Since these issues are associated with each other, Chen *et al.* suggested that they should be jointly optimized in their guideline for designing an SSVEP-based BCI [5]. Indeed, their SSVEP-based BCI speller, which combined the novel stimulus coding based on a joint frequency-phase modulation (JFPM) and the individual template-based target identification, achieved the highest ITR of 267 bits/min in the ones which had ever been reported [5]. However, its performance is still affected by the interference from the spontaneous EEG activities and artifacts. Therefore, the performance of an SSVEP-based BCI still has room for improvement.

Scalp EEG signals are generally considered as a linear instantaneous mixture of multiple time series from different cortical sources [17]. In other words, the cortical source activities can be estimated by applying weight coefficients to scalp EEG recordings from multiple electrodes. Based on this idea, several approaches, which extract optimal spatial filters (i.e., weight coefficients) to reconstruct source activities from scalp recordings, have been proposed to improve the signal-to-noise ratio (SNR) of SSVEPs [6,15,18–21]. For instance, independent component analysis (ICA) and principal component analysis (PCA) were applied for improving the accuracy of SSVEP detection [15,18]. These methods are based only on statistics between decomposed source signals. In contrast, model-based approaches such as canonical correlation analysis (CCA), minimum energy combination (MEC), and maximum contrast combination (MCC), in which computer-generated SSVEP models are employed, have shown reliable performance for detecting SSVEPs [19–21]. More recently, a novel data-driven method based on the task-related component analysis (TRCA) was demonstrated to enhance the SNR of SSVEPs by increasing the reproducibility of SSVEPs across multiple trials [6]. The TRCA-based spatial filtering has been successfully combined with the individual template-based target-identification algorithm [6]. However, to the best of our knowledge, it still remains unknown which spatial-filtering techniques make a better combination with the individual template-based method.

This chapter provides a comprehensive review of the spatial-filtering techniques in SSVEP detection. This chapter first provides the framework of the individual template-based target-identification algorithm. The filter bank analysis proposed by Chen *et al.* will then be introduced for improving the accuracy of target identification [22]. This chapter will then introduce the methods for deriving spatial filters including (1) simple average across electrodes (AVG), (2) MEC, (3) MCC, (4) CCA, (5) PCA, (6) ICA, and (7) TRCA. This chapter also performs a quantitative comparison of the spatial-filtering methods combined with the individual template-based method. Finally, this chapter discusses the characteristics of each spatial filter, future direction, and applications in real-life environments.

11.2 Individual template-based SSVEP detection

11.2.1 Basic framework

The individual template-based target-identification algorithm was first proposed by Nakanishi *et al.* [4]. The effectiveness of the method has been proven in the study of frequency-coded SSVEPs [22] and hybrid frequency- and phase-coded SSVEPs [4,5]. In the target identification, the goal is to take an input (i.e., multichannel EEG signal) and assign it to one of the labels for each class (i.e., visual stimulus). The diagram of the feature extraction in this method is illustrated in Figure 11.2. When using this method, users first need to collect training data for each individual to construct spatial filters and templates before the online operation of the BCI system. The individual training data for the nth stimulus are denoted by a three-way tensor $\chi_n = (\chi_n)_{jkh} \in \mathbb{R}^{N_c \times N_s \times N_t}, n = 1, 2, \ldots, N_f$. Here, n indicates the stimulus index, N_f is the number of visual stimuli, j indicates the channel index, N_c is the number of channels, k indicates the index of sample points, N_s is the number of sampling points in each trial, h indicates the index of training trials, and N_t is the number of training trials. For each visual stimulus, individual templates $\overline{\chi}_n \in \mathbb{R}^{N_c \times N_s}$ can be obtained

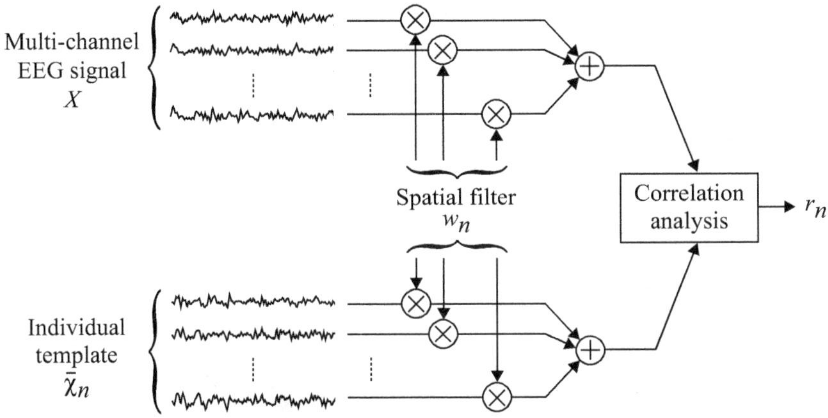

Figure 11.2 Framework of feature extraction including spatial filtering in the individual template-based method for detecting SSVEPs

by averaging multiple training trials as $(\overline{\chi}_n)_{jk} = 1/N_t \sum_{h=1}^{N_t} (\chi_n)_{jkh}$. Spatial filters $\boldsymbol{w}_n \in \mathbb{R}^{N_c}$ are derived from the training data for each stimulus, and then they are applied to the individual templates and input EEG data $\boldsymbol{X} \in \mathbb{R}^{N_c \times N_s}$ to enhance the SNR of SSVEPs. The detailed procedures of constructing spatial filters will be described in Section 11.3. After spatial filtering, the correlations r_n between the input data $\boldsymbol{X}^T \boldsymbol{w}_n$ and templates $\overline{\boldsymbol{\chi}}_n^T \boldsymbol{w}_n$ are calculated as feature values as follows:

$$r_n = \rho\left(\boldsymbol{X}^T \boldsymbol{w}_n, \overline{\boldsymbol{\chi}}_n^T \boldsymbol{w}_n\right) \tag{11.1}$$

where $\rho(a, b)$ indicates the Pearson's correlation analysis between two signals a and b. The template that maximizes the correlation is selected as the one corresponding to the target stimulus. Target class L_τ can be identified by the following equation:

$$\tau = \arg\max_n r_n, n = 1, 2, \ldots, N_f \tag{11.2}$$

In an SSVEP-based BCI, it is important to find an optimal spatial-filtering approach to increase the correlation difference between the target and nontarget stimuli, leading to a higher accuracy of target identification.

11.2.2 Ensemble strategy

To further improve the accuracy, an ensemble spatial-filtering strategy was introduced [6]. Since there are N_f individual training data corresponding to all visual stimuli, N_f different spatial filters can be obtained. Ideally, they should be similar to each other because the mixing coefficients from the sources of SSVEPs to scalp EEG signals could be considered similar within a narrow frequency range [23,24], which indicates the possibility of further improvements by integrating all spatial filters. An ensemble spatial filter $\boldsymbol{W} \in \mathbb{R}^{N_c \times N_f}$ is obtained as follows:

$$\boldsymbol{W} = \left[\boldsymbol{w}_1, \boldsymbol{w}_2, \cdots, \boldsymbol{w}_{N_f}\right] \tag{11.3}$$

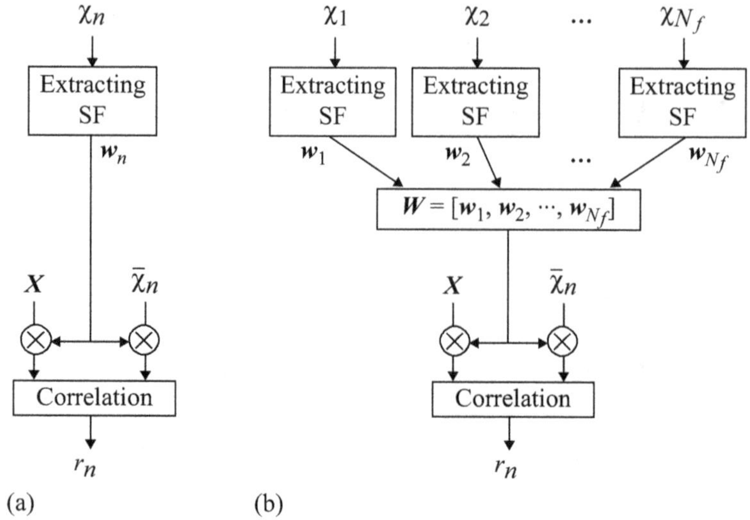

Figure 11.3 Two types of frameworks of template-based method. (a) Basic approach and (b) the ensemble strategy. The "SF" stands for spatial filter

Then (11.1) can be modified as follows:

$$r_n = \rho\left(\boldsymbol{X}^T \boldsymbol{W}, \overline{\boldsymbol{\chi}}_n^T \boldsymbol{W}\right) \tag{11.4}$$

where $\rho(\boldsymbol{a}, \boldsymbol{b})$ indicates the two-dimensional correlation analysis between $\boldsymbol{a}$ and $\boldsymbol{b}$. After obtaining all the features r_n, the target can be identified through (11.2). Figure 11.3 shows the flowchart of target identifications based on basic and ensemble strategies.

11.2.3 Filter bank analysis

The filter bank analysis is another way to improve the accuracy of SSVEP detection by decomposing SSVEPs into subband components so that independent information embedded in the harmonic components can be extracted more efficiently [22]. According to the study [22], the lower and upper cutoff frequencies of the mth subband were set to $m \times 8$ and 90 Hz, respectively. After applying mth zero-phase Chebyshev Type I infinite impulse response (IIR) filters, the training data and the test data are denoted as $\boldsymbol{\chi}_n^{(m)} \in \mathbb{R}^{N_c \times N_s \times N_t}$ and $\boldsymbol{X}^{(m)} \in \mathbb{R}^{N_c \times N_s}$, respectively. In the template-based method, correlation-based feature values for the mth subband and nth stimuli can be calculated as $r_n^{(m)} = f\left(\boldsymbol{\chi}_n^{(m)}, \boldsymbol{X}^{(m)}\right)$, where $f(\cdot)$ indicates the function for the feature extraction described in (11.1) or (11.4). A weighted sum of squares of the combined correlation values corresponding to all subband components was calculated as the feature for target identification:

$$\tilde{r}_n = \sum_{m=1}^{N_m} a(m) \cdot \left(r_n^{(m)}\right) \tag{11.5}$$

where N_m is the total number of subbands, and $a(m)$ was defined as $a(m) = m^{-1.25} + 0.25$ according to [22]. Target identification described in (11.2) can be modified as follows:

$$\tau = \arg\max_n \tilde{r}_n, n = 1, 2, \ldots, N_f \tag{11.6}$$

11.3 Spatial-filtering techniques

As mentioned in Section 11.2, spatial-filtering techniques have great potentials to improve the performance of the individual template-based SSVEP detection. Until the early 2000s, a target stimulus in an SSVEP-based BCI had been detected by finding the peak amplitude of an SSVEP in its spectrum [14]. In most cases, the amplitude spectra of SSVEPs were estimated from a single electrode placed at the occipital area (e.g., Oz). However, due to the interference from spontaneous EEG activities, the SNR of SSVEPs had been not sufficient to implement a practical BCI system. In the last decade, researchers have developed novel approaches to enhance the SNR of SSVEPs by taking advantages of multichannel EEG recording as listed in Table 11.1. This chapter introduces six spatial-filtering approaches for SSVEP.

11.3.1 Average combination

The simplest approach is taking an average across the signals from all electrodes [21]. The AVG-based spatial filters can be constructed as follows:

$$\boldsymbol{w}_n^{[j]} = \frac{1}{N_C}, j = 1, 2, \ldots, N_c. \tag{11.7}$$

Due to the individual difference in the structure of the brain, the best electrode to detect SSVEPs for each individual is unknown. This approach might average out the accuracy of detecting SSVEPs. In addition, when a few electrodes are much more affected by artifacts than others, the AVG-based spatial filtering might achieve better performance than a single-channel analysis.

11.3.2 Minimum energy combination

The MEC-based SSVEP detection was proposed by Friman *et al.* [21]. In their study, several different methods including the MEC and MCC-based methods were compared. The result showed that the MEC-based method achieved the classification accuracy of 84% using short data epochs (i.e., 1 s) and six electrodes, which was significantly higher than a bipolar approach (71%).

Table 11.1 List of previous studies of SSVEP-based BCIs using spatial-filtering techniques

Previous works	Spatial filter	Experimental design	Target identification			Results
			Data length	Electrodes	Algorithm	
Wang *et al.* [15]	ICA	*Paradigm*: Online *No. of subjects*: 16 *No. of stimuli*: 13 (frequency only)	3.4–5.7 s	13 (After ICA, 2 optimal electrodes were selected)	PSDA-based method	*Accuracy*: >95% *ITR*: 43 bits/min
Friman *et al.* [21]	AVG MEC MCC	*Paradigm*: Offline *No. of subjects*: 10 *No. of stimuli*: 6 (frequency only)	1.0 s	6 (Occipital)	Model-based method	*Accuracy*: AVG: 59% MEC: 84% MCC: 81%
Pouryazdian *et al.* [18]	PCA	*Paradigm*: Offline *No. of subjects*: 3 *No. of stimuli*: 26 (frequency only)	3.0 s	7 (Occipital)	PSDA-based method	*Accuracy*: 83.4%
Bin *et al.* [20]	CCA	*Paradigm*: Online *No. of subjects*: 12 *No. of stimuli*: 6 (frequency only)	2.0 s	9 (Occipital)	Model-based method	*Accuracy*: 95.3% *ITR*: 58 bits/min
Wang *et al.* [7]	CCA	*Paradigm*: Online *No. of subjects*: 4 *No. of stimuli*: 12 (frequency only)	Not provided	4 (Occipital)	Model-based method	*ITR*: 46 bits/min
Zhang *et al.* [26]	CCA	*Paradigm*: Offline *No. of subjects*: 10 *No. of stimuli*: 4 (multiple frequency-coding)	3 s	4 (Occipital)	Model-based method	*Accuracy*: 90.1% *ITR*: 24 bits/min

Kimura *et al.* [12]	CCA	*Paradigm*: Online *No. of subjects*: 10 *No. of stimuli*: 10 (FSK)	5.68 s	5 (Occipital)	Model-based method	*Accuracy*: 81.8% *ITR*: 29 bits/min
Chen *et al.* [11]	CCA	*Paradigm*: Online *No. of subjects*: 10 *No. of stimuli*: 45 (frequency only)	2.0 s	9 (Occipital)	Model-based method	*Accuracy*: 84.1% *ITR*: 105 bits/min
Nakanishi *et al.* [4]	CCA	*Paradigm*: Offline *No. of subjects*: 13 *No. of stimuli*: 32 (mixed frequency and phase coding)	1.0 s	8 (Occipital)	Individual template-based method	*Accuracy*: 91.35% *ITR*: 167 bits/min
Nakanishi *et al.* [16]	CCA	*Paradigm*: Offline *No. of subjects*: 10 *No. of stimuli*: 12 (JFPM)	1.0 s	8 (Occipital)	Individual template-based method	*Accuracy*: 92.78% *ITR*: 92 bits/min
Chen *et al.* [5]	CCA	*Paradigm*: Online *No. of subjects*: 12 *No. of stimuli*: 40 (JFPM)	0.5–1.0 s	9 (Occipital)	Individual template-based method	*Accuracy*: 98.64% *ITR*: 270 bits/min
Wang *et al.* [27]	CCA	*Paradigm*: Online *No. of subjects*: 5 *No. of stimuli*: 12 (JFPM)	4.0 s	8 (Behind-the-ears)	Individual template-based method	*Accuracy*: 84.08% *ITR*: 30 bits/min
Nakanishi *et al.* [6]	TRCA	*Paradigm*: Online *No. of subjects*: 20 *No. of stimuli*: 40 (JFPM)	0.3 s	9 (Occipital)	Individual template-based method	*Accuracy*: 89.83% *ITR*: 325 bits/min

The MEC-based spatial filters are derived to minimize as much energy of nuisance signals as possible. To this end, the first step is to remove potential SSVEP components from all the scalp EEG signals using computer-generated models of SSVEPs. In general, SSVEPs can be modeled by sinusoidal waveforms at fundamental and harmonic frequencies [25]. The models $\boldsymbol{Y}_n \in \mathbb{R}^{2N_h \times N_s}$ for nth stimulus can be formulated as

$$\boldsymbol{Y}_n = \begin{bmatrix} \sin(2\pi f_n t) \\ \cos(2\pi f_n t) \\ \vdots \\ \sin(2\pi N_h f_n t) \\ \cos(2\pi N_h f_n t) \end{bmatrix}, t = \left[\frac{1}{f_s}, \frac{2}{f_s}, \ldots, \frac{N_s}{f_s}\right] \tag{11.8}$$

where f_n is the stimulation frequency, f_s is the sampling frequency, and N_h is the number of harmonics. The nuisance signals $\tilde{\boldsymbol{\chi}}_n \in \mathbb{R}^{N_c \times N_s}$ can be estimated by subtracting the SSVEP models $\boldsymbol{Y}_n$ from individual template $\overline{\boldsymbol{\chi}}_n$ as

$$\tilde{\boldsymbol{\chi}}_n = \overline{\boldsymbol{\chi}}_n - \overline{\boldsymbol{\chi}}_n\left(\boldsymbol{Y}_n^T\left(\boldsymbol{Y}_n\boldsymbol{Y}_n^T\right)^{-1}\boldsymbol{Y}_n\right)^T \tag{11.9}$$

Then, a weight vector can be found so as to minimize the resulting energy of its projected signals $\tilde{\boldsymbol{\chi}}_n\boldsymbol{w}$ by solving the following problem as

$$\boldsymbol{w}_n = \underset{\boldsymbol{w}}{\operatorname{argmin}} \|\tilde{\boldsymbol{\chi}}_n^T\boldsymbol{w}\|^2 = \underset{\boldsymbol{w}}{\operatorname{argmin}}\, \boldsymbol{w}^T\tilde{\boldsymbol{\chi}}_n\tilde{\boldsymbol{\chi}}_n^T\boldsymbol{w} \tag{11.10}$$

The solution of the minimization problem is given by the eigenvector corresponding to the smallest eigenvalue. Note that, not only the eigenvector corresponding to the smallest eigenvalue but also some of the other eigenvectors might contain nuisance signals. Therefore, multiple eigenvectors should be included in the spatial filters so as to discard 90% of the energy in nuisance signals.

11.3.3 Canonical correlation analysis

The CCA-based spatial filtering was first proposed by Lin *et al.* [19]. After that, CCA has become the most popular method in detecting SSVEPs because of its effectiveness, robustness, and simple implementation. [4,5,7,11,12,16,20,26,27]. Bin *et al.* first employed the CCA-based spatial filtering in an online SSVEP-based BCI [20]. In their study, the BCI system achieved the accuracy of 95.3% using the CCA-based method with 2-s-long data and nine electrodes. Chen *et al.* implemented a similarly designed system with an increased number of stimuli [11]. The CCA-based method has also been used to detect SSVEPs elicited by different types of visual stimulus design [4,12,26]. Zhang *et al.* and Kimura *et al.* succeeded to classify the SSVEPs tagged with multiple frequencies using the CCA-based method [12,26]. Nakanishi *et al.* developed the BCI system based on SSVEPs elicited by the mixed frequency and phase coding, in which 32 visual stimuli were modulated by the combination of 8 frequencies and 4 phases [4]. In their study, the CCA-based spatial filtering was first combined with the template-based target identification. Despite a large number of stimuli, the recorded SSVEPs could be accurately classified using the combined method with 1-s-long data. In subsequent studies, the joint frequency and phase modulation, which finds the optimal phase parameters for the template-based method, was proposed toward the highest ITRs [5,16]. Indeed, the online BCI speller developed by Chen *et al.* achieved an ITR of 270 bits/min [5].

CCA is a statistical way to measure the underlying correlation between two sets of multidimensional variables. CCA finds weight coefficients, which maximize the correlation between the two variables projected onto the coefficients. In the template-based SSVEP detection, the weight vectors $\boldsymbol{w}_x$ and $\boldsymbol{w}_y$ can be obtained by calculating the canonical correlation between individual templates $\overline{\boldsymbol{\chi}}_n$ and the computer-generated SSVEP models $\boldsymbol{Y}_n$ described in (11.8). This process is formulated as follows:

$$\max_{\boldsymbol{w}_x,\boldsymbol{w}_y} \frac{E\left[\boldsymbol{w}_x^T \overline{\boldsymbol{\chi}}_n \boldsymbol{Y}_n^T \boldsymbol{w}_y\right]}{\sqrt{E\left[\boldsymbol{w}_x^T \overline{\boldsymbol{\chi}}_n \overline{\boldsymbol{\chi}}_n^T \boldsymbol{w}_x\right] E\left[\boldsymbol{w}_y^T \boldsymbol{Y}_n \boldsymbol{Y}_n^T \boldsymbol{w}_y\right]}} \tag{11.11}$$

In the template-based method, the optimized weight coefficient for the templates $\boldsymbol{w}_x$ is used as $\boldsymbol{w}_n$ in the testing stage.

11.3.4 Independent component analysis

ICA, which decomposes multidimensional signals into statistically independent components, has been widely used in EEG signal processing and neuroimaging [28,29]. Wang *et al.* applied ICA to 13-channel scalp EEG signals to decompose them into the components corresponding to SSVEPs and artifacts [15]. In this method, the best reference electrode, which maximizes/minimizes the correlation with noise/SSVEP components, was selected to remove the artifact from the signals in target electrodes, resulting in improved accuracy over 95%. Another approach based on multivariate statistical analysis was proposed by Pouryazdian *et al.* [18]. They demonstrated that PCA could extract the component highly associated with SSVEPs, leading to the higher classification accuracy than the PSDA-based method with a single scalp electrode. However, a comparison study showed that the ICA-based spatial filters were more efficient than the PCA-based one [30].

The ICA was first combined with the template-based SSVEP detection by Nakanishi *et al.* [31]. In this approach, training trials for the nth stimulus $\boldsymbol{\chi}_n \in \mathbb{R}^{N_c \times N_s \times N_t}$ are concatenated as $\dot{\boldsymbol{\chi}}_n \in \mathbb{R}^{N_c \times N_s \cdot N_t}$. By applying the informax ICA algorithm implemented in EEGLAB [29] to the concatenated data, unmixing matrix $\boldsymbol{A} \in \mathbb{R}^{N_c \times N_c}$ and activities of ICs $\boldsymbol{U} = (U)_{jk} = \boldsymbol{A}^T \dot{\boldsymbol{\chi}}_n \in \mathbb{R}^{N_c \times N_s \times N_t}$ can be obtained. The unmixing matrix is formed as

$$\boldsymbol{A} = \left[\boldsymbol{a}_1, \boldsymbol{a}_2, \ldots, \boldsymbol{a}_{N_c}\right] \tag{11.12}$$

where $\boldsymbol{a}_j \in \mathbb{R}^{N_c}$ is an unmixing vector to decompose original EEG signals into the jth IC. Although each IC is considered to be generated from independent sources, the correspondence between ICs and source activities is unknown. If an IC contains SSVEPs, its correlation to the computational models of SSVEPs should be higher than those with ICs unrelated to SSVEPs. Based on this idea, canonical correlations between each IC and the SSVEP models described in (11.8) were calculated, and the unmixing vectors were sorted by the values of the canonical correlations in descending order. Assume that $\boldsymbol{I}(j) \in \mathbb{N}^{N_c}$ is an index vector of original unmixing vectors, the sorted unmixing vectors $\tilde{\boldsymbol{A}} \in \mathbb{R}^{N_c \times N_c}$ can be described as

$$\tilde{\boldsymbol{A}} = \left[\boldsymbol{a}_{\boldsymbol{I}(1)}, \boldsymbol{a}_{\boldsymbol{I}(2)}, \ldots, \boldsymbol{a}_{\boldsymbol{I}(N_c)}\right]. \tag{11.13}$$

After sorting the unmixing vectors, the lth unmixing vector $\boldsymbol{a}_{\boldsymbol{I}(l)}$ is extracted and sorted as a spatial filter $\boldsymbol{w}_n$ for nth visual stimuli.

11.3.5 Task-related component analysis

TRCA is the method that extracts task-related components efficiently by maximizing their reproducibility during task periods [32]. Since SSVEPs are known as time-locked photic-driving responses to repetitive visual stimuli [1,33], the TRCA-based spatial filtering has great potential to enhance the SNR of SSVEPs. Nakanishi *et al.* demonstrated the effectiveness of the TRCA-based spatial filtering to improve the performance of an SSVEP-based BCI, in which the highest ITR of 325 bits/min was achieved using only 0.3-s-long data epochs [6].

The TRCA finds a linear coefficient vector $\boldsymbol{w} \in \mathbb{R}^{N_c}$ to maximize intertrial correlation of its projections. The hth trial in the training data for nth stimulus and task-related components are described as $\boldsymbol{\chi}_n^{(h)} \in \mathbb{R}^{N_c \times N_s}$ and $\boldsymbol{y}_n^{(h)} = \boldsymbol{w}^T \boldsymbol{\chi}_n^{(h)} \in \mathbb{R}^{N_s}$, respectively. The covariance C_{h_1,h_2} between h_1th and h_2th trials of $\boldsymbol{\chi}_n^{(h)}$ is described as

$$\begin{aligned} C_{h_1,h_2} &= \mathrm{Cov}\left(\boldsymbol{y}_n^{(h_1)}, \boldsymbol{y}_n^{(h_2)}\right) \\ &= \sum_{j_1,j_2=1}^{N_c} w_{j_1} w_{j_2} \mathrm{Cov}\left(\boldsymbol{\chi}_{nj_1}^{(h_1)}, \boldsymbol{\chi}_{nj_2}^{(h_2)}\right) \end{aligned} \tag{11.14}$$

All possible combinations of N_t trials are summed as

$$\begin{aligned} \sum_{\substack{h_1,h_2=1\\h_1\neq h_1}}^{N_t} C_{h_1,h_2} &= \sum_{\substack{h_1,h_2=1\\h_1\neq h_1}}^{N_t} \sum_{j_1,j_2=1}^{N_c} w_{j_1} w_{j_2} \mathrm{Cov}\left(\boldsymbol{\chi}_{nj_1}^{(h_1)}, \boldsymbol{\chi}_{nj_2}^{(h_2)}\right) \\ &= \mathbf{w}^T \boldsymbol{S}_n \boldsymbol{w} \end{aligned} \tag{11.15}$$

Here, the matrix $\boldsymbol{S} \in \mathbb{R}^{N_c \times N_c}$ is defined as

$$(S_n)_{j_1,j_2} = \sum_{\substack{h_1,h_2=1\\h_1\neq h_1}}^{N_t} \mathrm{Cov}\left(\boldsymbol{\chi}_{nj_1}^{(h_1)}, \boldsymbol{\chi}_{nj_2}^{(h_2)}\right) \tag{11.16}$$

To obtain a finite solution, the variance of $\boldsymbol{y}$ is constrained using concatenated training trials $\dot{\boldsymbol{\chi}}_n \in \mathbb{R}^{N_c \times N_s \cdot N_t}$ as

$$\begin{aligned} \mathrm{Var}(\boldsymbol{y}) &= \sum_{j_1,j_2=1}^{N_c} w_{j_1} w_{j_2} \mathrm{Cov}\left(\dot{\boldsymbol{\chi}}_{nj_1}, \dot{\boldsymbol{\chi}}_{nj_2}\right) \\ &= \mathbf{w}^T \boldsymbol{Q}_n \boldsymbol{w} \\ &= 1 \end{aligned} \tag{11.17}$$

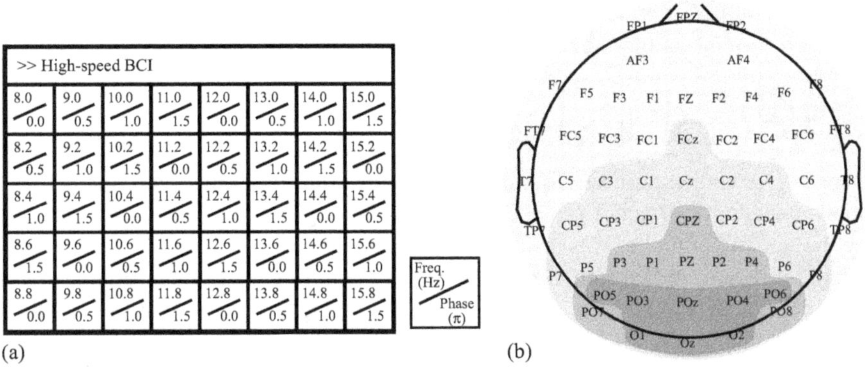

Figure 11.4 (a) Forty target visual stimuli designed with the joint frequency-phase modulation (JFPM), and (b) electrode location used for data recording. Four different electrodes setting used in the performance evaluation are marked with gray scale shades with different intensities

The constrained optimization problem can be solved as

$$w_n = \underset{w}{\operatorname{argmax}} \frac{\mathbf{w}^T \boldsymbol{S}_n \boldsymbol{w}}{\mathbf{w}^T \boldsymbol{Q}_n \boldsymbol{w}} \tag{11.18}$$

The optimal coefficient vector is obtained as the eigenvector of the matrix $\boldsymbol{Q}_n^{-1}\boldsymbol{S}_n$. The eigenvector corresponding to the largest eigenvalue is selected as the spatial filter to extract task-related components.

11.4 Material and methods

11.4.1 Dataset

This study performs a comparative analysis using a benchmark dataset publicly available on the Internet [34]. The visual stimulator designed for a text speller was used in this study. Forty visual flickers coded by the JFPM method were presented on a 23.6-in liquid-crystal display monitor with the resolution of 1,920 × 1,080 pixels and the refresh rate of 60 Hz. As illustrated in Figure 11.4(a), the user interface is a 5 × 8 matrix of the stimuli containing 40 characters (26 English alphabet letters, 10 digits, and 4 other symbols). Specifically, the 40 stimuli were tagged with linearly increasing frequencies and phases, of which the increments are both proportional to the stimulus index. The frequency range was selected from 8 to 15.8 Hz with an interval of 0.2 Hz. The phase values started from 0 and the phase interval was 0.5 π. The stimulus parameters were implemented using a sampled-sinusoidal stimulation method [11,35]. At frequency f and phase $\emptyset$, the stimulus sequence $s(f, \emptyset, i)$ can be generated by modulating the luminance of the screen:

$$s(f, \emptyset, i) = \frac{1}{2}\left\{\sin\left[2\pi f\left(\frac{i}{\text{RefreshRate}}\right) + \emptyset\right] + 1\right\} \tag{11.19}$$

where sin() generates a sine wave and i indicates the frame index in the sequence. In the stimulus sequence, 0 and 1 represent dark and the highest luminance, respectively. The size of the stimulus was 140 × 140 pixel square with 50 pixels vertical and horizontal distances between two neighboring stimuli. The stimulus program was developed under MATLAB® (MathWorks, Inc.) using the Psychophysics Toolbox Version 3 [36].

The EEG data were recorded from 35 healthy adults (17 females, mean age of 22 years) with normal or corrected-to-normal vision. Among the subjects, eight of them had experience of participating in experiments using SSVEPs. All the participants were asked to read and sign an informed consent form before the experiment. The Research Ethics Committee of Tsinghua University approved this experiment. The subjects were seated in a comfortable chair 70 cm away from the monitor. They performed a simulated online experiment to record data for offline analysis. The experiment consisted of 6 blocks containing 40 trials corresponding to 40 characters. In the beginning of each trial, a red square was presented at the target location as a target cue for 0.5 s. The subjects were asked to shift their gaze to the target as soon as possible. After the cue, all stimuli started to flicker on the screen simultaneously for 5 s. Then, the screen was blank for 0.5 s before the next trial began. Subjects were instructed to avoid eye blinks during the 5-s stimulation duration. There was a rest for several minutes between two consecutive blocks. The EEG signals were recorded by 64 electrodes according to the international 10–20 system using a Synamps2 EEG system (Neuroscan, Inc.) at a sample rate of 1,000 Hz as shown in Figure 11.4(b). The reference electrode was placed at the vertex (Cz). Electrode impedances were kept below 10 kΩ during recording. Event triggers generated by the stimulus program were recorded on an event channel of the amplifier synchronized to the EEG data.

11.4.2 Performance evaluation

Data epochs comprising 64-channel SSVEPs were first extracted according to the event triggers. All data epochs were down-sampled to 250 Hz and then band-pass filtered from 7 to 90 Hz with an IIR filter. Zero-phase forward and reverse filtering was implemented using the filtfilt() function in MATLAB. Considering a latency delay in the visual pathway, the data epochs were extracted in [0.14 s 0.14+d s], where the time 0 indicated the stimulus onset and d indicated the data length used in the analysis. The target-identification performance was evaluated using different conditions: (1) basic or ensemble strategies, (2) the number of subbands in the filter bank analysis ($N_m = 1, 2, \ldots, 5$ in this study), and (3) the layout of electrodes. Figure 11.4(b) shows four different electrode settings marked with gray scale shades with different intensities.

The classification accuracy was estimated using a leave-one-out cross-validation. In each of six iterations, the cross-validation was performed using five blocks (i.e., $N_t = 5$) for training and one block for testing. In addition to classification accuracy, the online BCI performance was also evaluated by ITR [14]:

$$\text{ITR} = \left(\log_2 N_f + P \log_2 P + (1 - P) \log_2 \left[\frac{1 - P}{N_f - 1} \right] \right) \times \left(\frac{60}{T} \right) \quad (11.20)$$

where P is the classification accuracy and T (seconds/selection) is the average time for a selection. This study calculated classification performance using different T (target gazing time d: 0.1–1.0 s with an interval of 0.1 s; gaze shifting time: 0.5 s).

11.5 Results and discussions

11.5.1 Signal features of SSVEPs after spatial filtering

Figure 11.5 shows ERP images of SSVEPs from all the subjects after applying the aforementioned spatial-filtering methods. The images were constructed by vertically stacking color-coded horizontal bars representing each single data epoch and applying a vertical moving averaging. The plots and the edges of gray areas below the ERP images indicate the average and standard deviation of EEG signals across all the trials, respectively. The time 0 indicates the stimulus onset. The spatial filters were constructed by 1-s-long data epochs ($d = 1.0$) with eight electrodes over the occipital areas. The ERP images obtained by the CCA-, ICA-, and TRCA-based spatial filtering showed clear positive and negative peaks in the amplitude, resulting in sinusoidal-like averaged waveform across trials [Figure 11.4(c)–(e)]. This result suggests that these spatial-filtering techniques could extract the components corresponding to SSVEPs from the scalp EEG signals. In the averaged trials after the TRCA-based spatial filtering, the amplitude is increasing gradually after the stimulus onset. On the other hand, the CCA-based spatial filter produced an averaged SSVEP with consistent amplitude. It is well known that the amplitude of SSVEPs can be modulated by the users' attention [37], leading to the deteriorated performance in detecting SSVEPs. Therefore, model-based spatial-filtering methods require components, which well fit in the computational models. The performance of the TRCA-based spatial filtering might also be affected by the shift of visual attention. Since TRCA is a data-driven algorithm, it would perform well as long as the waveforms are consistent across trials. In contrast, the AVG- and MEC-based spatial filters seem not better at extracting SSVEP components than others [Figure 11.4(a) and (b)]. Note that, ICA might decompose SSVEPs into several components. In addition, the MEC requires multiple weight coefficients to remove noise components. Therefore, the SNR of SSVEPs could be improved by integrating multiple spatial filters when using the MEC- or ICA-based methods.

11.5.2 A comparison of frameworks for SSVEP detection

Figure 11.6 shows the averaged accuracy across subjects with different data length from 0.1 to 1.0 s with an interval of 0.1 s and a different number of subbands in the filter bank analysis from 1 to 5 with an interval of 1. The basic framework and

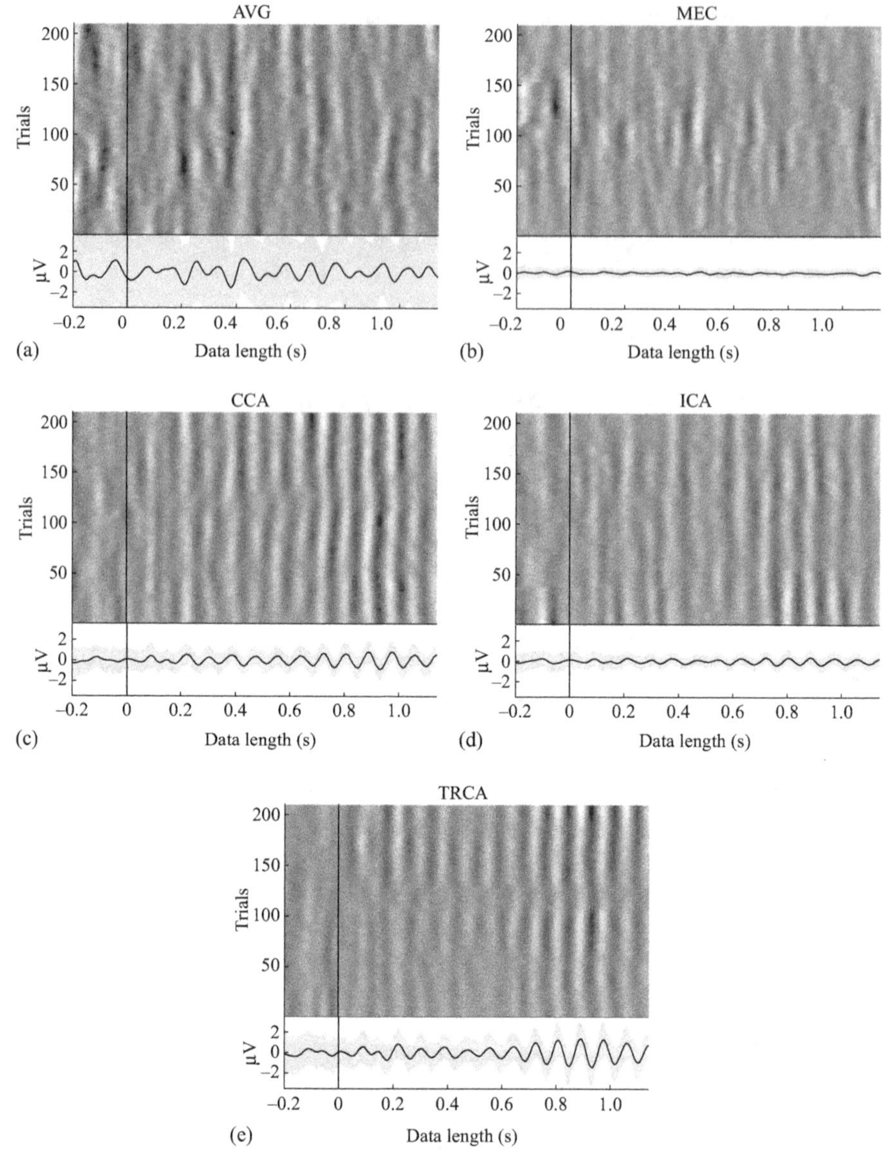

Figure 11.5 ERP images of data epochs from all subjects after spatial filtering based on (a) AVG, (b) MEC, (c) CCA, (d) ICA, and (e) TRCA. The color scales are normalized for each method. The traces below each ERP image show the averaged ERPs with standard deviation across trials. The solid lines at time 0 indicate stimulus onsets

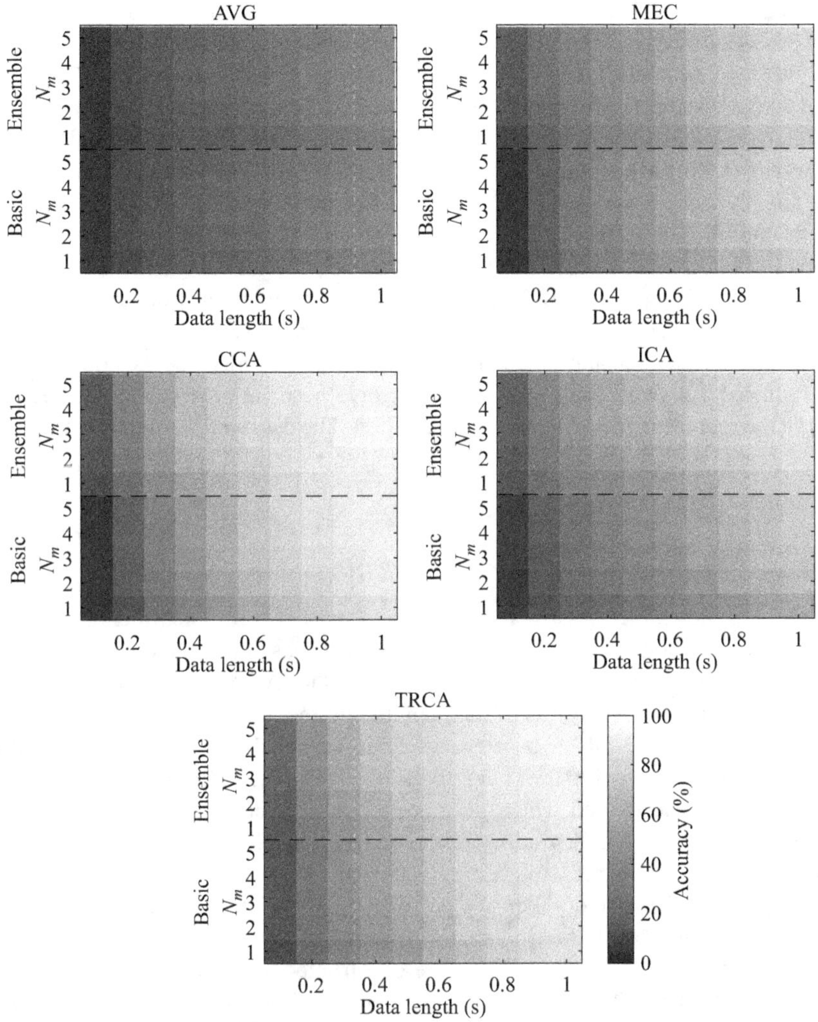

Figure 11.6 Heat maps illustrating averaged target-identification accuracy across subjects using the template-based methods with basic and ensemble strategies and different number of subbands as a function of data length

the ensemble strategy were also compared using eight electrodes over occipital areas. Empirical results suggest that the ensemble strategy achieved better performance than the basic framework in all the spatial-filtering approaches except for the AVG-based method. Note that, the accuracy of the AVG-based method was identical between the basic and ensemble frameworks. In addition, the accuracy increased as the number of subbands increased. The averaged accuracy across subjects was submitted to a four-way analysis of variance (ANOVA), showing significant main effects of frameworks [$F(1{,}376) = 2{,}587.80$, $p < 0.001$], the number of subbands [$F(4{,}376) = 371.99$, $p < 0.001$], spatial-filtering methods [$F(4{,}376) = 8{,}320.73$, $p < 0.001$], and data length [$F(9{,}376) = 6{,}896.25$, $p < 0.001$]. The performance of the individual template-based methods can be further improved by finding the best combination of the parameters. The four-way ANOVA also showed significant interactions between frameworks and the number of subbands [$F(4{,}376) = 3.45$, $p = 0.008$], frameworks and algorithms [$F(4{,}376) = 380.45$, $p < 0.001$], and the number of subbands and algorithms [$F(16{,}376) = 13.46$, $p < 0.001$], respectively. The previous study of the TRCA-based SSVEP detection suggested that the number of subbands should be five or more [6]. However, the ensemble approach and filter bank analysis increase the computational cost required in a single trials analysis. Therefore, these parameters can be selected so as to fit the requirements of the target application. The data length should be optimized toward maximal ITRs for each spatial-filtering approach.

11.5.3 A comparison of electrodes settings

Figure 11.7 shows the classification accuracy and ITRs for each method with different electrode setting (i.e., N_c). For all the methods, the ensemble strategy was employed with five subbands (i.e., $N_m = 5$). The TRCA-based method achieved the highest performance regardless of data length and the number of electrodes. Except for the MEC- and the TRCA-based method, all approaches had a tendency that the accuracy was higher with fewer numbers of electrodes. The classification accuracy might be deteriorated when there are more electrodes that might not contain SSVEPs. Especially, the degrading in the accuracy of the CCA-based method was more furious than the others. Interestingly, the MEC-based method achieved a higher accuracy curve when the number of electrodes was 16. The possible explanation for this result is that the MEC-based spatial filtering is more robust to noisy channels than others because the spatial filters are constructed so as to minimize the power of nuisance signals. The TRCA-based method also obtained better accuracy when the number of electrodes was 16. Indeed, the highest ITR of 229.40 bits/min was obtained when using 0.4 s-long data epochs with 16 electrodes. Two-way repeated measures ANOVAs revealed there were significant main effects of the electrode settings for each method [AVG: $F(3{,}102) = 45.64$, $p < 0.001$; MEC: $F(3{,}102) = 32.91$, $p < 0.001$; CCA: $F(3{,}102) = 195.86$, $p < 0.001$; ICA: $F(3{,}102) = 78.32$, $p < 0.001$; TRCA: $F(3{,}102) = 10.77$, $p < 0.001$]. This result suggests that the optimal channel setting might be different for each method, but electrodes should be placed only around the occipital area. The difference in the TRCA-based method among channel settings was smaller than the others. In a real environment, scalp EEG signals might be contaminated by various artifacts. Moreover, some electrodes could have much higher impedance. In this case, the TRCA-based spatial filtering has great potential to ignore bad channels and only extract SSVEP components.

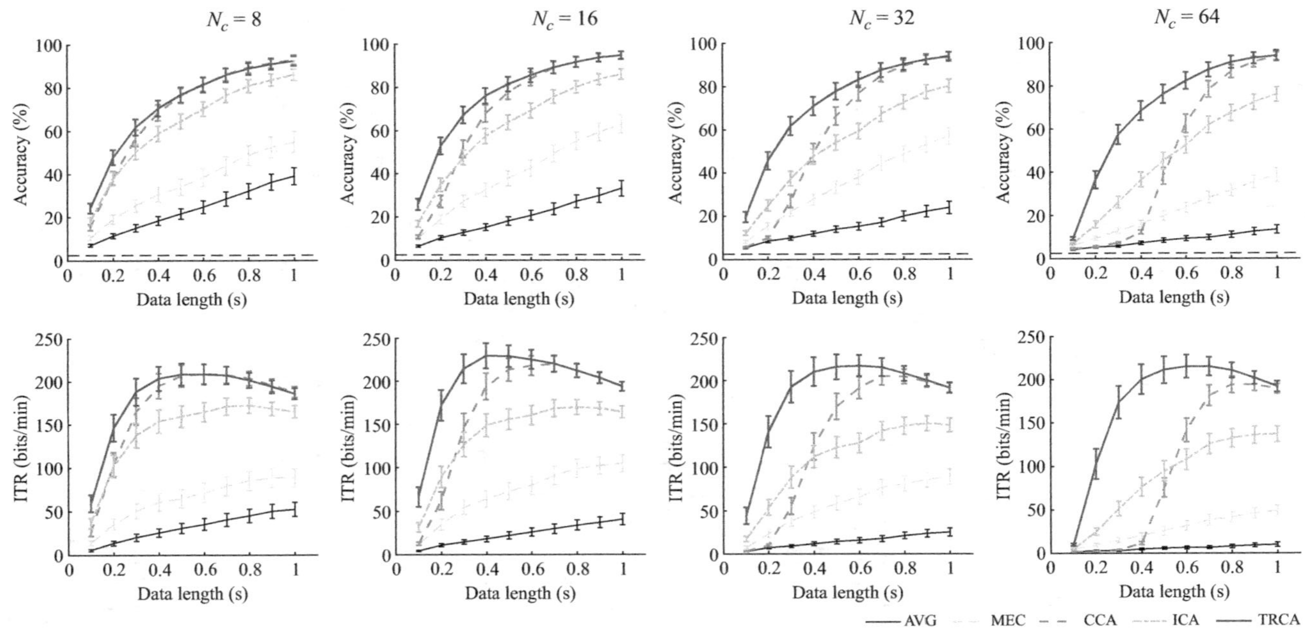

Figure 11.7 Averaged target-identification accuracy and ITRs across subjects using the ensemble template-based method with five subbands with different number of channels. Error bars indicate standard errors. The dashed line indicates the chance-level accuracy (i.e., 2.5%)

11.5.4 Toward further improvement

In the template-based SSVEP detection, the number of training data is one of the most important parameters to improve its performance. Previous studies, which showed the classification accuracy increased with the number of training trials, suggested collecting five or more trials for deriving effective CCA-based spatial filters [4,16]. Another study also showed significant improvements in the accuracy of the TRCA-based method when using more training data [6]. Therefore, further investigations on the performance limitation caused by the number of training trials and the optimal number are required for a practical BCI system. In addition, an online adaptation algorithm, which updates template signals and spatial filters in online operation, might enhance the performance. Yuan *et al.* proposed an adaptation algorithm for the CCA-based SSVEP detection [38]. This algorithm can increase the number of training trials over time and alleviate the nonstationarity of brain states. This approach can also be applied to the other types of spatial-filtering algorithms.

Redesigning visual stimuli is another way to optimize the performance. For each spatial-filtering method, the optimal phase interval in the JFPM coding might be different. As mentioned, visual stimuli and target-identification algorithms should be jointly optimized according to the guideline for designing an SSVEP-based BCI [5]. To this end, the joint frequency-phase optimization method, which runs a grid-search of phase interval and data length using an offline dataset, was proposed in Chen *et al.*'s study [5]. In this way, they found the best phase interval of 0.35π when using the individual template-based algorithm with the CCA-based spatial filters, leading to a high ITR of 267 bits/min. In contrast, it has been proven that frequency resolution does not affect the accuracy of classifying SSVEPs when using the individual template-based methods [39]. Therefore, stimulus frequencies can be selected flexibly according to the number of commands required by target applications.

The ITRs obtained in this study were much lower than the ones reported in previous studies of the high-speed BCIs (i.e., 267 bits/min in Chen *et al.*'s study [5], and 325 bits/min in Nakanishi *et al.*'s study [6]). This might be because large parts of subjects were naïve for the experiment using SSVEPs. Wang *et al.* pointed out that the naïve subjects have much lower performance than experienced ones in their research [34]. Unlike other BCI paradigms such as motor imagery or P300 [35,36], researchers have not attempted to investigate the way to train subjects. Therefore, proposing effective user training methods is important toward further improvement of BCI performance.

11.5.5 Challenges and future direction

Although collecting a large amount of training data is crucial to guarantee the performance of the template-based methods, this process might be time consuming and decrease usability. Assuming 10 trials for each of the 40 visual stimuli, the training data with a trial length of 1.5 s (target gazing time: 1.0 s; gaze shifting time: 0.5 s) can be acquired within approximately 10 min excluding break time. In addition, since spatial-filtering algorithms generally assume the consistency in the electrode layout, users need to collect training data every time when they change electrode positions. Several template-transferring approaches (e.g., transfer across different subjects [38], different recording sessions [40], or different visual stimuli [41]) have been proposed to reduce the training time. Stimulus design might also play an important role to reduce training time. For example, multiple frequency coding including FSK requires fewer frequencies than the number of visual stimuli [12]. Therefore, a well-designed FSK modulation can significantly reduce the training time without decreasing classification accuracy.

The individual template-based SSVEP detection algorithms require precise time-synchronization between a stimulation program and EEG recordings to maximize the advantage of the JFPM. In conventional experiments, event triggers were generated by a stimulation program and sent to an EEG amplifier through a physical cable. Recently, reliable synchronization can be established using wireless communication technologies including the lab streaming layer [42]. In addition, the stability of stimulus presentation might impact the performance of the individual template-based algorithm. Although large variabilities in the latency of SSVEPs were observed when using mobile platforms (e.g., a tablet and a cell phone) as a stimulation device, the feasibility of a mobile stimulation for SSVEP-based BCIs has been proven [7,43]. That is, the major barriers to implementing real-world applications of BCIs based on a truly portable platform have already been eliminated.

11.6 Conclusions

This chapter described the framework of template-based SSVEP-detection algorithms and spatial-filtering techniques. This framework significantly contributed to the performance improvement of an SSVEP-based BCI reported in recent years. The systematic comparisons showed the efficacy of integrating the ensemble strategy and filter-bank analysis. The robustness of an electrode layout was also investigated, which validated the effectiveness of the TRCA-based spatial-filtering method. We hope this chapter will encourage more real-life BCI applications for communication and control.

References

[1] Vialatte F.B., Maurice M., Dauwels J., Cichocki A. 'Steady-state visual evoked potentials: Focus on essential paradigms and future perspectives'. *Prog. Neurobiol.* 2010, vol. 90, pp. 418–438.

[2] Regan D. 'Some characteristics of average steady-state and transient responses evoked by modulated light'. *Electroencephalogr. Clin. Neurophysiol.* 1966, vol. 20(3), pp. 238–248.

[3] Wang Y., Gao X., Hong B., Jia C., Gao S. 'Brain–computer interfaces based on visual evoked potentials: Feasibility of practical system design'. *IEEE Eng. Med. Biol. Mag.* 2008, vol. 27, pp. 64–71.

[4] Nakanishi M., Wang Y., Wang Y.T., Mitsukura Y., Jung T.P. 'A high-speed brain speller using steady-state visual evoked potentials'. *Int. J. Neural Syst.* 2014, vol. 24(6), p. 1450019.

[5] Chen X., Wang Y., Nakanishi M., Gao X., Jung T.P., Gao S. 'High-speed spelling with a noninvasive brain–computer interface'. *Proc. Natl. Acad. Sci. U.S.A.* 2015, vol. 112(44), pp. E6058–E6067.

[6] Nakanishi M., Wang Y., Chen X., Wang Y.T., Gao X, Jung T.P. 'Enhancing detection of SSVEPs for a high-speed brain speller using task-related component analysis'. *IEEE Trans. Biomed. Eng.* 2018, vol. 65(1), pp. 104–112.

[7] Wang Y.T., Wang Y., Jung T.P. 'A cell-phone-based brain–computer interface for communication in daily life'. *J. Neural Eng.* 2011, vol. 8(2), p. 025018.

[8] Martinez P., Bakardjian H., Cichocki A. 'Fully online multicommand brain–computer interface with visual neurofeedback using SSVEP paradigm'. *Comput. Intell. Neurosci.* 2007, vol. 2007, p. 94561.

[9] Nakanishi M., Wang Y., Wang Y.T., Mitsukura Y., Jung T.P. 'Generating visual flickers for eliciting robust steady-state visual evoked potentials at flexible frequencies using monitor refresh rate'. *PLoS One.* 2014, vol. 9(6), p. e99235.

[10] Wang Y., Wang Y.T., Jung T.P. 'Visual stimulus design for high-rate SSVEP BCI'. *Electron. Lett.* 2010, vol. 46(15), pp. 1057–1058.

[11] Chen X., 'A high-ITR SSVEP-based BCI speller', *Brain–Comput. Interf.* 2014, vol. 1(3/4), pp. 181–191.

[12] Kimura Y., Tanaka T., Higashi H., Morikawa N. 'SSVEP-based brain–computer interfaces using FSK-modulated visual stimuli'. *IEEE Trans. Biomed. Eng.* 2013, vol. 60(10), pp. 2831–2838.

[13] Jia C., Gao X., Hong Bo., Gao S. 'Frequency and phase mixed coding in SSVEP-based brain–computer interface', *IEEE Trans. Biomed. Eng.* 2011, vol. 58(1), pp. 200–206.

[14] Cheng M., Gao X., Gao X., Xu D. Design and implementation of a brain–computer interface with high transfer rates. *IEEE Trans. Biomed. Eng.* 2002, vol. 49(10), pp. 1181–1186.

[15] Wang Y., Wang R., Gao X., Hong B., Gao S. 'A practical VEP-based brain–computer interface'. *IEEE Trans. Neural Syst. Rehabil. Eng.* 2006, vol. 14(2), pp. 234–239.

[16] Nakanishi M., Wang Y., Wang Y.T., Jung T.P. 'A comparison study of canonical correlation analysis based methods for detecting steady-state visual evoked potentials'. *PLoS One.* 2015, vol. 10(10), p. e0140703.

[17] Onton J., Westerfiled M., Townsend J., Makeig S. 'Imaging human EEG dynamics using independent component analysis'. *Neurosci. Biobehav. Rev.* 2006, vol. 30(6), pp. 808–822.

[18] Pouryazdian S., Erfanian A. 'Detection of steady-state visual evoked potentials for brain–computer interfaces using PCA and high-order statistics'. In Dossel O., Schlegel W.C. (eds.). World Congress on Medical Physics and Biomedical Engineering. Berlin, Heidelberg: Springer; 2009. pp. 480–483.

[19] Lin Z., Zhang C., Wu W., Gao X. 'Frequency recognition based on canonical correlation analysis for SSVEP-based BCIs'. *IEEE Trans. Biomed. Eng.* 2007, vol. 54(6), pp. 1172–1176.

[20] Bin B., Gao X., Yan Z., Hong B., Gao S. 'An online multi-channel SSVEP-based brain–computer interface using a canonical correlation analysis method'. *J. Neural Eng.* 2009, vol. 6(4), p. 046002.

[21] Friman O., Volosyak I., Graser A. 'Multiple channel detection of steady-state visual evoked potentials for brain–computer interfaces'. *IEEE Trans. Biomed. Eng.* 2007, vol. 54(4), pp. 742–790.

[22] Chen X., Wang Y., Gao S., Jung T.P., Gao X. 'Filter bank canonical correlation analysis for implementing a high-speed SSVEP-based brain–computer interface'. *J. Neural Eng.* 2015, vol. 12, p. 046008.

[23] Srinivasan R., Bibi F.A., Nunez P.L. 'Steady-state visual evoked potentials: distributed local sources and wave-like dynamics are sensitive to flicker frequency'. *Brain Topogr.* 2006, vol. 18(3), pp. 167–187.

[24] Ales J.M., Norcia A.M. 'Assessing direction-specific adaptation using the steady-state visual evoked potential: Results from EEG source imaging'. *J. Vis.* 2009, vol. 9(7), pp. 1–13.

[25] Wang Y., Gao X., Gao S. 'Computational modeling and application of steady-state visual evoked potentials in brain–computer interfaces'. *Sci. Suppl.* 2015, vol. 350(6256), pp. 43–46.

[26] Zhang Y., Xu P., Liu T., Hu J., Zhang R., Yao D. 'Multiple frequencies sequential coding for SSVEP-based brain–computer interface'. *PLoS One* 2012, vol. 7(3), p. e29519.

[27] Wang Y.T., Nakanishi M., Wang Y., Wei C.S., Cheng C.K., Jung T.P. 'An online brain–computer interface based on SSVEPs measured from non-hair-bearing areas'. *IEEE Trans. Neural Syst. Rehabil. Eng.* 2017. vol. 25(1), pp. 11–18.

[28] Jung T.P., Makeig S., Humphries C., *et al.* 'Removing electroencephalographic artifacts by blind source separation'. *Psychophysiology.* 2000, vol. 37(2), pp. 163–178.

[29] Delorme A., Makeig S. 'EEGLAB: An open source toolbox for analysis of single-trial EEG dynamics including independent component analysis'. *J. Neurosci. Methods* 2004, vol. 134, pp. 9–21.

[30] Garcia-Molina G., Zhu D. 'Optimal spatial filtering for the steady-state visual evoked potential: BCI application'. *Proceedings of the 5th Int. IEEE EMBS Conf. Neural Eng.*; Cancun, Mexico, Apr 2011. pp. 156–160.

[31] Nakanishi M., Wang Y., Hsu S.H., Wang Y.T., Jung T.P. 'Independent component analysis-based spatial filtering improves template-based SSVEP detection'. *Proceedings of the 39th Ann. Int. Conf. IEEE Eng. Med. Biol. Soc.*; Jeju Island, Korea, Jul 2017, pp. 3620–3623.

[32] Tanaka H., Katsura T., Sato H. 'Task-related component analysis for functional neuroimaging and application to near-infrared spectroscopy data'. *NeuroImage* 2013, vol. 64, pp. 308–327.

[33] Regan D. 'Human brain electrophysiology: Evoked potentials and evoked magnetic fields in science and medicine'. *Br. J. Ophthalmol.* 1990, vol. 74(4), p. 255.

[34] Wang Y., Chen X., Gao X., Gao S. 'A benchmark dataset for SSVEP-based brain–computer interfaces'. *IEEE Trans. Neural Syst. Rehabil. Eng.* 2017, vol. 25(10), pp. 1745–1752.

[35] Manyakov N.V., Chumerin N., Robben A., Combaz A., van Vliet M., Van Hulle M.M. 'Sampled sinusoidal stimulation profile and multichannel fuzzy logic classification for monitor-based phase-coded SSVEP brain–computer interfacing'. *J. Neural Eng.* 2013, vol. 10(3), p. 036011.

[36] Brainard D.H. 'The psychophysics tooblox'. *Spat. Vis.* 1997, vol. 10, pp. 433–436.

[37] Xu M., Wang Y., Nakanishi M., *et al.* 'Fast detection of covert visualspatial attention using hybrid N2pc and SSVEP features', *J Neural Eng.* 2016, vol. 13, p. 066003.

[38] Yuan P., Chen X., Wang Y., Gao X., Gao S. 'Enhancing performance of SSVEP-based brain–computer interfaces via exploiting inter-subject information'. *J. Neural Eng.* 2015, vol. 12, p. 046006.

[39] Nakanishi M., Wang Y., Wang Y.T., Jung T.P., 'Does frequency resolution affect the classification performance of steady-state visual evoked potentials?'. *Proceedings of the 8th Int. IEEE EMBS Conf. Neural Eng.*; Shanghai, China, May 2017, pp. 341–344.

[40] Nakanishi M., Wang Y., Jung T.P. 'Session-to-session transfer in detecting steady-state visual evoked potentials with individual training data'. In Schmorrow D.D., Fidopiastis C.M. (eds.). *Foundations of Augmented Cognition: Neuroergonomics and Operational Neuroscience.* Lecture Notes in Computer Science, vol. 9743. Cham, Switzerland: Springer: 2016, pp. 253–260.

[41] Suefusa K., Tanaka T. 'Reduced calibration by efficient transformation of teplates for high speed hybrid coded SSVEP brain–computer interfaces', *Proceedings of the 42nd IEEE Int. Conf. Acoust. Speech Sig. Process.*; New Orleans, USA, Mar 2017. pp. 929–933.

[42] Kothe C.A., Makeig S., 'BCILAB: a platform for brain–computer interface development', *J. Neural Eng.* 2013, vol. 10. p. 056014.

[43] Wang Y.T., Wang Y., Cheng C.K., Jung T.P., 'Developing stimulus presentation on mobile devices for a truly portable SSVEP-based BCI', *Proceedings of the 35th Ann. Int. Conf. IEEE Eng. Med. Biol. Soc.*; Osaka, Japan, Jul 2013, pp. 5271–5274.

Chapter 12
A review of feature extraction and classification algorithms for image RSVP-based BCI

Zhengwei Wang[1], *Graham Healy*[1], *Alan F. Smeaton*[1], *and Tomas E. Ward*[1]

Abstract

In this chapter, we introduce an architecture for rapid serial visual presentation (RSVP)-based brain–computer interface (BCI) systems that use electroencephalography (EEG). Hereafter, we will refer to the coupling of the RSVP protocol with EEG to support a target-search BCI as RSVP-EEG. Our focus in this chapter is on a review of feature extraction and classification algorithms applied in RSVP-EEG development. We briefly present the commonly deployed algorithms and describe their properties based on the literature. We conclude with a discussion on the future trajectory of this exciting branch of BCI research.

12.1 Introduction

The rapid serial visual presentation (RSVP) is a method that can be used to extend the brain–computer interface (BCI) approach to enable high throughput target image recognition applications [1–3]. Using electroencephalography (EEG) signals to label or rank images is of practical interest as many types of images cannot be automatically labeled by a computer [2]. A common example here is to enhance the performance of satellite imagery analysts, by performing selection to get a smaller number of images for later and more detailed inspection [1]. In the RSVP target-search paradigm (see Figure 12.1), there is a rapid succession of images presented on screen, in which only a small percentage contain target images. Images are typically presented to participants at a very fast speed on a monitor (5–12 images per second). These infrequent target images are known to elicit the P300 event-related potential (ERP), a type of brain response that has a well-established history of study [4]. The idea is that the participant is unaware when a target stimulus is going to appear; hence, its presentation on screen elicits the P300 ERP reflecting the orientation of participant's attention to the

[1]Insight Centre for Data Analytics, School of Computing, Dublin City University, Ireland

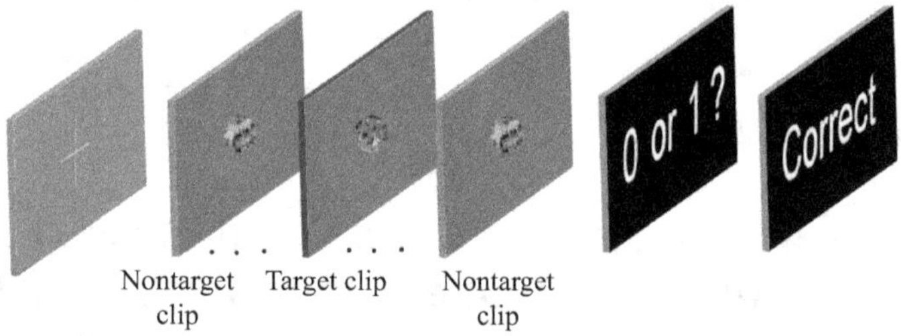

Figure 12.1 RSVP paradigm protocol in Section 12.1 [1]

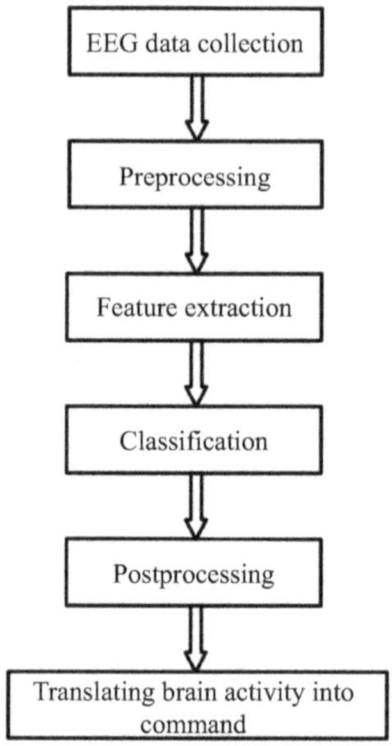

Figure 12.2 Block diagram of a typical BCI system in Section 12.1

stimulus. These brain activity related responses, when extracted, can be used through application of signal processing and machine-learning techniques to enable labeling and/or ranking of images.

In order to use an RSVP-EEG BCI in this way, a user must be capable of responding with brain activity patterns that can be identified automatically. In this regard, the use of an "oddball" paradigm to elicit P300 ERP responses is ideal as targets searched for tend to be infrequent in many datasets and the response has characteristic features.

Figure 12.2 shows the stages of a typical BCI system. Preprocessing, feature extraction, classification and postprocessing are classification system components. Changes in any one of these components can alter the performance of a BCI system. Preprocessing refers to denoising signals, i.e., filtering, artifact rejection, normalization, etc. Feature extraction and classification both belong to the machine-learning section and are essential elements of BCI systems. Postprocessing refers to the use of context information to eliminate outliers which can improve the performance of a classifier.

To date, a number of thorough reviews of classification techniques for BCI have been published [5,6], but none have been specifically dedicated to the review of feature extraction and classification algorithms used for RSVP-BCI research. More broadly, the RSVP target-detection problem is part of a wider field of study that investigates single-trial detection methods [1,2,7].

In Section 12.2, we give a brief introduction to the RSVP-EEG experimental setup. We then show several spatiotemporal signals that are typically present in RSVP-EEG, which have discriminative properties, e.g., the P300 and N200. It should be noted that the common objective of all BCI systems is to maximize classification accuracy rather than providing an interpretation of the underlying neurophysiology.

Finally, we describe the preprocessing step and the problems of RSVP-EEG data availability in the literature which has impact on algorithm development, reproducibility and benchmarking.

In Section 12.3, we outline common strategies used to extract useful features from RSVP-EEG data, namely, spatial filters achieved using (un)supervised techniques such as independent component analysis (ICA) which aims to find a linear representation of non-Gaussian data so as to maximize a statistical independence metric, time-frequency representation which decomposes RSVP-EEG to the time-frequency domain and some other feature extraction methods. Spatial filtering allows for dimensionality reduction by transforming high spatial dimension EEG to a subspace according to different optimization objectives, e.g., improving the signal-to-noise ratio (SNR). Reducing data dimensionality in this way is often essential to overcoming issues with having relatively fewer training examples than there are a high number of features—a scenario commonly referred to as the "curse of dimensionality."

In Section 12.4, we explore a number of commonly used classification strategies which covers both linear and nonlinear techniques. Linear classification techniques include linear discriminant analysis (LDA), Bayesian linear regression (BLR), logistic regression (LR) and support vector machine (SVM). Nonlinear classification approaches are mainly focused on artificial neural networks (NNs).

Therefore, and overall, this paper aims to survey the different feature extraction and classification techniques used in RSVP-based BCI research and to identify their critical properties, shortcomings and advantages. It also provides newcomers to the RSVP-BCI area with an introduction—a framework within which an analysis of RSVP-EEG data can be understood.

12.2 Overview of RSVP experiments and EEG data

12.2.1 RSVP experiment for EEG data acquisition

Data acquisition for RSVP-EEG experiments is typically carried out using two computers. One computer is used for stimulus presentation and the other for recording and monitoring of EEG data from participants. A typical setup is shown in Figure 12.3.

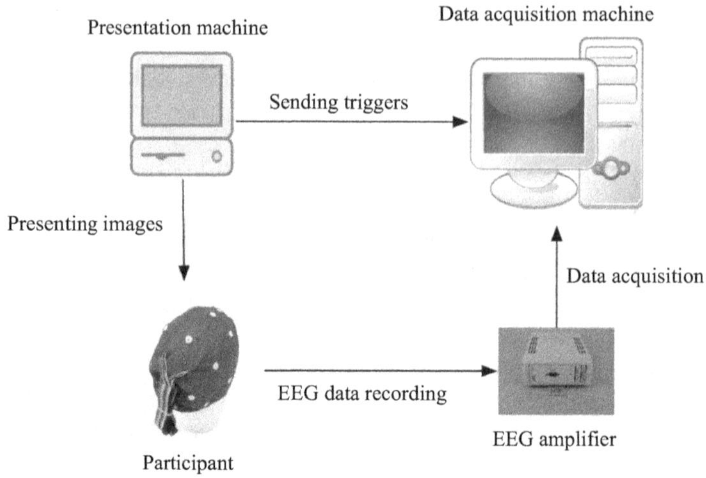

Figure 12.3 RSVP experiment setup in Section 12.2.1

The EEG amplifier is used for recording the EEG signals measured from the participant. When displaying the image sequence to participants, a timestamp for each image must be recorded and aligned with the multichannel time-series EEG captured on the acquisition computer. These are commonly referred to as triggers.

In RSVP-based BCI research, triggers are normally sent from the presentation software (e.g., PsychoPy, E-prime) either to the EEG acquisition device directly [7,8] via a physical port or to the acquisition software [9]. Due to the fast presentation speeds involved with RSVP-EEG, careful attention should be given to ensure that stimulus presentation timings are as expected. We suggest the validation of software triggers in an RSVP experiment against triggers captured using an optical sensor exposed to the presentation screen as this can help to resolve subtle timing issues that may be present [10].

12.2.2 Brief introduction to RSVP-EEG pattern

The most widely used pattern in the EEG signals acquired during RSVP-BCI is the P300 ERP. The P300 is a complex endogenous response that can be subdivided into a novelty-related P3a component and a posterior occurring component commonly encountered in RSVP-search referred to as the P3b. The discovery of the P300 arose from the confluence of increased technological capability for signal averaging applied to human neuroelectric measures and the impact of information theory on psychological research [11]. The P300 is often characterized by its amplitude and latency, where it is defined as the largest positive-going peak in the time range of 300–800 ms following a stimulus presentation. Its latency and amplitude can vary depending on stimulus modality, task conditions, subject age and other factors [4]. Figure 12.4 shows an example of a P300 (P3b) response at channel Pz in one RSVP search

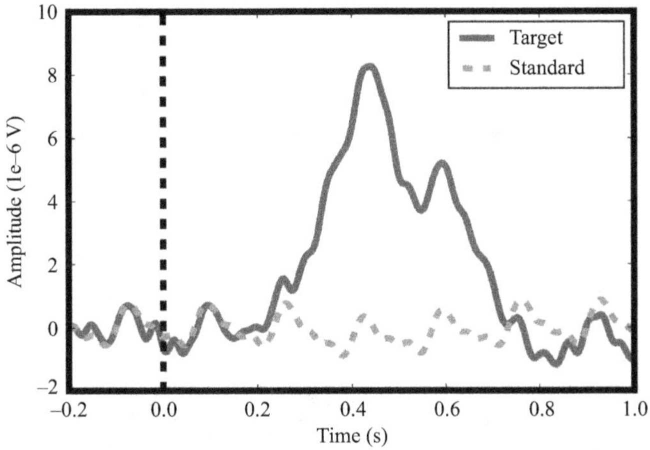

Figure 12.4 P300 response example at the Pz channel in an RSVP experiment in Section 12.2.2, the EEG signal has been band-passed between 0.1 and 30 Hz

task. It can be seen that the P300 peak occurs at around 480 ms. The periodic oscillation that can be seen in the ERP average for standard images is due to a steady-state visual evoked potential response [12]. To facilitate presentation of the concepts involved in RSVP-EEG, we make use of the neurally augmented image labeling strategies (NAILS) dataset [13]. This EEG dataset is part of an open data challenge carried out in 2017 [14].[1]

It is worth noting that not all participants display the stereotyped P300 response, with some displaying characteristics such as low amplitude components leading to unfavorable SNR properties. Reasons for this will not be explored here, but further information can be found in [15].

The P300 is not the only ERP that is commonly encountered when using an RSVP target search paradigm. Earlier, ERPs (notably the N200) are often present alongside the P3 [16] and can be useful in providing discriminative information for classification. In Figure 12.5, it can be seen that both early- and later-time regions generate discriminative ERP-related activity across participants [13].

[1]EEG data from 1 to 9 participants in NAILS was used in this work. Data collection was carried out with approval from Dublin City University's Research Ethics Committee (DCU REC/2016/099). Each participant completed six different tasks (INSTR, WIND1, WIND2, UAV1, UAV2 and BIRD). For each task, participants were asked to search for specific target images from the presented images (i.e., an airplane has the role of target in UAV1 and UAV2 tasks, a keyboard instrument is the target for the INSTR task, while a windfarm is the target in WIND1 and WIND2 tasks, parrot being the target in BIRD task). Each task was divided into 9 blocks, where each block contains 180 images (9 targets/171 standards); thus, there were 486 target and 9,234 standard images available for each participant. Images were presented to participants at a 6-Hz presentation rate. EEG data was recorded using a 32-channel BrainVision actiCHamp at 1,000 samples/s sampling frequency, using electrode locations as defined by the 10–20 system.

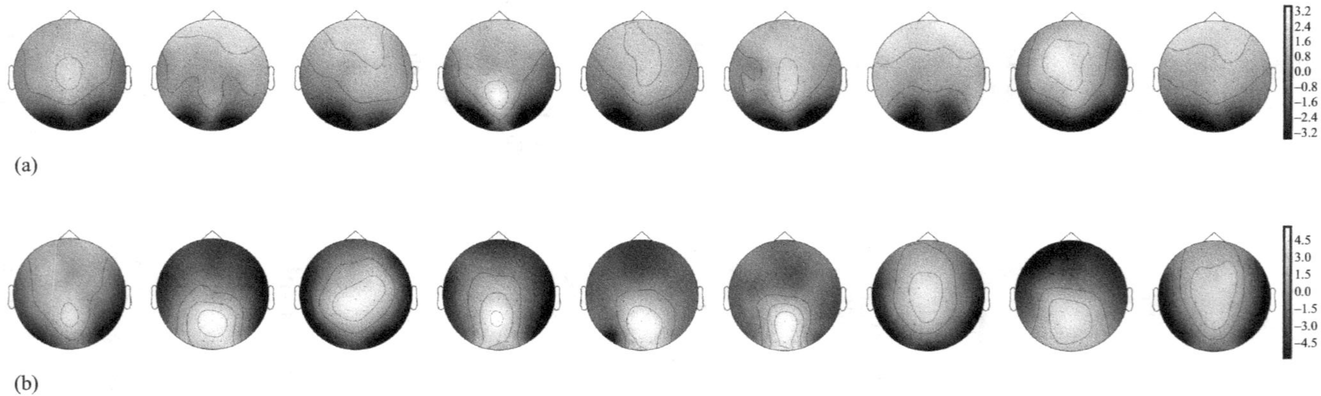

Figure 12.5 *ERPs for each participant corresponding to time regions of 200–340 and 370–700 ms in Section 12.2.2. These are selected for presentation to emphasize the presence of discriminative ERP-related activity in these time regions across participants, namely, time regions coinciding with P200, N200 and P300 ERP activity. (a) Early time region ERP for nine participants from left to right and (b) later time region ERP for nine participants from left to right*

Single-trial detection methods are not restricted solely to BCI-related contexts and are often found as part of a researcher's toolkit in developing an analysis pipeline when working with neuroscientific data. What such methods typically strive to accomplish is striking a balance between the neurophysiological interpretability of a model and its complexity. Transparent models are more likely to lend themselves to meaningful interpretations. While a strategy of enforcing simplistic models with few parameters may aid in interpretability, the purpose of RSVP-EEG is to maximize information throughout as defined by some performance metric.

In reality, a BCI can make use of nonneural signal sources present in the EEG. For example, some participants may, without realizing, blink their eyes upon seeing a target in an RSVP experiment, where the eye-blink will impart a large voltage deflection in the EEG. We are primarily concerned, however, with direct neural signal sources in this chapter and strategies to utilize these.

12.2.3 RSVP-EEG data preprocessing and properties

Preprocessing of some kind is generally a required step before any meaningful interpretation or use of the EEG data can be realized. Preprocessing typically involves re-referencing (changing the referencing channel), filtering the signal (by applying a bandpass filter to remove environmental noise or to remove activity in nonrelevant frequencies), epoching (extracting a time epoch typically surrounding the stimulus onset), trial/channel rejection (to remove those containing artifacts), etc. See [17] for further information. In RSVP-EEG, a common average reference or mastoid reference is often used. A bandpass filter (e.g. 0.1–30 Hz) is commonly applied in RSVP-EEG. The EEG signal is preferably analyzed as epochs (i.e., the whole EEG data is cut by using a fixed time window (e.g., 0–1,000 ms) corresponding to each trigger onset) and each segment is named as an epoch. These epochs can then be used for analysis (e.g., feature extraction, classification).

The presence of many artifacts such as those related to muscle movements in the EEG signal can be sometimes removed by using a bandpass filter as the frequencies of interest in RSVP-EEG do not always detrimentally overlap. During RSVP-EEG experiments, it can be very common for eye-blink behavior to occur in response to target images. This perhaps arises as a result of the participant withholding eye-blinks until a target is seen. While this may be favorable for improving the detection rate of targets, without inspection of the data it may lead to erroneous conclusions on what discriminative information is actually driving the performance of a classifier. One common strategy to investigate this involves identifying a spatial component in the EEG signal related to eye-blinks via ICA to determine if it is trial-locked to targets in any way. Additionally, ICA allows for such activity to be in part attenuated. An investigation of commonly used strategies (and subtle pitfalls) can be found in [18].

Before extracting features from RSVP-EEG data, some critical properties of EEG signals have to be considered concerning the design of an RSVP-based BCI system:

- Low SNR: EEG in noninvasive BCI has an inherently poor SNR and task-related ERPs are typically overwhelmed by strong ongoing EEG background activity in single trials;

- Curse of dimensionality: In RSVP-based BCI system, EEG data can have high-dimensionality spanning space and time. However, typically limited training sets are available especially considering the target image class which are usually infrequent;
- Overlapping epochs: There is substantial overlap between adjacent target epochs and standard epochs because of the short interstimulus interval used in the RSVP paradigm;
- Imbalanced datasets: Target images are overwhelmed by standard images in an RSVP application which leads to an imbalanced classification problem.

These critical properties have to be considered before feature extraction. The last one can be overcome through cost-sensitive learning [19], while the first three are inherent challenges in the design of an RSVP-based BCI system.

Two main differences between the RSVP-EEG paradigm and other ERP paradigms are that the former requires single-trial detection in the presence of overlapping epochs. Traditional ERP analysis typically computes a grand average ERP where phase-locked activity in the signal remains after averaging, whilst other non-locked background activity increasingly attenuates as more trials are averaged. For example, the P300 speller is an ERP paradigm that has been a benchmark for P300 BCI systems. In this paradigm, each desired symbol is spelt several consecutive times by a participant where the epochs corresponding to each row/column are averaged over the trials. This averaging process is able to improve the EEG SNR for the system because averaging reduces the contribution of random background EEG oscillations [20]. This repetition of an image stimulus is not always applicable in the RSVP-EEG paradigm because it can introduce unintentional behaviors such as a participant attending to an image due it being a salient repetition rather than it being a target. In single trial detection, low SNR is a challenge for the detection of discriminating ERP activity. Furthermore, the overlapping epoch's problem may contribute to overfitting when training a machine-learning model [21]. In summary, low SNR and overlapping epochs are two challenging problems for RSVP-EEG when compared to other ERP paradigms.

12.2.4 Performance evaluation metrics

A machine-learning model's performance can be evaluated by a variety of evaluation metrics. Area under the receiver operating characteristic (ROC) curve (AUC) is widely used as it illustrates the discriminative ability of a binary classifier system as its discrimination threshold is varied [22]. One may want to adjust this threshold for example to optimize for fewer false positives at the cost of more false negatives. ROC-AUC is the most widely used evaluation metric in RSVP-EEG research [1,2,7]. However, ROC-AUC score may not be suitable when evaluating some real-world systems because it gives an unified measure of the performance of a classifier across all potential thresholds and in effect sidesteps the issue of the impact of threshold selection.

Balanced accuracy (BA) is well suited for evaluating RSVP-EEG systems that utilize binary classifications [23]. BA can be calculated as below:

$$BA = \frac{1}{2}\,(sensitivity + specificity) \tag{12.1}$$

where $sensitivity = (TP/(TP + FN))$ and $specificity = (TN/(TN + FP))$.

Choosing evaluation metrics critically depends on the application. For example, if the classification system is used to rank target above standard images, then AUC would be the preferred evaluation metric. If the classification system is designed to give a binary classification (target vs standard), then BA can be a good evaluation metric. Both ROC-AUC and BA are robust to targets/standards ratio imbalances in dataset.

12.3 Feature extraction methods used in RSVP-based BCI research

The challenge for feature extraction methods is to find intrinsic characteristics of the EEG signals that relate to certain cognitive responses. Feature extraction in BCI systems plays an important role since it can greatly affect the SNR and the classification strategy used, which in turn determines the performance of the BCI.

This section focuses on extracting RSVP-EEG from three aspects: (1) spatial filtering (supervised and unsupervised); (2) time-frequency representation; (3) other feature extraction methods.

12.3.1 Spatial filtering

12.3.1.1 Supervised spatial filtering

As mentioned in the previous section, RSVP-EEG data suffers from low SNR and often high spatial dimensionality. Spatial filtering is an efficient technique for mitigating these concerns. In the area of BCI research, xDAWN [24], beamformer [25] and common spatial pattern (CSP) [26] are widely used for generating supervised spatial filters. In this chapter, we focus on three methods for generating spatial filters: xDAWN, CSP and LDA beamformer. For xDAWN, the goal is to maximize the signal-to-signal-plus-noise ratio (SSNR), whereas for CSP, the goal is to maximize the ratio between the discriminative activity and the common activity, leading to optimal variances for the discrimination of two types of stimulus EEG signals. LDA beamformer is used for source signal reconstruction where it maximizes the SNR.

Problem formulation: Let $\mathbf{X} \in \mathbb{R}^{C\times T}$ be EEG epochs corresponding to each image stimulus, where C is channel number and T is epoch time length. The problem of spatial filtering is to find a set of projection vectors (each comprised of weights for each channel) $\mathbf{w} \in \mathbb{R}^{C\times n}$ (n is the number of components) to project $\mathbf{X}$ to a subspace, where $\mathbf{w}$ is calculated by different algorithms, i.e., xDAWN, beamformer, CSP, etc.

$$\mathbf{X}_{\text{sub}} = \mathbf{w}'\mathbf{X} \tag{12.2}$$

CSP: CSP generates sets of channel weightings that can be used to project multichannel EEG data into a low-dimensional subspace, where this transformation can maximize the variance of two-class signal matrices. Let $X_+(i) \in \mathbb{R}^{C\times T}$ and $X_-(i) \in \mathbb{R}^{C\times T}$ be the ith event locked EEG epochs (C is the channel number and T is the time length) in two experimental conditions, i.e., $X_+(i)$ for the target image condition and $X_-(i)$ for the standard image condition. Covariance matrices in the two conditions can be estimated as

$$\mathbf{\Sigma}_c = \frac{1}{n}\sum_{i=1}^{n} \frac{\mathbf{X}_c(i)\mathbf{X}'_c(i)}{trace(\mathbf{X}_c(i)\mathbf{X}'_c(i))} \quad (c \in \{+,-\}) \tag{12.3}$$

where "′" denotes the matrix transposition and $\mathbf{\Sigma}_c \in \mathbf{R}^{\mathbf{C}\times\mathbf{C}}$. The CSP optimization problem can be formulated as

$$\{\max, \min\}_{\mathbf{w}\in\mathbb{R}^C} \frac{\mathbf{w}'\mathbf{\Sigma}_+\mathbf{w}}{\mathbf{w}'\mathbf{\Sigma}_-\mathbf{w}} \tag{12.4}$$

This optimization problem is given by the simultaneous digitalization of the two covariance matrices. This can be achieved by solving the generalized eigenvalue problem:

$$\mathbf{\Sigma}_+\mathbf{w} = \lambda\mathbf{\Sigma}_-\mathbf{w} \tag{12.5}$$

> **Note:** The objective of CSP is to maximize the variance in one class while minimizing the variance in the other class. Maximizing the variance in this way corresponds to maximizing the frequency-power of target-related activity in the signal. When using CSP, multiple spatial filters will be obtained and crossvalidation is normally deployed to choose a spatial filter(s). This strategy can be adapted to use multiple different band-passed versions of the same EEG signal epoch to leverage different sources of discriminative information present across different frequencies (that often also differ in spatial characteristics). CSP is widely applied in motor imagery-based BCI [27]. In Yu's work, CSP has been applied for producing spatial filters for RSVP-based BCI [8].

xDAWN: The xDAWN algorithm has been successfully applied in the P300 speller BCI application [24]. The basic goal of xDAWN is to enhance the SSNR of the responses corresponding to the target stimulus. Let recorded signals be $\boldsymbol{X} \in \mathbb{R}^{N_t\times N_s}$, where N_t is the time length of recorded EEG signals and N_s is the number of channels. It considers the following model:

$$\mathbf{X} = \mathbf{DA} + \mathbf{H} \tag{12.6}$$

where $\mathbf{D} \in \mathbb{R}^{N_t \times N_e}$ (N_e is the number of temporal samples of ERP corresponding to the target stimulus) is the real Toeplitz matrices and $\mathbf{A}$ is the ERP response to the target. $\mathbf{D}$ has its first column elements set to zero except for those that correspond to a target stimulus onset and $\mathbf{H}$ is the on-going EEG activity.

The problem statement for xDAWN becomes how to estimate the spatial filter for (12.6) such that the synchronous response is enhanced by spatial filtering:

$$\mathbf{XU} = \mathbf{DAU} + \mathbf{HU} \tag{12.7}$$

where $\mathbf{U} \in \mathbb{R}^{N_s \times N_f}$ (N_f is the number of spatial filters). The optimized solution can be achieved by

$$\hat{\mathbf{U}} = \arg\max_{\mathbf{U}} \frac{Trace(\mathbf{U}'\hat{\mathbf{A}}'\mathbf{D}'\mathbf{D}\hat{\mathbf{A}}\mathbf{U})}{Trace(\mathbf{U}'\mathbf{X}'\mathbf{X}\mathbf{U})} \tag{12.8}$$

where $\hat{\mathbf{A}}$ is the least squares estimation of response $\mathbf{A}$. More details about the computation method can be found in [24].

> **Note:** Separately from CSP, the numerator in xDAWN in (12.8) is the ERP response rather than target EEG epochs, i.e., ERP response being the mean value of target EEG epochs. xDAWN aims to enhance the SSNR of the response corresponding to the target stimulus and it is originally designed for enhancing the P300 evoked potential for the P300 speller BCI [24]. Similar to CSP, xDAWN generates multiple spatial filters as well. It is suggested to use crossvalidation to determine the number of spatial filters. In recent published work, xDAWN has been applied to RSVP-based BCI for spatial filtering [28].

LDA beamformer: LDA beamformer has been proposed to maximize the SNR of EEG in a way which is robust to correlated sources [29]. The generation of a spatial filter using LDA beamformer is comprised of three steps: (1) spatial pattern estimation; (2) covariance matrix estimation; (3) spatial filter optimization. Let column vectors $\mathbf{p}_1 \in \mathbb{R}^{C\times 1}$ and $\mathbf{p}_2 \in \mathbb{R}^{C\times 1}$ be the spatial pattern of a specific component in two different experimental conditions, where C is the number of channel. We denote the difference pattern as $\mathbf{p} := \mathbf{p}_1 - \mathbf{p}_2$ and the covariance matrix $\mathbf{\Sigma} \in \mathbb{R}^{C\times C}$. The optimization problem for the LDA beamformer can be stated as

$$\begin{aligned} &\underset{\mathbf{w}}{\text{minimize}} \quad \mathbf{w}'\mathbf{\Sigma}\mathbf{w} \\ &\text{s.t.} \quad \mathbf{w}'\mathbf{p} = 1 \end{aligned} \tag{12.9}$$

and the optimized solution can be determined as

$$\mathbf{w} = \mathbf{C}^{-1}\mathbf{p}\left(\mathbf{p}'\mathbf{C}^{-1}\mathbf{p}\right)^{-1} \tag{12.10}$$

> **Note:** Separately from the previous two methods, the LDA beamformer method generates an optimal spatial filter (only one spatial filter) that maximizes the SNR. One optimal projection vector may not be able to fully capture all available information from the original EEG epoch due to the sources of variability such as different spatial characteristics of early and late target-discriminative ERPs across tasks and participants. Therefore, multiple time window LDA beamformers (where the researcher trains the LDA beamformer in multiple time windows) are often applied for improved performance with RSVP-based EEG data.

So far we have introduced three supervised approaches to generate useful spatial filters. After applying the spatial filter(s) to the original epoch, it can be appreciated that the dimensionality of the projected subspace has been reduced significantly and that this subspace signal may have different properties (optimized SSNR for xDAWN, optimized SNR for LDA beamformer, maximum difference of variance between two classes for CSP) depending on which algorithm has been used for generating the spatial weightings. This projected subspace can then be used as the basis of a feature set for training a practical classifier. It is worth noting that the overall effect of employing spatial filtering methods is an improvement in the SNR, a reduction in computation cost and a more favorable situation for many classification algorithms that suffer issues with high-dimensional feature vectors (particularly when few training examples are available). These are desirable properties of a signal processing pipeline for RSVP-based BCI.

12.3.1.2 Unsupervised spatial filtering

ICA: EEG source activity refers to the time-varying far-field potentials arising within an EEG source and volume-conducted to the scalp electrodes. The recorded EEG signals are then, according to this interpretation, the summation of neural activity, contributions of nonbrain sources such as scalp muscle, eye movement and cardiac artifacts, plus (ideally small) electrode and environmental noise [30]. Successful separation of contributions from these nonneural activity related sources can improve the SNR of the signals of interest. ICA is a technique that can aid here and can be used to find linear representations such that time-series signals obtained via its components' projections are statistically independent from each other (or as independent as possible). Such a representation is capable of capturing the essential structure of the data in many applications, including feature extraction and signal separation [31]. Essentially, ICA produces a matrix of spatial filters. ICA has been widely applied to EEG signal fields for denoising [32] and artifact removal [33]. Figure 12.6 illustrates the characteristics of three independent components (ICs) corresponding to eyeblinks, P300 activity and other trial locked ERP activity, respectively. From the IC plots (left), it is noticeable that different signal sources have different IC localizations, i.e., eyeblink is distributed frontally while the P300 is localized posteriorly. From an ERP image and

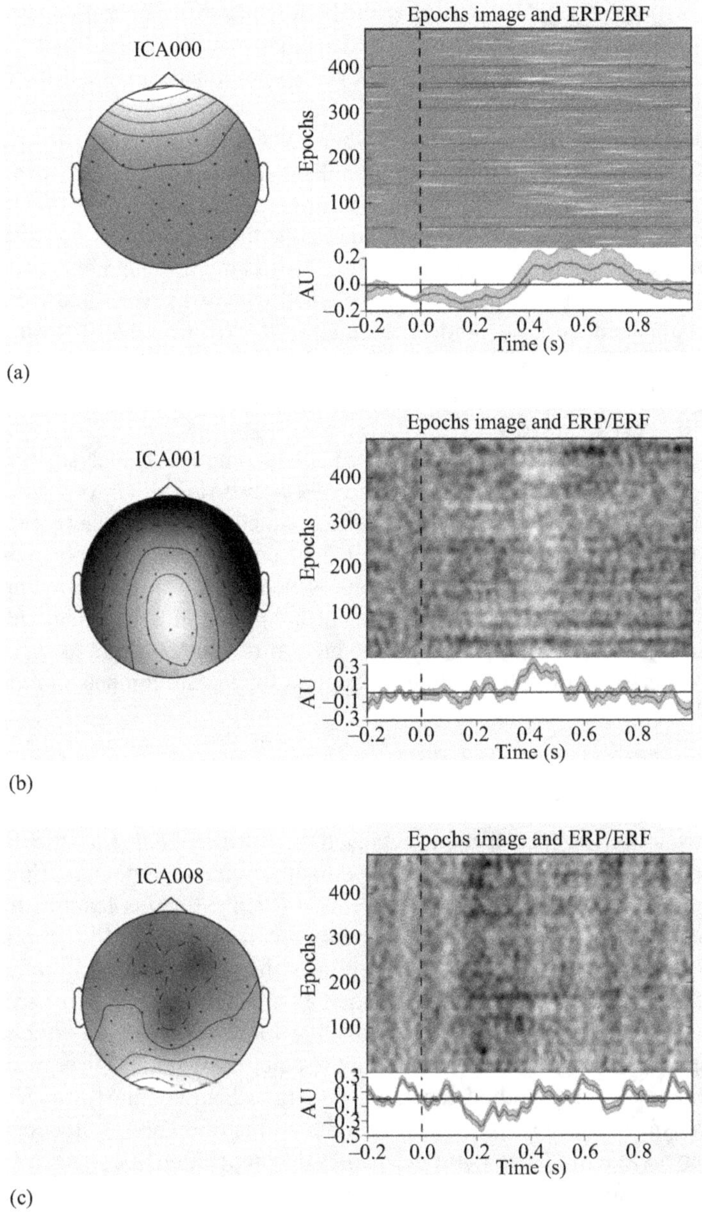

Figure 12.6 Examples of ICA components (left) and ERP images (right) in Section 12.3.1.2 for eye blink related activity (a), posterior P300 activity (b) and other trial-locked EEG activity (c). The results are generated using one participant's dataset (from NAILS) and only includes only RSVP target related trials

ERP time series plot (right hand side top and bottom plots), it can be seen that eyeblink activity and P300-related activity are locked to target stimuli. Moreover, it can be seen that the eyeblink time-locked activity is noticeably prominent at around 400 ms which is close to the time region where we also see P300 activity. This indicates that this participant sometimes blinked their eyes when presented with a target image likely as a result of an active effort to suppress eyeblinks up to that point in case they missed a target. Eyeblink artifacts in such instances can potentially be beneficial for the classification process. However, we are going to remove these as we only consider signals of direct neural origin in this exposition. It is worth emphasizing here that ICA successfully resolved the signal into a distinct physiologically interpretable source in this instance. Published work by Bigdely-Shamlo *et al.* demonstrated ICA successfully applied to RSVP-based BCI generating ICs and independent time course templates (ITs) for each IC. These ITs were selected as features for training the classifier [1] quite successfully.

Principal component analysis: Principal component analysis (PCA) is a statistical technique which uses eigenvalue decomposition to convert a set of correlated variables into a set of linearly uncorrelated variables where each of the resultant variables is referred to as a principal component [34]. For multivariate datasets, notably data in a high-dimensional space, PCA can be particularly effective for dimensionality reduction. PCA has been applied in EEG signal analysis for dimensionality reduction [35] and the production of spatial filters [36]. In RSVP-based BCI literature, PCA has only been applied for feature dimension reduction to date [1,37].

12.3.2 Time-frequency representation

Feature extraction in BCI can be achieved in the time domain, the frequency domain and the combined time-frequency domain. In the time domain, time regions coinciding with ERPs such as the P300 are used when extracting features for single-trial event detection [3]. Frequency domain features such as the amplitudes of μ (8–13 Hz) and β (14–26 Hz) are widely used in sensorimotor control BCI as it has been shown changes occur in these when a participant imagines (or engages) in certain types of movements. However, frequency domain features have not been used in the image RSVP paradigm in the literature to date. Time-frequency representations can be generated using methods such as short-time Fourier transform (STFT) and wavelet transforms. The wavelet transform method is often preferred over STFT in many instances as it produces a time-frequency decomposition of a signal over a range of characteristic frequencies that separates individual signal components more effectively than the STFT method does. A number of other properties of the source signal such as stationarity assumptions should be considered when utilizing one approach over another [38].

A set of common types of time-frequency features found in RSVP-based EEG has been proposed in Meng's work [39]. In Figure 12.7, we show the time-frequency mean power generated using target EEG epochs from the NAILS dataset. It can be seen that both high power and strong ITC appear in low frequencies (0–6 Hz) and in two time

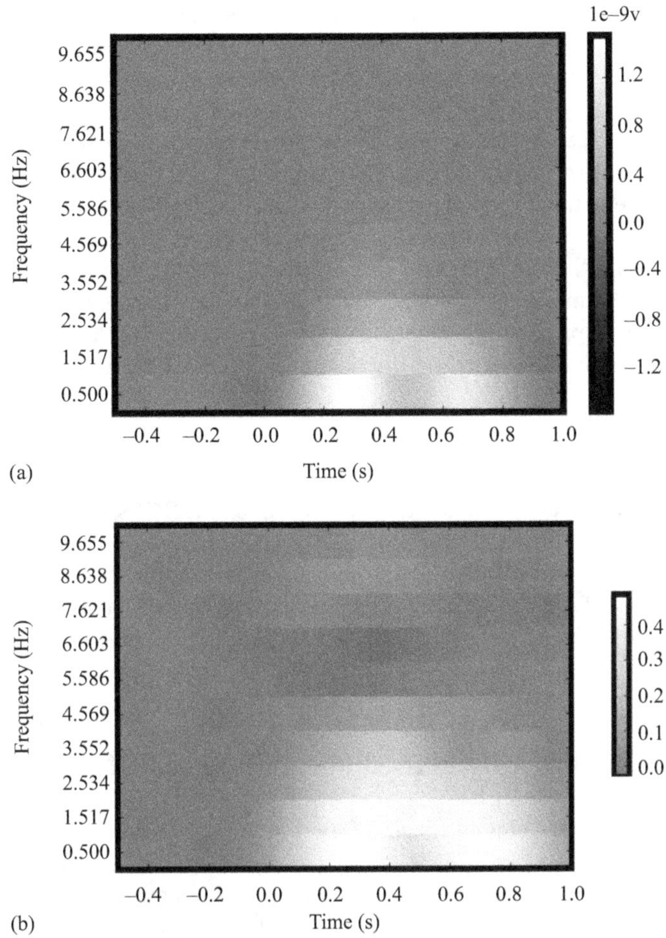

Figure 12.7 Time-frequency representation example corresponding to target images at the Pz channel for averaging nine participants in an RSVP experiment using Morlet wavelet transform in Section 12.3.2, the EEG signal has been band-passed between 0.1 and 30 Hz. (a) Time-frequency mean power and (b) intertrial coherence.

regions (200–400 and 600–800 ms). The discriminative ERSP-related activity appears in both an early time region and a later time region. The time-frequency representation strongly depends on the type of the image stimulus and the experimental environment.

12.3.3 Other feature extraction methods

In Huang's work, EEG signals from the stimulus onset to 500 ms poststimulus were extracted for each channel and concatenated to form a feature vector [7]. This resulted in each trial containing 32 × 129 features (where 32 is the number of channels and 129 is the number of time points). This strategy of building feature vectors as a concatenation over time regions (and channels) of interest is commonly found in the RSVP literature and often yields good results. Hierarchical discriminant component analysis has been proposed in [40], where this method estimates EEG signatures of target detection events using multiple linear discriminators, each trained at a different time window relative to the image onset. Since EEG signals contain both spatial and temporal information, a spatiotemporal representation for RSVP-based EEG data has been proposed by Alpert [37]. This representation is divided into two steps: (1) LDA is applied at each timestamp to produce the spatial weights and a spatial weight matrix is then used for mapping original epoch to a new space, and (2) PCA is then used for dimensionality reduction based on the temporal domain, i.e., for each independent channel.

12.3.4 Summary

Feature extraction is an essential step when designing a BCI system because pertinent features can significantly improve performance of the resulting classifier and additionally it can significantly reduce the computational cost. In RSVP-based BCI, discriminative ERP-related activity often occurs in both early and late time region. Feature extraction for RSVP-based BCI is best designed by considering these ERPs' properties.

12.4 Survey of classifiers used in RSVP-based BCI research

This section surveys the classifiers used for recognition of target and standard events in RSVP-based BCI systems. Due to the fact that the nonlinear problem has not been well explored in RSVP-based EEG data in the literature so far, this section is divided into linear and NN classifiers. Since deep-learning technology is very popular currently in other application domains such as computer vision and natural language processing, we introduced some deep-learning methods in the NN section.

12.4.1 Linear classifiers

Linear classifiers are widely used for designing BCI applications due to their good performance, often simple implementation and low computational complexity. Four main linear classifiers will be in introduced in this section, namely, LDA, BLR, LR and SVM. In this section, we consider our model as

$$y = \mathbf{w}'\mathbf{x} + b \tag{12.11}$$

where y is classifier output, $\mathbf{x}$ is the feature vector and b is the threshold.

12.4.1.1 Linear discriminant analysis

LDA is a supervised subspace learning method which is based on the Fisher criterion, and it is equivalent to least squares regression (LSR) if the regression targets are set

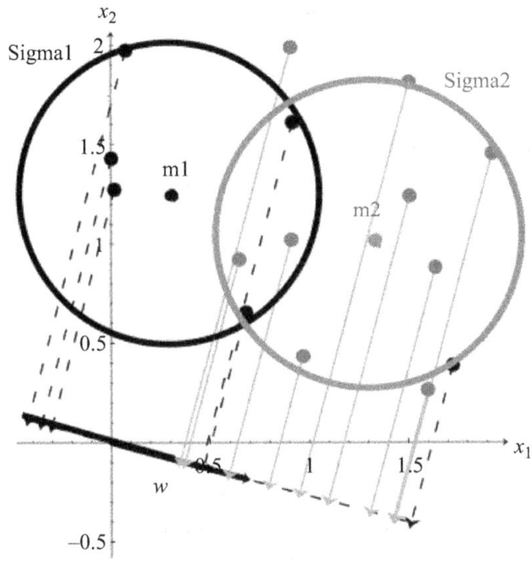

Figure 12.8 Projection of two different classes onto a line by LDA in Section 12.4.1.1 [43]

to N/N_1 for samples from class 1 and $-(N/N_2)$ for samples from class 2 (where N is the total number of training samples, N_1 is the number of samples from class 1 and N_2 is the number of samples from class 2) [41]. It aims to find an optimal linear transformation $\mathbf{w}$ that maps $\mathbf{x}$ to a subspace in which the between-class scatter is maximized while the within-class scatter is minimized in that subspace. The optimization problem for LDA is to maximize the cost function as below:

$$J = \frac{\mathbf{w}'\mathbf{S}_B\mathbf{w}}{\mathbf{w}'\mathbf{S}_W\mathbf{w}} \tag{12.12}$$

where $\mathbf{S}_B$ is the between-class scatter and $\mathbf{S}_W$ is the within-class scatter. Regularization is often applied in order to avoid the singular matrix problem of $\mathbf{S}_W$ [42]. Figure 12.8 shows the LDA implementation on two different classes with equal covariance and different mean values. The solid line is the projected subspace $\mathbf{w}$ where these two classes will be projected by LDA. This transformation enables the best separation between two classes on the subspace $\mathbf{w}$. Details of the method can be found in Duda's book [43].

LDA has very low computational complexity which makes it suitable for online BCI systems. As mentioned earlier, classification of RSVP-based EEG data suffers from the imbalanced data set problem. In Xue's work [44], he showed that there is no reliable empirical evidence to support that an imbalanced data set has a negative effect on the performance of LDA for generating the linear transformation vector. Consequently, LDA is suitable and has been successfully used in RSVP-based BCI

[1,39]. LDA typically suffers from issues particularly when dealing with data sources that contain issues such as outliers. For this reason, regularization strategies are typically employed but are not covered here [42].

12.4.1.2 Bayesian linear regression

BLR, also named Bayesian linear discriminant analysis, can be seen as an extension of LDA or LSR. In BLR, regularization for parameters is used for preventing overfitting caused by high dimensional and noisy data. BLR assumes the parameter distribution and target distribution are both Gaussian [41]. We introduce LSR as a starting point for the description of BLR. Given the linear model in (12.11), the solution of LSR can be stated as

$$\mathbf{w} = (\mathbf{X}\mathbf{X}^{\mathrm{T}})^{-1}\mathbf{X}\mathbf{y} \tag{12.13}$$

Note that $y = (N/N_1)$ for class 1 and $y = -(N/N_2)$ for class 2 here (threshold can be determined by adding a column with all one as the first column in $\mathbf{X}$). LSR does not consider the parameter distribution in this case, and it maximizes the likelihood. For BLR, it considers the parameter distribution and maximizes the posterior. Given the prior target distribution $p(y) \sim \mathcal{N}(\mu, \beta^{-1})$ and parameter distribution $p(\mathbf{w}) \sim \mathcal{N}(0, \alpha^{-1}\mathbf{I})$ (where β and α are the inverse variance), BLR gives the optimized estimation for the parameter:

$$\mathbf{w} = \beta(\beta\mathbf{X}\mathbf{X}^{\mathrm{T}} + \alpha\mathbf{I})^{-1}\mathbf{X}\mathbf{y} \tag{12.14}$$

It can be seen that the optimization of BLR is added with the prior information of parameter and data. Hence, the optimization depends on the hyperparameters β and α. In real-world applications, the hyperparameters can be tuned using crossvalidation or the maximum likelihood solution with an iterative algorithm [41,45]. BLR has been proven to have very good performance in BCI research [46,47].

12.4.1.3 Logistic regression

LR models the conditional probability as a linear regression of feature inputs. Considering the linear regression model of RSVP-based EEG data in (12.11), the logistic model can be constructed as

$$p(\mathbf{x}) = \frac{1}{1 + e^{-\mathbf{w}'\mathbf{x}+b}} \tag{12.15}$$

The optimization problem of an LR can be constructed by minimizing the cost function as below:

$$J(\mathbf{w}, b) = -\frac{1}{m}\sum_{i=1}^{m}\left[y_i \log(p(\mathbf{x}_i)) + (1 - y_i)\log(1 - p(\mathbf{x}_i))\right] \tag{12.16}$$

where $y \in \{0, 1\}$ and m is the sample number of two classes. Similar to SVM mentioned earlier, LR can be modified by penalizing different cost terms to each class and the cost function can be modified as below:

$$J(\mathbf{w}, b) = -\frac{1}{n_0 + n_1} \sum_{i=1}^{n_0+n_1} [n_0 y_i \log(p(\mathbf{x}_i)) + n_1(1 - y_i) \log(1 - p(\mathbf{x}_i))] \quad (12.17)$$

where n_0 and n_1 are the numbers of standard and target image clips, respectively.

LR is part of a broader family of generalized linear models (GLMs), where the conditional distribution of the response falls in some parametric family, and the parameters are set by the linear predictor. LR is the case where the response is binomial and it can give the prediction of the conditional probability estimation. LR is easily implemented and has been successfully applied for RSVP-based BCI research [48,49].

12.4.1.4 Support vector machine

An SVM aims to select the hyperplane which maximizes the margins (i.e., the distance from the nearest training samples). In order to overcome the imbalanced classification problem, a weighted support vector machine is proposed [50]. An SVM can be used for linear and nonlinear classification by using the "kernel trick." This consists of mapping data to other spaces using a kernel function $\kappa(x_i, x_j)$. For linear classification, kernel function can be chosen as $\kappa(x_i, x_j) = \langle x_i, x_j \rangle$. For nonlinear classification, Gaussian kernel, $\kappa(x_i, x_j) = \exp(-(\|x_i - x_j\|^2/2\delta^2))$, is widely used in the classification area.

SVM has a small number of hyperparameters which need to be tuned manually and this is often done by using crossvalidation. There is already considerable use of SVMs in RSVP-based BCI research [8,51,52].

12.4.1.5 Other machine-learning algorithms

Here, we have introduced four linear classification algorithms that are widely used in the RSVP-based BCI research area. There are numerous machine-learning algorithms available currently, and we encourage readers to experiment with them. Scikit-learn is a machine-learning library in Python, and it provides lots of machine-learning algorithm implementations. Linear classification methods can be found in the GLMs list.

> **Note:** For the four classifiers above, every method involves hyperparameters except LDA (unless using a regularized version that relies on some parameter selection). We suggest using grid search or maximum likelihood estimation to determine these hyperparameters. K-fold validation can be used for the evaluation of each set of tuned hyperparameters. Importantly, classifier performance should be evaluated on a withheld testset such as an experimental block that does not overlap with the data used for model training or hyperparameter selection.

12.4.1.6 Summary

To summarize, we have introduced four types of linear classifier from the viewpoints of different optimization objectives. LDA aims to find the subspace which gives the best separation between two classes after projection. BLR uses prior information about data distribution and weights, constraining the estimated weights closer to zeros, which helps to stop overfitting. SVM finds the optimal hyperplane maximizing the margins and it can be applied to nonlinear cases by using the "kernel trick." LR can predict the conditional probability. We recommend using LDA (a regularized form) and BLR for RSVP-based BCI because of their low computational cost and good performance. Recent work in RSVP-based BCI research shows that BLR outperforms LDA and SVM [47].

12.4.2 Neural networks

NNs is yet another category of classifiers that are increasingly used in BCI research. NN comprises several artificial neurons that can enable nonlinear decision boundaries.

This section is divided into two parts. The first describes the multilayer perceptron (MLP), the most widely used NN, and then some deep-learning techniques are introduced.

12.4.2.1 Multilayer perceptron

An MLP is minimally comprised of three layers of neurons, namely an input layer, one or several hidden layers and an output layer [53]. In each hidden layer, each neuron is connected to the output of each neuron in the previous layer and its output is the input of each neuron in the next layer. Considering RSVP-based BCI, there is only one output in the output layer. Parameters in MLP can be updated by the BackPropagation algorithms which involve computing partial derivatives to update parameters in the direction of a gradient which decrease the overall loss function [54]. A number of what are called gradient descent optimization algorithms exist for this purpose [55].

NN and MLP are very flexible classifiers that can be applied to a great variety of problems because they are universal approximators [5]. Hence, MLP can be applied for almost all machine-learning problems including binary classification or multi-class classification or modeling. The main disadvantage of MLP is overfitting [56] due to the limited training samples especially for targets in RSVP-based EEG data. Therefore, one has to be careful when designing a MLP architecture and regularization is often required [57].

12.4.2.2 Some deep-learning techniques

Modern deep learning provides a powerful framework for supervised learning [58]. With more layers and more neurons in layers, a deep network can represent increasingly complex nonlinear patterns, and it has been successfully applied to many fields including computer vision [59], natural language processing [60], etc.

Since deep-learning implementations in the area of EEG is still rare, we will introduce three representative methods in the deep-learning field, namely, convolutional neural networks (CNNs), recurrent neural networks (RNN) and deep belief nets (DBN).

Convolutional neural networks

CNN is a type of NN that employs a mathematical operation called convolution specialized for processing a grid of values where the arrangement of the values is not arbitrary such as is the case with an image where pixels near to any one pixel tend to be correlated in a meaningful way [61]. Convolutional networks are simply NNs that use convolutions in place of general matrix multiplications in at least one of their layers [58]. CNN has been tremendously successful in many practical applications [62–64]. CNN leverages three properties that improve learning, namely sparse interactions, parameter sharing and equivariant representations [58]. Sparse interactions are accomplished by the convolution operation while choosing the kernel smaller than the input size. This property enables meaningful features to be extracted from input data. Parameter sharing refers to the fact that each member of the kernel is used at every position of the input with the same parameters. Equivariant representations means that if the input changes, the output changes in the same way [58].

Since a CNN is capable of extracting features from the input data automatically, CNN has become the most widely used deep-learning architecture in RSVP-EEG research [28,65–67]. All of these works have shown that CNN is effective in combining the spatial filtering and the classification steps in an unified way. CNN is also the winning solution to the NAILS competition in the 13th NTCIR conference [21] which shows better performance than traditional methods.

Recurrent neural networks

Different from CNN specialized for processing a grid of values, an RNN is a family of NNs which is designed for processing sequential data such as speech signals [68]. RNN processes the sequence which contains vectors $\mathbf{x}(t)$ with the time step index t ranging from 1 to τ [58].

There are several implementations of RNN in EEG signal analysis and classification [69–71]. However, RNN implementations on RSVP-based BCI have not been stated in the literature thus far. Since EEG is sequential and the P300 has a temporal property, it is an open question if an RNN can be used effectively in RSVP-based BCI.

Deep belief nets

A deep belief network (DBN) is a generative graphical model which comprises multiple layers of latent variables ("hidden units"), with connections between the layers but not between units within each layer [72]. It provides an efficient way to learn a multiple-layered restricted Boltzmann machine [73].

A DBN has shown efficacy in RSVP-based BCI from Ahmed's work and it can extract discriminant features from RSVP-based EEG data as well [74].

12.4.2.3 Summary

In this section, we have introduced the use of NN techniques for classification. The first part introduced an MLP framework which is the most classically used NN framework. The MLP has a simple implementation and it is flexible but easily suffers from issues related to overfitting.

Deep learning is very popular and has had great success in many real-world applications but often requires very large volumes of data. With better data acquisition and more advanced generative models [75], it is possible to train better deep network models for RSVP-EEG leveraging very large datasets.

The main difficulty with EEG signals is the potentially very large feature dimensionality when considering all combinations of channel, frequency and time features. This makes feature extraction a very complex step that must cater to inter-subject and intertask variability. In traditional RSVP-based BCI systems, feature extraction and classification are always separated. Deep learning provides a potentially unified way to accomplish this.

12.5 Conclusion

In this chapter, we have introduced the main parts in a typical RSVP-based BCI system including RSVP data acquisition, data preprocessing, feature extraction and classification. We focused on the machine-learning architecture (feature extraction and classification) as it is the most important part of a typical BCI system.

We have shown that discriminative patterns in RSVP-based EEG data can appear in different time regions (both early and late), i.e., the P300 is not the only ERP being elicited in the RSVP paradigm. Therefore, we suggest designing a feature extraction method that takes into account the properties of the discriminative ERPs for a given task. A good feature extraction method can not only improve the classifier's performance in the later stage but also reduce the computational cost. In this chapter, we introduce existing feature extraction methods used in the literature so far. We stated the objective of each method and a direct comparison between those methods will be part of future work.

The other part of the machine-learning architecture in an RSVP-EEG BCI system is the classifier. We divided our discussion on classifiers into both linear classifiers and NNs. The choice of the classifier remains difficult and depends mainly on the number of available trials and feature vector dimensionality. Linear classifiers remain popular as they have low computational complexity, are easy to implement and have good performance in classification accuracy. NNs possibly outperform linear classifiers with a large number of trials as deep NNs are able to capture high level features related to the variability of the EEG signals across participants and over time. However, acquiring EEG data is time consuming and the variability in the EEG of a specific participant can change over time, which indicates that the number of trials in RSVP-based BCI is limited. Therefore, the choice of classification method should be capable of training a model with a limited amount of available data. In this aspect, linear classifiers are more preferable than NNs due to fewer parameters in the model which in turn can help to prevent overfitting on the noisy and limited RSVP-based EEG data. Here, we suggest to use LDA and BLR in RSVP-based BCI research as they are easy to implement, efficient and have good performance.

The area of RSVP-EEG stretches back well over a decade and there has been significant progress in this time. With the emergence of deep-learning approaches, computer vision recognition applications are able to perform at or even above a human level raising questions about whether people are still needed to perform image labeling tasks. We believe that when labeled image datasets are limited, these computer vision system may not perform very well as a typical component to their success is the availability of very large labeled image datasets. In this way, RSVP-EEG may assist in more efficiently labeling large datasets of image content to support this process. Similarly, many image labeling tasks may require subjective (or expert) knowledge about the image that cannot be easily learned by a deep-learning architecture but that may be readily detected when using an RSVP-EEG system. We see these systems as being able to work in a synergistic manner rather than competitively.

Acknowledgment

This work is funded as part of the Insight Centre for Data Analytics which is supported by Science Foundation Ireland under Grant Number SFI/12/RC/2289.

References

[1] Nima Bigdely-Shamlo, Andrey Vankov, Rey R Ramirez, and Scott Makeig. Brain activity-based image classification from rapid serial visual presentation. *IEEE Transactions on Neural Systems and Rehabilitation Engineering*, 16(5):432–441, 2008.

[2] Adam D Gerson, Lucas C Parra, and Paul Sajda. Cortically coupled computer vision for rapid image search. *IEEE Transactions on Neural Systems and Rehabilitation Engineering*, 14(2):174–179, 2006.

[3] Paul Sajda, Adam Gerson, and Lucas Parra. High-throughput image search via single-trial event detection in a rapid serial visual presentation task. In *Neural Engineering, 2003. Conference Proceedings. First International IEEE EMBS Conference on*, pages 7–10. IEEE, 2003.

[4] John Polich. Updating p300: an integrative theory of p3a and p3b. *Clinical Neurophysiology*, 118(10):2128–2148, 2007.

[5] Fabien Lotte, Marco Congedo, Anatole Lécuyer, Fabrice Lamarche, and Bruno Arnaldi. A review of classification algorithms for EEG-based brain–computer interfaces. *Journal of Neural Engineering*, 4(2):R1, 2007.

[6] Klaus-Robert Müller, Matthias Krauledat, Guido Dornhege, Gabriel Curio, and Benjamin Blankertz. Machine learning techniques for brain–computer interfaces. *Biomedical Technology*, 49(1):11–22, 2004.

[7] Yonghong Huang, Deniz Erdogmus, Misha Pavel, Santosh Mathan, and Kenneth E Hild Ii. A framework for rapid visual image search using single-trial brain evoked responses. *Neurocomputing*, 74(12–13):2041–2051, 2011.

[8] Ke Yu, Kaiquan Shen, Shiyun Shao, Wu Chun Ng, Kenneth Kwok, Xiaoping Li. Common spatio-temporal pattern for single-trial detection of event-related potential in rapid serial visual presentation triage. *IEEE Transactions on Biomedical Engineering*, 58(9):2513–2520, 2011.

[9] Swartz Center for Computational Neuroscience UCSD. Lab streaming layer [online], available from http://github.com/sccn/labstreaminglayer/wiki. 2016.

[10] Zhengwei Wang, Graham Healy, Alan F Smeaton, and Tomas E Ward. An investigation of triggering approaches for the rapid serial visual presentation paradigm in brain computer interfacing. In *Signals and Systems Conference (ISSC), 2016 27th Irish*, pages 1–6. IEEE, 2016.

[11] Henri Begleiter. *Evoked brain potentials and behavior*, volume 2. New York: Springer Science & Business Media, 2012.

[12] Anthony M Norcia, Lawrence Gregory Appelbaum, Justin M Ales, Benoit R Cottereau, and Bruno Rossion. The steady-state visual evoked potential in vision research: a review. *Journal of Vision*, 15(6):4–4, 2015.

[13] Graham Healy, Zhengwei Wang, Cathal Gurrin, Tomás E Ward, and Alan F Smeaton. *An EEG image-search dataset: a first-of-its-kind in IR/IIR*. In Proceedings of CHIIR Workshop on Challenges in Bringing Neuroscience to Research in Human-Information Interaction, Oslo, Norway, March 2017.

[14] Graham Healy, Tomás Ward, Cathal Gurrin, and Alan F Smeaton. Overview of NTCIR-13 NAILS task. In *The Thirteenth NTCIR Conference (NTCIR-13)*, Tokyo, Japan, 5–8 Dec 2017.

[15] Carmen Vidaurre and Benjamin Blankertz. Towards a cure for BCI illiteracy. *Brain Topography*, 23(2):194–198, 2010.

[16] Brian F Odonnell and Ronald A Cohen. *The N2-P3 complex of the evoked potential and human performance*. NASA. Langley Research Center, Mental-State Estimation, pages 269–286, 1987.

[17] Steven J Luck. *An introduction to the event-related potential technique*. Cambridge, MA: MIT Press, 2014.

[18] Michael Plöchl, José P Ossandón, and Peter König. Combining EEG and eye tracking: identification, characterization, and correction of eye movement artifacts in electroencephalographic data. *Frontiers in Human Neuroscience*, 6, 2012.

[19] Charles Elkan. The foundations of cost-sensitive learning. In *International Joint Conference on Artificial Intelligence*, volume 17, pages 973–978. Lawrence Erlbaum Associates Ltd, 2001.

[20] Reza Fazel-Rezai, Brendan Z Allison, Christoph Guger, Eric W Sellers, Sonja C Kleih, and Andrea Kübler. P300 brain computer interface: current challenges and emerging trends. *Frontiers in Neuroengineering*, 5, 2012.

[21] Amelia J Solon, Stephen M Gordon, Brent J. Lance and Vernon J. Lawhern. Deep learning approaches for p300 classification in image triage: Applications to the NAILS task. In *Proceedings of the 13th NTCIR Conference on Evaluation of Information Access Technologies, NTCIR-13, Tokyo, Japan*, pages 5–8. 2017.

[22] James A Hanley and Barbara J McNeil. The meaning and use of the area under a receiver operating characteristic (ROC) curve. *Radiology*, 143(1):29–36, 1982.

[23] Kay Henning Brodersen, Cheng Soon Ong, Klaas Enno Stephan, and Joachim M Buhmann. The balanced accuracy and its posterior distribution. In *Pattern recognition (ICPR), 2010 20th International Conference on*, pages 3121–3124. IEEE, 2010.

[24] Bertrand Rivet, Antoine Souloumiac, Virginie Attina, and Guillaume Gibert. xDAWN algorithm to enhance evoked potentials: application to brain–computer interface. *IEEE Transactions on Biomedical Engineering*, 56(8):2035–2043, 2009.

[25] Yaqub Jonmohamadi, Govinda Poudel, Carrie Innes, Daniel Weiss, Rejko Krueger, and Richard Jones. Comparison of beamformers for EEG source signal reconstruction. *Biomedical Signal Processing and Control*, 14:175–188, 2014.

[26] Benjamin Blankertz, Ryota Tomioka, Steven Lemm, Motoaki Kawanabe, and Klaus-Robert Muller. Optimizing spatial filters for robust EEG single-trial analysis. *IEEE Signal Processing Magazine*, 25(1):41–56, 2008.

[27] Yijun Wang, Shangkai Gao, and Xiaornog Gao. Common spatial pattern method for channel selection in motor imagery based brain–computer interface. In *Engineering in Medicine and Biology Society, 2005. IEEE-EMBS 2005. 27th Annual International Conference of the*, pages 5392–5395. IEEE, 2006.

[28] Hubert Cecotti, Miguel P. Eckstein, and Barry Giesbrecht. Single-trial classification of event-related potentials in rapid serial visual presentation tasks using supervised spatial filtering. *IEEE Transactions on Neural Networks & Learning Systems*, 25(11):2030–2042, 2014.

[29] Matthias S Treder, Anne K Porbadnigk, Forooz Shahbazi Avarvand, Klaus-Robert Müller, and Benjamin Blankertz. The LDA beamformer: optimal estimation of ERP source time series using linear discriminant analysis. *NeuroImage*, 129:279–291, 2016.

[30] Scott Makeig and Julie Onton. ERP features and EEG dynamics: an ICA perspective. *Oxford Handbook of Event-Related Potential Components. New York, NY: Oxford*, 2009.

[31] Aapo Hyvärinen and Erkki Oja. Independent component analysis: algorithms and applications. *Neural Networks*, 13(4):411–430, 2000.

[32] Marzieh Mohammadi, Sepideh Hajipour Sardouie, and Mohammad Bagher Shamsollahi. Denoising of interictal EEG signals using ICA and time varying AR modeling. In *Biomedical Engineering (ICBME), 2014 21th Iranian Conference on*, pages 144–149. IEEE, 2014.

[33] Tzyy-Ping Jung, Scott Makeig, Colin Humphries, *et al.* Removing electroencephalographic artifacts by blind source separation. *Psychophysiology*, 37(2):163–178, 2000.

[34] Jonathon Shlens. A tutorial on principal component analysis. *arXiv preprint arXiv:1404.1100*, 2014.

[35] Muhammad Naeem, Clemens Brunner, and Gert Pfurtscheller. Dimensionality reduction and channel selection of motor imagery electroencephalographic data. *Computational Intelligence and Neuroscience*, 2009, 2009.
[36] T Zanotelli, SA Santos Filho, and Carlos J Tierra-Criollo. Optimum principal components for spatial filtering of EEG to detect imaginary movement by coherence. In *Engineering in Medicine and Biology Society (EMBC), 2010 Annual International Conference of the IEEE*, pages 3646–3649. IEEE, 2010.
[37] Galit Fuhrmann Alpert, Ran Manor, Assaf B Spanier, Leon Y Deouell, and Amir B Geva. Spatiotemporal representations of rapid visual target detection: a single-trial EEG classification algorithm. *IEEE Transactions on Biomedical Engineering*, 61(8):2290–2303, 2014.
[38] Mike X Cohen. *Analyzing neural time series data: theory and practice*. Cambridge, MA: MIT Press, 2014.
[39] Jia Meng, Lenis Mauricio Meriño, Nima Bigdely Shamlo, Scott Makeig, Kay Robbins, and Yufei Huang. Characterization and robust classification of EEG signal from image RSVP events with independent time-frequency features. *PLoS One*, 7(9):e44464, 2012.
[40] Adam D Gerson, Mads Dyrholm, An Luo, *et al.* Spatiotemporal linear decoding of brain state. *Signal Processing Magazine IEEE*, 25(1):107–115, 2008.
[41] Christopher M. Bishop. *Pattern recognition and machine learning*. New York: Springer. 2007.
[42] Jerome H Friedman. Regularized discriminant analysis. *Journal of the American Statistical Association*, 84(405):165–175, 1989.
[43] Richard O Duda, Peter E Hart, and David G Stork. *Pattern classification*. New York: John Wiley & Sons, 2012.
[44] Jing Hao Xue and Michael Titterington. Do unbalanced data have a negative effect on LDA? *Pattern Recognition*, 41(5):1558–1571, 2008.
[45] David JC MacKay. Bayesian interpolation. *Neural Computation*, 4(3):415–447, 1992.
[46] Ulrich Hoffmann, Jean-Marc Vesin, Touradj Ebrahimi, and Karin Diserens. An efficient P300-based brain–computer interface for disabled subjects. *Journal of Neuroscience Methods*, 167(1):115–125, 2008.
[47] Hubert Cecotti and Anthony J Ries. Best practice for single-trial detection of event-related potentials: application to brain–computer interfaces. *International Journal of Psychophysiology*, 111:156–169, 2017.
[48] Paul Sajda, Adam D Gerson, Marios G Philiastides, and Lucas C Parra. Single-trial analysis of EEG during rapid visual discrimination: enabling cortically-coupled computer vision. *Towards Brain–Computer Interfacing*, pages 423–44, 2007.
[49] Yonghong Huang, Deniz Erdogmus, Santosh Mathan, and Misha Pavel. Boosting linear logistic regression for single trial ERP detection in rapid serial visual presentation tasks. In *Engineering in Medicine and Biology Society, 2006. EMBS'06. 28th Annual International Conference of the IEEE*, pages 3369–3372. IEEE, 2006.

[50] Edgar E Osuna. *Support vector machines: training and applications*. Available from http://hdl.handle.net/1721.1/7290. 1997.

[51] Graham Healy and Alan F Smeaton. Optimising the number of channels in EEG-augmented image search. *Proceedings of the 25th BCS Conference on Human-Computer Interaction*. British Computer Society, 2011.

[52] Eva Mohedano, Kevin Mcguinness, Graham Healy, Noel E O'Connor, Alan F Smeaton, Amaia Salvador, Sergi Porta, and Xavier I-Nieto. Exploring EEG for object detection and retrieval. In *ACM on International Conference on Multimedia Retrieval*, pages 591–594, 2015.

[53] Albert Nigrin. *Neural networks for pattern recognition*. Oxford: Oxford University Press, 1995.

[54] Hecht-Nielsen. Theory of the backpropagation neural network. In *International Joint Conference on Neural Networks*, volume 1, pages 593–605, 1989.

[55] Sebastian Ruder. An overview of gradient descent optimization algorithms. *arXiv preprint arXiv:1609.04747*, 2016.

[56] Dineshbalu Balakrishnan and Sadasivan Puthusserypady. Multilayer perceptrons for the classification of brain computer interface data. In *Bioengineering Conference, 2005. Proceedings of the IEEE Northeast*, pages 118–119, 2005.

[57] Anil K Jain, Robert PW Duin, and Jianchang Mao. Statistical pattern recognition: a review. *IEEE Transactions on Pattern Analysis & Machine Intelligence*, 22(1):4–37, 2000.

[58] Ian Goodfellow, Yoshua Bengio, and Aaron Courville. *Deep learning*. Cambridge, MA: MIT Press, 2016.

[59] Gustavo Carneiro, Jacinto C. Nascimento, and António Freitas. The segmentation of the left ventricle of the heart from ultrasound data using deep learning architectures and derivative-based search methods. *IEEE Transactions on Image Processing*, 21(3):968–982, 2012.

[60] Jianfeng Gao, Xiaodong He, and Li Deng. Deep learning for web search and natural language processing. Available from https://www.microsoft.com/en-us/research/publication/deep-learning-for-web-search-and-natural-language-processing/, 2015.

[61] Sergey Zagoruyko and Nikos Komodakis. Learning to compare image patches via convolutional neural networks. In *Computer Vision and Pattern Recognition*, pages 4353–4361, 2015.

[62] Steve Lawrence and C Lee Giles, Ah Chung Tsoi, and Andrew D Back. Face recognition: a convolutional neural-network approach. *IEEE Transactions on Neural Networks*, 8(1):98–113, 1997.

[63] Baotian Hu, Zhengdong Lu, Hang Li, and Qingcai Chen. Convolutional neural network architectures for matching natural language sentences. *Advances in Neural Information Processing Systems*, 3:2042–2050, 2015.

[64] Ming Liang and Xiaolin Hu. *Recurrent convolutional neural network for object recognition. Proceedings of the IEEE Conference on Computer Vision and Pattern Recognition*. 2015.

[65] Hyungtae Lee and Heesung Kwon. Single-trial EEG RSVP classification using convolutional neural networks. In *SPIE Defense + Security*, page 983622, 2016.

[66] Vernon J Lawhern, Amelia J Solon, Nicholas R Waytowich, Stephen M Gordon, Chou P Hung, and Brent J Lance. EEGnet: a compact convolutional network for EEG-based brain–computer interfaces. *arXiv preprint arXiv:1611.08024*, 2016.

[67] Ran Manor and Amir B. Geva. Convolutional neural network for multi-category rapid serial visual presentation BCI. *Frontiers in Computational Neuroscience*, 9:146, 2015.

[68] Tomas Mikolov, Martin Karafiát, Lukas Burget, Jan Cernocký, and Sanjeev Khudanpur. Recurrent neural network based language model. In *INTERSPEECH 2010, Conference of the International Speech Communication Association, Makuhari, Chiba, Japan, September*, pages 1045–1048, 2010.

[69] Elif Derya Übeyli. Analysis of EEG signals by implementing eigenvector methods/recurrent neural networks. *Digital Signal Processing*, 19(1):134–143, 2009.

[70] Nihal Fatma Güler, Elif Derya Übeyli, and İnan Güler. Recurrent neural networks employing Lyapunov exponents for EEG signals classification. *Expert Systems with Applications*, 29(3):506–514, 2005.

[71] Elliott M Forney and Charles W Anderson. Classification of EEG during imagined mental tasks by forecasting with Elman recurrent neural networks. In *International Joint Conference on Neural Networks*, pages 2749–2755, 2011.

[72] Geoffrey E Hinton. Deep belief networks. *Scholarpedia*, 4(6):5947, 2009.

[73] Geoffrey E Hinton. A practical guide to training restricted Boltzmann machines. *Momentum*, 9(1):599–619, 2012.

[74] Shaheen Ahmed, Lenis Mauricio Merino, Zijing Mao, Jia Meng, Kay Robbins, and Yufei Huang. A deep learning method for classification of images RSVP events with EEG data. In *IEEE Global Conference on Signal and Information Processing*, pages 33–36, 2013.

[75] Ian J Goodfellow, Jean Pouget-Abadie, Mehdi Mirza, *et al.* Generative adversarial networks. *Advances in Neural Information Processing Systems*, 2014.

Chapter 13

Decoding music perception and imagination using deep-learning techniques

Sebastian Stober[1] *and Avital Sternin*[2]

Abstract

Deep learning is a sub-field of machine learning that has recently gained substantial popularity in various domains such as computer vision, automatic speech recognition, natural language processing, and bioinformatics. Deep-learning techniques are able to learn complex feature representations from raw signals and thus also have potential to improve signal processing in the context of brain–computer interfaces (BCIs). However, they typically require large amounts of data for training – much more than what can often be provided with reasonable effort when working with brain activity recordings of any kind. In order to still leverage the power of deep-learning techniques with limited available data, special care needs to be taken when designing the BCI task, defining the structure of the deep model, and choosing the training method.

This chapter presents example approaches for the specific scenario of music-based brain–computer interaction through electroencephalography – in the hope that these will prove to be valuable in different settings as well. We explain important decisions for the design of the BCI task and their impact on the models and training techniques that can be used. Furthermore, we present and compare various pre-training techniques that aim to improve the signal-to-noise ratio. Finally, we discuss approaches to interpret the trained models.

13.1 Introduction and motivation

13.1.1 Evidence from research on auditory perception and imagination

There are several related works that support the hypothesis that music imagery can successfully be used as a robust electroencephalography (EEG)-based brain–computer interface (BCI) task by exploiting the strong temporal characteristics of

[1]Research Focus Cognitive Sciences, University of Potsdam, Germany

[2]Department of Psychology, Brain and Mind Institute, University of Western Ontario, Canada

Colour versions of the figures in this chapter are available on IET Digital Library – see http://digital-library.theiet.org/

music and the salience of musical features in the corresponding brain activity. Music is inherently rhythmic, so we will begin with evidence related to rhythmic patterns. Oscillatory neural activity is sensitive to accented tones in a rhythmic sequence [1], and neural oscillations entrain to rhythmic sequences [2,3] and increase in anticipation of strong tones in a non-isochronous, rhythmic sequence [4–6]. When subjects hear rhythmic sequences, the magnitude of the oscillations changes for frequencies related to the metrical structure of the rhythm [7,8]. EEG studies [9] have further shown that perturbations of the rhythmic pattern lead to distinguishable electrophysiological responses – commonly referred to as event-related potentials (ERPs). This effect appears to be independent of the listener's level of musical proficiency. Furthermore, [10] showed that imagined auditory accents imposed on top of a steady metronome click can be recognised from ERPs. However, to deal with noise in the EEG signal and reduce the impact of unrelated brain activity, an ERP analysis requires averaging brain responses recorded over many events. Collecting many repetitions is not feasible for a BCI. In order to obtain an acceptable communication speed (bit rate), a BCI should work with only a few short trials. A stimulus that may work in such a BCI setting is a short clip of music.

EEG has been successfully used to distinguish perceived melodies. In a study by Schaefer *et al.* [11], ten participants listened to seven short melody clips with a length between 3.26 and 4.36 s. For single-trial classification, each stimulus was presented 140 times in randomised back-to-back sequences of all stimuli. Using a quadratically regularised linear logistic-regression classifier with 10-fold cross-validation, the accuracy varied between 25% and 70% within subject and between 35% and 53% across subjects. Combining all trials from all subjects and stimuli into a grand average ERP, a fronto-central component was identified through singular-value decomposition. This component explained 23% of the total signal variance with the corresponding time course showing significant differences between stimuli that were strong enough to allow cross-participant classification. Furthermore, a correlation with the stimulus envelopes of up to 0.48 was observed with the highest value over all stimuli at a time lag of 70–100 ms. Although the above results are promising, the classification was done on perceived music which is not useful in a BCI setting where imagination of music is needed.

Functional magnetic resonance imaging (FMRI) studies [12,13] have shown that similar brain structures and processes are involved during music perception and imagination. A recent review of the literature on auditory imagery concludes that '*auditory imagery preserves many structural and temporal properties of auditory stimuli*' and '*involves many of the same brain areas as auditory perception*' [14]. This is also underlined by Schaefer [15, p. 142] whose '*most important conclusion is that there is a substantial amount of overlap between the two tasks* [music perception and imagination], *and that "internally" creating a perceptual experience uses functionalities of "normal" perception.*' Thus, brain signals recorded while listening to a music piece could serve as reference data to detect salient elements expected during imagination. A recent meta-analysis [16] summarised evidence that EEG is capable of detecting brain activity during the imagination of music. Most notably, preliminary results for recognising imagined music fragments from EEG recordings were reported in [17] in which four out of eight participants produced imagery that was classifiable (in a binary comparison) with an accuracy between 70% and 90% after 11 trials. These results are already very encouraging. However, much improvement is still needed for a robust BCI.

Another closely related field of research is the reconstruction of auditory stimuli from EEG recordings. In [18], EEG recorded during listening to natural speech was reported to contain traces of the speech amplitude envelope. Independent component analysis (ICA) and a source localisation technique were applied to enhance the strength of this signal. Applying this technique to imagined speech, single-sentence classification performance was statistically significant for two of eight subjects with better performance when several sentences were combined for a longer trial duration. In [19], a method for decoding attentional selection in a cocktail party environment from single-trial EEG recordings has been proposed. In an experiment, 40 subjects were presented with two classic works of fiction at the same time – each one to a different ear – for 30 one-minute trials. To determine which of the two stimuli a subject attended to, both stimulus envelopes were reconstructed from the recorded EEG using decoders trained with linear regression. The stimulus with the best reconstruction from the EEG was predicted – using the Pearson correlation of the envelopes as measure of quality. A within-subject accuracy of 89% was reported for the attended stimulus. This technology could be used in a BCI for real-time decoding of music imagination from an EEG signal.

13.1.2 Existing auditory and music-based BCIs

There is evidence that auditory P300-ERP paradigms are less accurate and slower than similar visual ERP paradigms (e.g. [20]). However, as pointed out in [21], an auditory BCI paradigm may be one of the only remaining communication channels in some BCI application domains. Furthermore, visual BCI paradigms may not function well when eye-gaze control is limited. In [20], six environmental sounds were used to operate a 6×6 P300 speller where users separately selected the row and the column to type one of 36 characters. A similar 5×5 design was studied in [22]. In [23], a T9-like 3×3 design was used where a combination of pitch (high/medium/low) and direction (left/middle/right) was employed for a two-dimensional encoding scheme that only requires a single character selection step in combination with word prediction. A similar approach with additional redundancy was used in [24]. In [25], a three-stimulus oddball paradigm with one frequent stimulus and two target stimuli was used for binary communication. Subjects attended to one specified target stimulus, masking the other two non-target events. Target stimuli differed either in loudness, pitch or direction with pitch resulting in the best performance for 14 of 20 healthy subjects.

Another attention-driven auditory BCI approach involves the simultaneous presentation of two or more auditory streams. In [26], two concurrent sequences were presented – each with individual target events and stimulus onset asynchronies, which elicited different averaged ERPs when attended to. A similarly designed BCI described in [27] used tone burst trains with different modulation rates (37 and 43 Hz) resulting in distinguishable steady-state auditory evoked potentials (SSAEPs). In a comparison of ERPs and SSAEPs in a two-stream design using 5-s stimulation intervals for single-trial BCI communication, attention modulation of ERPs showed clearly superior performance [28].

There is evidence that using natural stimuli instead of artificial sounds improves auditory BCIs with respect to ergonomics and performance [29]. A musical oddball paradigm was described in [30] using a minimalistic version of *Just can't get enough* (Depeche Mode) consisting of synthesised bass, keyboard, and drums. From this, randomised 40-s stimuli were composed by repeating a characteristic 1-bar pattern for each instrument (standard stimulus) and varying it infrequently (deviant stimulus) without violating the characteristics of the musical idiom. Subjects were instructed to count the number of deviants for the attended instrument. This resulted in characteristic ERP responses which allowed to identify the focus of attention for 10 of 11 healthy subjects with an accuracy above 80%. In [31], this study was extended to a second music stimulus – a Jazz piece using more natural sounding acoustic samples of flute, bass and piano, manually adjusted loudness and micro timing for more expressivity, different pattern lengths of 3, 4, and 5 beats per instrument, and stereo panning. Using three instrument-specific classifiers and posterior probabilities, a mean accuracy above 90% across 11 healthy participants was achieved for both stimuli. The stronger independence of the different voices in the Jazz piece did not lead to a significant effect.

A special sub-class of BCIs, Brain–Computer Music Interfaces (BCMIs) allow their users to control music – either voluntarily (hard BCMIs) or passively (soft BCMIs). An overview of this field is provided in [32]. Applications range from artistic tools for music composers and performers to medical application such as for music therapy.

13.2 Deep learning for EEG analysis – the state of the art

13.2.1 Challenges

Deep learning has gained a lot of popularity in general over the course of the last decade. However, it is still far from being widely adopted for anything but the 'mainstream' – i.e. application domains with large and often publicly available datasets such as computer vision or natural language processing. Compared to these, for several reasons, EEG analysis appears particularly unattractive for deep-learning researchers who are looking for new challenges.

1. Deep learning usually requires vast amounts of data but public access to EEG data is very limited and typical EEG datasets are small with respect to the number of subjects and trials per subject.
2. EEG data are complex and high dimensional. Recordings often comprise many channels and have moderately high sampling rates (much higher than video frame rates but lower than typical audio sampling rates). This results in a high number of values (input dimensions) per trial. For instance, a 6-s recording of 64 EEG channels sampled at 512 Hz has 196,608 dimensions. This is the same dimensionality

as for a colour image in the ImageNet dataset [33] at the typically used resolution of 256×256 pixel. As an equivalent of the more than 14 million labelled images in ImageNet, roughly 1,000 days of EEG recordings would be required.
3. EEG data are noisy. There are many unrelated brain processes that may dominate the recorded signal and there are many sources that can easily introduce signal artefacts such as eye blinks, muscle activity, or electrical noise from nearby devices. Experimental design, choice of equipment, and preprocessing techniques can improve signal quality to some extent but usually results in an increase in cost per recording.

The combination of these three points results in rather adverse conditions for the application of deep learning. Hence, successfully analysing EEG with deep-learning techniques is by all means not 'low-hanging fruit' that could be attained with little effort. Instead, special strategies have to be devised that address each of the above challenges specifically. Instead of an immediate breakthrough, rather incremental progress by little steps at a time should be expected. In the following, an overview of the state of the art in deep learning for EEG analysis is provided that points out potential factors contributing to a working solution.

13.2.2 Deep learning applied to EEG analysis

Already in 2009, Mirowski *et al.* [34] successfully trained convolutional neural networks (CNNs) for epileptic seizure prediction in intracranial EEG. They did not use raw EEG data but computed bivariate features (from 15 pairs of 6 channels) related to EEG synchronisation comprising cross-correlation, non-linear interdependence, dynamical entrainment, and wavelet-based phase synchrony for several frequencies. Their CNNs classified 5-s windows of EEG sampled at 256 Hz. The CNN architecture was similar to LeNet5 [35] with five layers: three convolutional layers with two subsampling layers in between. Subsampling was only performed in the time domain. The first and last convolutions were performed over the time axis only, whereas the second convolution considered all dimensions (including the feature axis). L1-norm regularisation was applied on the weights for feature selection – i.e. only features with non-zero (or non-negligible) weights were considered as relevant for the classification task. However, as pointed out later by Haufe *et al.* [36], model parameters for classification or decoding should not be directly interpreted in terms of the brain activity as they depend on all noise components in the data, too. In particular, they specifically showed that regularisation such as L1-sparsification does not improve interpretability of the model parameters. Instead, a forward model should be derived that explains how the measured signals were generated from the neural sources.

Wulsin *et al.* [37] used deep belief nets (DBNs) to detect anomalies related to epilepsy in EEG recordings by classifying individual 'channel-seconds', i.e. 1-s chunks from a single EEG channel without further information from other channels or about prior values. This approach helps to drastically reduce data dimensionality and at the same time results in more training examples. But it also reduces the complexity of patterns that can potentially be detected. The classifier was first pre-trained layer by layer as an auto-encoder on unlabelled data, followed by a supervised fine-tuning with back-propagation on a much smaller labelled dataset. Working on raw, unprocessed data sampled at 256 Hz led to a classification accuracy comparable to hand-chosen features. Längkvist *et al.* [38] similarly employed DBNs combined with hidden Markov models to classify different sleep stages. Again, the data was segmented into 1-s chunks. Here, a DBN on raw data showed a classification accuracy close to one using 28 hand-selected features.

Lin *et al.* [39] learned features for epileptic seizure detection from single-channel EEG segments (each with a duration of 23.6 s) using a 3-hidden-layer stacked sparse auto-encoder and applied a simple Softmax classifier. A similar approach using stacked denoising auto-encoders was described earlier in [40] to classify EEG recordings of rhythm perception and identify their ethnic origin. In a follow-up study, CNNs were applied to distinguish the individual rhythms [41]. Similarly, Tabar and Halici [42] investigated CNNs and stacked auto-encoders (SAEs) to classify EEG motor imagery signals. Another CNN for detecting P300 waves was described in [43]. Here, convolutions were applied in both, space and time domain. In [44], a CNN was pruned for feature selection.

Nurse *et al.* [45] proposed to use multi-layer perceptrons to discover suitable features for motor imagery BCI tasks in combination with a genetic algorithm for tuning hyper-parameters such as the number of hidden layers and neurons. In a follow-up experiment, Nurse *et al.* [46] followed a different strategy using CNNs for motor imagery classification. An *et al.* [47] demonstrated that DBNs are also suitable for classifying motor imagery in a BCI setting. There has also been early work on emotion recognition from EEG using deep neural networks such as described by Jirayucharoensak *et al.* [48] and Zheng *et al.* [49].

13.2.3 Custom solutions developed for EEG analysis

Cecotti and Gräser [50] used a special CNN for classification of steady-state visual evoked potentials – i.e. brain oscillation induced by visual stimuli. The network integrated the Fourier transform between convolutional layers, which transformed the data from the time domain to a time-frequency representation. The transform itself had no trainable parameters but its implementation still allowed layers before and afterwards to be trained in the usual fashion using back-propagation.

This demonstrates an important design pattern for custom solutions: custom elements can be integrated into the forward computation (producing an output for given inputs) between other trainable elements as long as there is a way to back-propagate gradients through them – independently of whether or not they introduce trainable parameters themselves. Generally, custom network elements need to produce well-behaved gradients to avoid negative impacts on the learning dynamics.[1]

Note: A similar transformation could be learned by a set of convolutional filters. But learning filters no longer guarantees a meaningful ordering such as from low to

[1] For functions that have no useful gradient such as step functions, it might still be possible to estimate gradients using reinforcement learning algorithms like REINFORCE [51].

high frequencies. As a result, subsequent application of convolution along the filter (frequency) axis would no longer be semantically justified.[2]

The approach proposed by Cecotti *et al.* [50] allowed convolution to be applied in the time-domain as well as in the time-frequency domain. Bashivan *et al.* [52] further described a way that convolution could be applied in the spatial domain. Similar to topographic EEG plots, they mapped EEG channels into 2D space according to the EEG montage and interpolated values in between accordingly. Further, three frequency bands (obtained using the fast Fourier transform) were mapped to the three colour channels (RGB). The resulting sequence of images was then processed using state-of-the-art video classification approaches. In their experiments for a mental load classification task, the authors found that a combination of CNNs and recurrent neural networks (RNNs) resulted in the best accuracy. They further argued that their proposed transformation allows researchers to combine data acquired with different EEG hardware and montages.

13.2.4 The need for open science

Jingwei *et al.* [53] proposed to learn so-called deep motor features for BCIs based on motor imagery using multi-scale deep CNNs. The authors claim that these features are 'dissimilar for different tasks and alike for the same tasks' and report '100% accuracy for 4 classes [of] imagined motor EEG.' Such unrealistically high accuracy is usually an indicator of a flawed experiment. For instance, the way the data was recorded and split into subsets for training and evaluation could have introduced artefacts that were exploited by the classifier. This so-called Horse phenomenon has, for instance, been described by Sturm [54]. Another common cause for flawed results are software bugs that cause mistakes in indexing trials or that accidentally duplicate trials resulting in overlapping training and evaluation data.

The above is only one example of several papers in neuroscience using deep learning with questionable results. Unfortunately, in many of these papers, data and code are usually not available for peer review and replication of the results. This is in contrast to the common practice in machine learning, and especially in deep-learning research, to share code and if possible also data. Making data and code accessible increases the credibility, reproducibility and impact of the results. It would be beneficial for the still young field of deep-learning research in neuroscience to adopt open science principles as well.

13.2.5 Summary

Overall, whilst largely out of fashion in mainstream deep-learning research, using unsupervised pre-training of DBNs, SAEs and the like to initialise lower layers and identify representative features is still a popular strategy for EEG analysis. It allows for training deep networks even with small datasets but requires extra care and effort

[2]With a given ordering of filter outputs resulting from a fixed filter bank, moving the receptive field of a convolutional filter along the newly created axis corresponds to shifting the frequency band in focus. Without an ordering of the filters, this would not be semantically meaningful.

for successful outcomes. DBNs in particular have the additional advantage of being generative models and as such can generate samples of what they have learned. This could be exploited for model introspection – i.e. to gain insights about the learned patterns – but has so far not been investigated for EEG to the knowledge of the authors.

In terms of network structure, convolution has become popular. In contrast, the potential benefits of recurrent connections have not been extensively explored yet in spite of promising early studies such as by Petrosian *et al.* [55]. A possible reason for this might be the high temporal resolution of EEG signals that result in long input sequences even for very short signal segments. Under these conditions, it can be problematic to learn patterns beyond a limited local context. As a workaround, the EEG signal can be transformed into a time-frequency representation with a reduced temporal resolution like in [50,52].

13.3 Experimental design

Our aim was to create a dataset that would facilitate interdisciplinary research towards music imagery information retrieval – i.e. retrieving music information from brain activity – and ultimately BCIs driven by music perception and imagination. We wanted the dataset to allow researchers to explore the music stimulus space and investigate which musical dimensions are best captured in EEG recordings and thus would be well suited for a BCI. The result of this effort is the OpenMIIR dataset [56].[3] This section highlights and explains important experimental design decisions we made when creating this dataset.

13.3.1 Stimulus selection

There are many characteristics of music that may play a role in how well the stimuli can be classified from EEG data such as time signature, key signature, lyrics, instrumentation and emotion. Based on the existing literature on music perception and imagination, we decided to control two particularly promising musical aspects: time signature (meter) and the presence/absence of lyrics. Other aspects like tempo, key, mood or genre were not systematically controlled but we generally strove to cover a broad spectrum with very differently sounding stimuli.

In order to make imagination of the stimuli as easy as possible, we specifically focused on over-learned, familiar pieces considering songs from existing lists of children's nursery rhymes, movie soundtracks, Christmas carols, etc. Stimuli were selected based on whether they were in 3/4 (waltz) or 4/4 (march) time and whether the songs had lyrics or not. They were kept as similar in length as possible with care taken to ensure that they all contained complete musical phrases. For the no-lyrics category, we specifically chose pieces that do not have any lyrics associated with them to avoid any automatic and unwanted lyric imagination. For each song with lyrics, we selected a version with and without a singing voice and tempo-matched the

[3]The OpenMIIR dataset is available at https://openmiir.github.io/.

Table 13.1 Overview of the stimuli used in the experiment

ID	Name	Meter	Length (s)	Tempo (BPM)	No. of bars	Bar length (s)
1	Chim Chim Cheree (lyrics)	3/4	13.3	212	16	0.85
2	Take Me Out to the Ballgame (lyrics)	3/4	7.7	189	8	0.95
3	Jingle Bells (lyrics)	4/4	9.7	200	8	1.20
4	Mary Had a Little Lamb (lyrics)	4/4	11.6	160	8	1.50
11	Chim Chim Cheree	3/4	13.5	212	16	0.85
12	Take Me Out to the Ballgame	3/4	7.7	189	8	0.95
13	Jingle Bells	4/4	9.0	200	8	1.20
14	Mary Had a Little Lamb	4/4	12.2	160	8	1.50
21	Emperor Waltz	3/4	8.3	178	8	1.01
22	Hedwig's Theme (Harry Potter)	3/4	16.0	166	15	1.08
23	Imperial March (Star Wars Theme)	4/4	9.2	104	4	2.30
24	Eine Kleine Nachtmusik	4/4	6.9	140	4	1.71

pair by applying (minimal) time stretching where necessary. The rationale behind this was to allow us to analyse the effect of a present or absent singing voice within the same song. As an additional requirement, stimuli had to start on the down beat (the beginning of the musical bar) because this would make it easier to start imagination on cue. Stimulus details can be found in Table 13.1.

13.3.2 Equipment and procedure

Based on the literature, we hypothesised that temporal features would play a key role in stimulus recognition as they might be captured well by the EEG. A problem for the music imagination recordings was that we had no control over the timing. It was not possible to use a background metronome click to keep participants in time during imagination because the stimuli had different tempi. Each click would result in an ERP, and the rate of the ERPs in the EEG recording would easily give away the identity of the stimulus during classification. Furthermore, hearing the clicks could have interfered with the imagination task. We therefore only used clicks to cue the tempo and the start of the stimulus. Additionally, we required participants to succeed in a tapping-along test prior to the EEG part of the study to make sure they were able to sense and keep the beat. We also checked their familiarity with the stimuli. The experiment included four conditions:

1. Stimulus perception preceded by metronome cue clicks
2. Stimulus imagination preceded by metronome cue clicks
3. Stimulus imagination without metronome cue clicks
4. Stimulus imagination without metronome cue clicks, with feedback

Cue clicks started approximately 2 s before the stimulus, began to fade out at the 1 s mark and stopped at the onset of the music (condition 1) or when imagination was

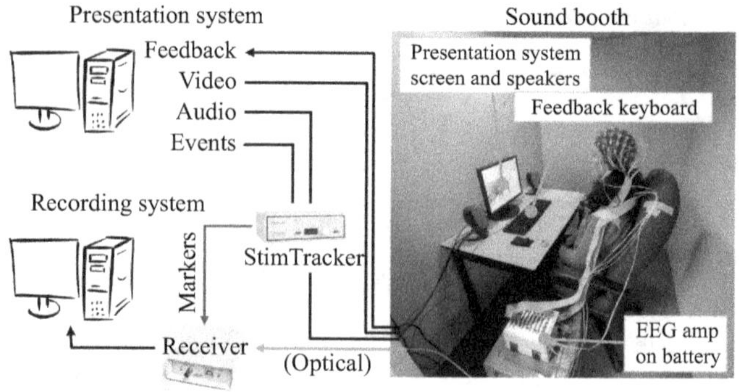

Figure 13.1 Setup for the EEG experiment. The presentation and recording systems were placed outside to reduce the impact of electrical line noise that could be picked up by the EEG amplifier

supposed to begin (condition 2). In conditions 3 and 4, we simulated spontaneous imagination like in a real BCI scenario by removing the metronome cue entirely.

For the EEG recording, participants sat in an audiometric room (Eckel model CL-13). A BioSemi Active-Two system with 64+2 EEG channels recorded EEG data at 512 Hz as shown in Figure 13.1. Horizontal and vertical electrooculography (EOG) channels recorded eye movements. The presented audio was routed through a Cedrus StimTracker connected to the EEG receiver, which allowed a high-precision synchronisation (<0.05 ms) of the stimulus onsets with the EEG data. The experiment was programmed and presented using PsychToolbox run in MATLAB 2014a. A computer monitor displayed the instructions and fixation cross for the participants to focus on during the trials to reduce eye movements. The stimuli and cue clicks were played through two tabletop speakers (Altec Lansing VS2121) at a comfortable level that was kept constant across participants. Headphones were not used because pilot participants reported headphones caused them to hear their heartbeat which interfered with the imagination portion of the experiment.

The EEG experiment was divided into two parts with five blocks each as illustrated in Figure 13.2. A single block comprised all 12 stimuli in randomised order. It is essential *not* to record trials for the same stimulus back to back in a block because machine-learning techniques could easily pick up stimuli-unrelated signal components to identify the block and thus the stimuli. Rather, test trials for evaluation should be taken from different blocks than the training trials. Our experimental design specifically supports this approach.

Between blocks, participants could take breaks at their own pace. Conditions 3 and 4, which simulate a more realistic query scenario during which the participant has not heard the stimulus immediately prior to imagining, were identical except for the trial context. While condition 1–3 trials were recorded directly back-to-back within the first part of the experiment, all condition 4 trials were recorded separately in the

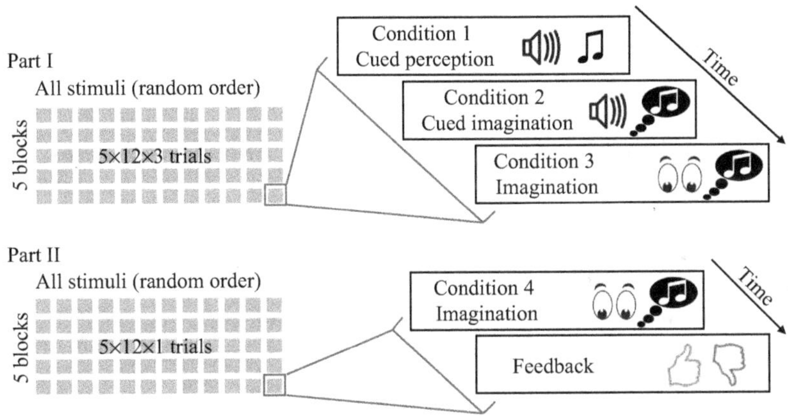

Figure 13.2 Illustration of the design for the EEG portion of the study

second part without any cue clicks or tempo priming by prior presentation of the stimulus. After each condition 4 trials, participants provided feedback by pressing one of two buttons indicating whether or not they felt they had imagined the stimulus correctly. In total, 240 trials (12 stimuli × 4 conditions × 5 blocks) were recorded per subject. With the explicit consent of all participants and the approval of the ethics board at the University of Western Ontario, all the data collected in this experiment were released under the Open Data Commons Public Domain Dedication and License.

13.3.3 Preprocessing

For all follow-up EEG analysis experiments, the raw EEG and EOG data were preprocessed using the MNE-Python toolbox [57]. Channels containing noise that could not be removed by simple filtering techniques (i.e. resulting from muscle movements or bad electrical contact with the scalp) were identified as 'bad' by visual inspection.[4] The bad channels were removed and interpolated (between 0 and 3 per subject). For interpolation, the spherical splines method described in Perrin *et al.* [58] was applied. The data were then filtered with an overlap-add FIR filter (filter length 10 s), keeping a frequency range between 0.5 and 30 Hz. The width of the transition band was 0.1 at 0.5 Hz and 0.5 at 30 Hz. The filtering removed unwanted high frequency information and any slow signal drift in the EEG. Removing unwanted noise (i.e. from external sources or muscle movements) restricts analyses to data within the frequency range of signals produced by the brain. We computed independent components using extended Infomax ICA [59] and removed components that had a high correlation with the EOG channels to reduce artefacts caused by eye blinks. This ensured that the final results could be attributed to brain responses and not other sources of electrical activity.

[4]Bad channels are highlighted for each participant in the topographic plots in Figure 13.4.

Finally, the data from the 64 EEG channels were reconstructed from the remaining independent components.

13.4 Representation learning techniques for pre-training

The techniques described in this section have been devised specifically to learn useful features from EEG signals that facilitate music stimulus classification for the Open-MIIR dataset. Beyond this scope, they are generally applicable to arbitrary encoder and decoder networks and in potentially many other domains.

13.4.1 Basic auto-encoder

No matter how much effort one puts into controlling the experimental conditions during EEG recordings, there will always be some individual differences between subjects and between recording sessions. This can make it hard to combine recordings from different subjects. One common way to address this issue is to average over many very short trials such that differences are cancelled out. When this is not feasible because of the trial length or a limited number of trials, an alternative strategy is to derive signal components from the raw EEG data hoping that these will be stable and representative across subjects. Here, principal component analysis (PCA) and ICA are well-established techniques. Both learn linear spatial filters, whose channel weights can be visualised by topographic maps as shown in Figure 13.3 (middle).

As a pilot experiment, we investigated whether basic convolutional auto-encoders (CAEs) can also be used to learn common EEG signal components similar to those obtained by PCA and ICA. CAEs are a special variant of CNNs that encode their input using convolution into a compressed internal representation which is then decoded using deconvolution into the original space. They are commonly trained to minimise the reconstruction error. Such encoder/decoder pairs can optionally be stacked as described in [60] to increase the complexity of the internal representations.

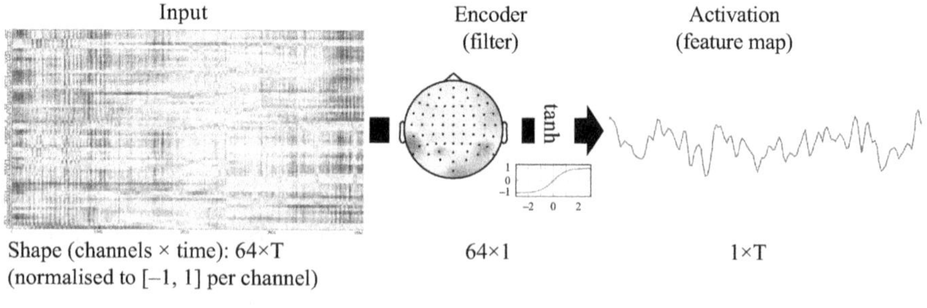

Figure 13.3 Example encoder pipeline with a single convolutional filter of width 1 (visualised as a topographic map instead of a column vector) that aggregates 64 EEG channels into a single waveform

We considered two measures of the reconstruction quality. The mean squared reconstruction error (MSRE) is the commonly used error measure for auto-encoding and is computed as the squared Euclidean distance between the input and the reconstruction averaged over all samples. We further defined the mean channel correlation (MCC) as the mean correlation of the individual input channels and their reconstructions. The correlation is computed as Pearson's r, which is in the range $[-1,1]$ where 1 is total positive correlation, 0 is no correlation, and -1 is total negative correlation.

The convolutional filters applied here were two-dimensional where the width is time (samples) and the height is space (channels). The filter height in the space dimension always equals the number of channels, i.e. weights are only shared along the time dimension. In the simplest case, the filter width in the time dimension is only 1. A convolutional layer with k filters then behaves like a conventional fully connected layer with k neurons that aggregates EEG data for a single time point across all channels into k output values – with the only difference being that a CNN processes the whole recording as one input. This is illustrated in Figure 13.3 for a single convolutional filter.

A CAE with filter width 1 and linear activation functions is strongly related to PCA [61] and consequently obtains similar results. Using an auto-encoder, however, opens up further possibilities such as non-linear activation functions, a filter width greater than 1, regularisation of channel weights and activations, and stacking multiple encoder/decoder layer pairs. This allows us to identify more complex components which can include the time dimension.

Moreover, a CAE can also learn individually adapted components that are linked between subjects. To this end, it is first trained on the combined recordings from the different subjects to identify the most common components. The learned filter weights are then used as initial values for individually trained auto-encoders for each subject. This leads to adaptations that reflect individual differences. Ideally, these are small enough such that the relation to the initial common components still remains. In this case, the filter activation of the individually adapted auto-encoders can be used as a common feature representation for further analysis.

Figure 13.4 shows the spatial signal filters learned by a basic CAE with 4 simple filters of time width 1 compared to components learned with PCA and extended Infomax ICA [59] on the concatenated trials for the perception and cued imagination conditions, respectively. All trials were preprocessed as described in Section 13.3.3. The deconvolution filters were set to share the weights with the corresponding convolution filters. We experimented with different activation functions (linear, tanh and rectified linear [62]) and obtained very similar topographies that only had small variations in intensity. We finally selected the tanh non-linearity because its output matches the value range of the network inputs ($[-1,1]$). Furthermore, we found that using bias terms did not lead to improvements and therefore chose network units without bias. Comparing the commonly used MSRE to the MCC measure showed that MSRE was the superior cost function in this setting. The training process was generally very stable such that a high learning rate with linear decay over time in combination with momentum could be used to quickly obtain results. As termination

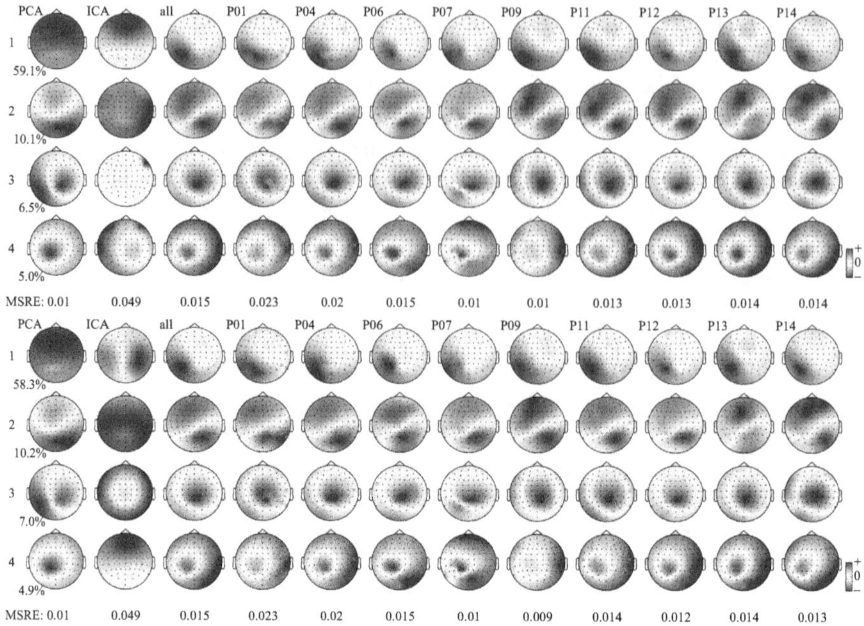

Figure 13.4 Topographic map visualisation of the EEG channel weights for the top-4 signal components derived from the perception (top) and cued imagination trials (bottom). The leftmost two columns show the top-4 principal components (explaining over 80% of the variance in total; individual values are shown next to the components) and the corresponding independent components computed using extended Infomax ICA. The remaining columns show the weights learned by a tied-weights CAE for all subjects combined (third column) and all individual subjects labelled P01 to P14 accordingly using four filters of width 1 (samples) with tanh non-linearity and without biases. Noisy EEG channels that were removed and interpolated are marked by Xs. Values at the bottom of each column refer to the mean squared reconstruction error (MRSE) per sample, which was used as cost function

criterion, we tracked the error on the training set for early stopping and specified a (rather defensive) maximum of 1,000 epochs.

We first trained an auto-encoder on all trials within a particular condition. The learned filter weights were then used as initialisation values for individual auto-encoders for each subject. Minor differences between subjects can be recognised in the example shown in Figure 13.4 but the component relations between subjects are still obvious. This was generally the case for different network structures that we tested. The individual components were very stable. Comparing results for perception and imagination, there was very little difference. The largest difference was seen in the ICA components between the two conditions. The MSRE was generally in the range of the value obtained with PCA and substantially lower than for ICA. This demonstrates the suitability for general dimensionality reduction with the additional benefit of obtaining a common data representation across subjects that can accommodate individual differences. However, the resulting representation might not be specific enough for a respective classification task. The derived techniques described in the following section address this potential shortcoming.

13.4.2 Cross-trial encoder

The basic auto-encoder scheme only aims to find general signal components. In EEG signal analysis, we are especially interested in finding features that are stable across trials of the same class and ultimately allow us to recognise different classes in a classification scenario. The straightforward approach in deep learning would be to directly train a classifier on the labelled trials in a supervised fashion. During the training process, respective features would naturally emerge in hidden network layers. The problem with this approach is that it requires a large amount of trials for training – especially given the high dimensionality of the EEG data and the expected complexity of the features to be learned. With too few trials, the network will likely fail to identify the relevant information thanks to the adverse signal-to-noise ratio. This will result in a poor generalisation performance. There is still the possibility of using regularisation to reduce model over-fitting to the training data but this usually does not address the lack of sufficient data in the first place. Therefore, we conceived a pre-training approach for small datasets that also acts as a regulariser.

The idea is to create a situation where the auto-encoder can only perform well (on the training data and in general) when it uses features that are stable across trials. Instead of trying to simply reconstruct the input trial, we attempt to reconstruct a different trial belonging to the same class.[5] This strategy can be considered as a special case of the generalised framework for auto-encoders proposed by Wang *et al.* [63].[6] Given n_C trials for a class C, n_C^2 or $n_C(n_C - 1)$ pairs of input and target trials can be formed depending on whether pairs with identical trials are included. This increases the number of training examples by a factor depending on the partitioning of the dataset with respect to the different classes. As for the basic auto-encoding scheme, the training objective is to minimise the reconstruction error. In this sense, it is unsupervised training, but the trials are paired for training using knowledge about their class labels. The technique can be extended as described in the following section.

[5]Under a strict perspective, trying to reconstruct a different trial may no longer be considered as *auto-encoding*. However, we rather see the term as a reference to the network architecture which is still the same as for regular auto-encoders. Only the training data has changed.

[6]They similarly define a 'reconstruction set' (all instances belonging to the same class) but use a different (and much simpler) loss function based on linear reconstruction and do not consider convolution.

13.4.3 Hydra-net cross-trial encoder

Paired trials for a cross-trial encoder can either belong to the same subject or to different subjects. The latter case is much harder as the auto-encoder has to deal with inter-subject differences. To allow the CAE to adapt to individual differences between the subjects, a modified network structure called *hydra-net* can be used as introduced in [64]. Hydra-nets allow the network to have separate processing pathways for subsets of a dataset. This makes it possible to have different weights – depending on trial meta-data – in selected layers of a deep network. Such a network can, for instance, adapt to each subject individually. With individual input layers, the structure can be considered as a network with multiple 'heads' – hence the reference to Hydra.

This approach can be applied at both ends of the CAE as illustrated in Figure 13.5, i.e. the encoder filters can be selected based on the input trial's subject whereas the decoder filters can be chosen to match the target trial's subject. Encoder and decoder weights can optionally be tied within subjects. The training procedure can be split into three stages as follows:

1. Train a general model using within-subject trial pairs.
2. Starting from the general model as initialisation, train an individual model for each subject still using within-subject trial pairs exclusively.
3. Combine the individual models into a hydra-net model and train it using a mix of cross-subject and within-subject trial pairs.

This procedure ensures that individualised filters still correspond across subjects, which facilitates comparing patterns and is also important for potential subsequent layers that are not individualised. Such follow-up layers that process a mix of data coming from different hydra-net branches generally need to rely on consistent semantics of their inputs in order to learn higher-level features.

Figure 13.5 shows the final result learned by a hydra-net cross-trial CAE with a single (spatial) filter of width 1 after stage 3 using the perception trials. A further example with a single (spatio-temporal) filter of width 3 is shown in Figure 13.6 after stage 1 (left column) and 3 (remaining columns). In our experiments described in [64], we found that using the output of this pre-trained encoder as features instead of the raw EEG significantly improved the stimulus classification accuracy.[7]

13.4.4 Similarity-constraint encoder

One downside of the approaches proposed above is that there is no straightforward way to directly evaluate the learned feature representations. Essentially, we are looking for distinctive time series. Using the features in a classifier may give us some idea about their usefulness but this creates a dependency on the classifier. This way, it becomes hard to tell whether a classification failed because of a poorly designed classifier or because the features did not contain enough relevant information to allow classification.

[7]Two classifiers were considered: a linear support vector classifier (SVC) and a neural network that just added a fully connected layer and a Softmax output layer on top.

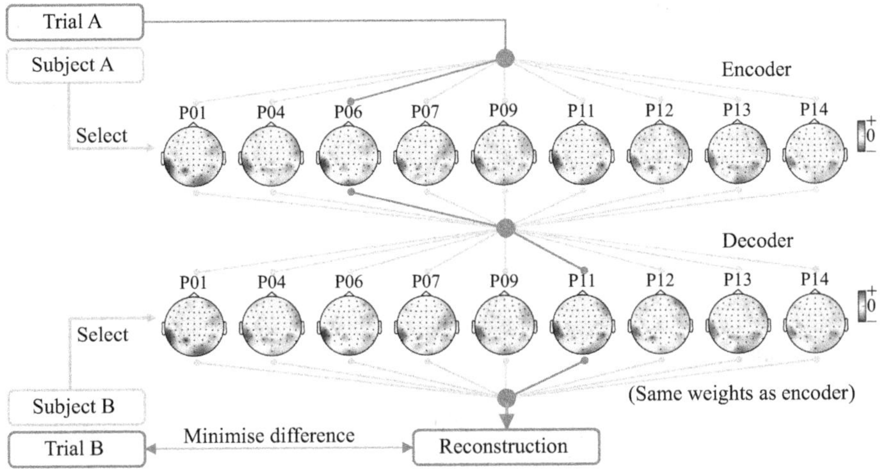

Figure 13.5 Training a hydra-net cross-trial encoder with cross-subject trial pairs (A,B). Both trials always belong to the same class. Two separate hydra-net layers with shared weights (within subjects) are used as encoder (convolution) and decoder (deconvolution). Each layer has a single convolutional filter (shape: 64 channels × 1 sample) per subject visualised as topographic map. Encoder weights are selected based on the subject meta-data of the input trial (A) whereas the decoder weights are selected based on the subject meta-data of the paired trial (B). In this example, a trial from subject P06 is paired with a trial from subject P11. The respective processing pathway for these selector values is highlighted. During back-propagation of the reconstruction error, only the filters along the selected pathway are adapted

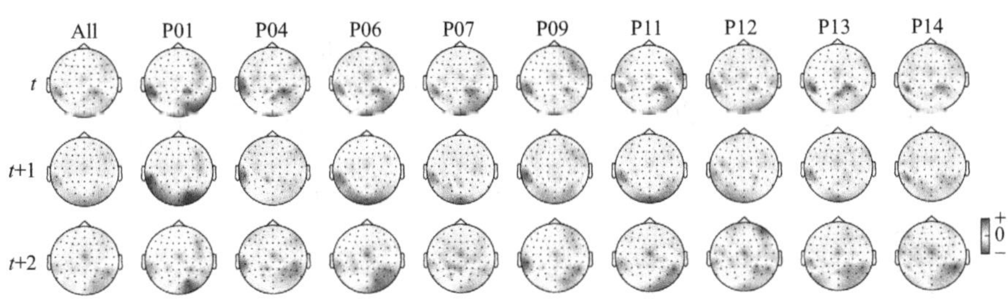

Figure 13.6 A spatio-temporal filter over all 64 EEG channels (visualised as topographic maps) and 3 samples (rows) after pre-training using cross-trial encoding. Left column: global filter after training on within-subject trial pairs (stage 1). Remaining columns: individual filters (hydra-net) after training with cross-subject trials (stage 3)

As an alternative to evaluating the classification performance, one can attempt to analyse the feature space. Obviously, the features should be similar for trials from the same class. Such an evaluation requires a similarity measure which creates a different kind of dependency. However, we consider the dependency on a similarity measure less strong than in the classification scenario. Moreover, the similarity measure could be learned along with the features in a slightly different encoding scheme.

For the similarity-constraint encoding scheme, we remove the decoder part of our original auto-encoder as we are only interested in the internal representations. Furthermore, we redesign the encoder such that it can process multiple trials at the same time as a single input tuple. To ensure that, all trials within a tuple are processed in the exactly same way, weights and biases need to be shared between the parallel processing pipelines. Finally, we need to compute similarities (or distances[8]) between the trials of a tuple. A straightforward choice would be the L2-norm (or Euclidean distance in correspondence with the MSRE) or the MCC as introduced in Section 13.4.1. But more complex computations are also possible as long as they allow training through back-propagation. In our experiments described in [64,65], we found that using the dot product resulted in the best features for the subsequent stimulus classification task. This observation can be explained by the fact that the same similarity measure is used throughout neural networks to compare inputs with learned patterns. If we construct a classifier by adding a fully connected layer to map the encoder output to the outputs (i.e. the stimulus label), this additional layer will learn a pattern for each output class and use the dot product to compare its input against these patterns. So the dot product appears like a natural choice. But still other similarity measures might, for instance, increase robustness.

Note that simply maximising the feature similarity for the paired trials derived in Section 13.4.2 would be an ill-posed learning problem. The optimum could easily be achieved by transforming all inputs into the same representation. This way, all trials would have maximum similarity – including those from different classes.[9] It would also be impractical to train the network with pre-defined target similarity values for all possible pairs of trials.

Instead, the network can be trained to encode relative similarity constraints. As introduced in [66], a relative similarity constraint (a, b, c) describes a relative comparison of the trials a, b and c in the form 'a is more similar to b than a is to c.' Here, a is the reference trial for the comparison. There exists a vast literature on using such constraints to learn similarity measures in various application domains. Here, we use them to define an alternative loss function for learning a feature encoding. To this end, we combine all pairs of trials (a, b) from the same class (as described in Section 13.4.2) with all trials c from other classes demanding that a and b are more similar.

[8] Distance measures like MSRE need to be transformed into a corresponding similarity measure.

[9] A way around this problem is to use a contrastive loss as in Siamese networks, which will be described in Section 13.4.5.

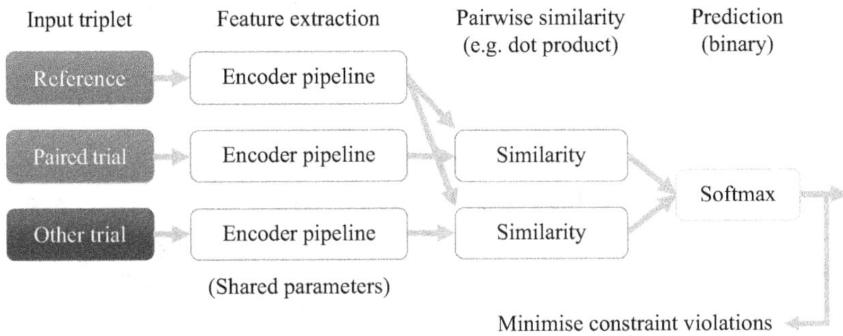

Figure 13.7 Processing scheme of a similarity-constraint encoder for triplets. Figure 13.3 shows an example of a simple encoder pipeline

Forming such trial triplets can quickly lead to a combinatorial 'explosion' of the training set. In this case, we can iterate only once over all pairs (a, b) in each epoch and draw the other trial c randomly from the different classes. This way, the training set differs in each iteration but still covers all pairs. For the computation of the training error, we add two layers to the network. The first one computes the pairwise similarity between the reference trial (at the first position in the triplet) and the other trials. The final layer directly applies the Softmax non-linearity to the two similarity scores (without affine transformation) producing a classification output. This is illustrated in Figure 13.7. The only trainable part is the shared encoder pipeline.

The network can be trained like a regular classifier trying to minimise the number of violated similarity constraints. When this value reaches zero, the different classes can be perfectly distinguished in the similarity space as the closest neighbours will belong to the same class. Each fulfilled constraint only makes a small local contribution to the global structure of the similarity space. This way, the network can more gradually adapt compared to a scenario where it would have to directly recognise the different classes based on a few training trials. Optionally, the triplets can be extended to tuples of higher order by adding more trials from other classes. This results in a gradually harder learning task because there are now more other trials to compare with. At the same time, each single training example comprises multiple similarity contains which might speed up learning.

The idea of hydra-nets can also be applied to similarity-constraint encoders (SCEs). However, both techniques already lead to an increase in computational complexity on their own. Therefore, their combination may easily become computationally infeasible for multi-layer encoder pipelines and thus should be carefully considered. The computational complexity also depends on the respective implementations and the deep-learning frameworks used. A detailed discussion is provided in the supplementary material of [64].

As pointed out in the following section, the SCE is closely related to two independently developed approaches.

13.4.5 Siamese networks and triplet networks

Siamese networks [67] and triplet networks [68] are two closely related techniques that can be applied alternatively to the SCE approach described above. Both methods process multiple input instances in parallel using identical encoder pipelines and then try to optimise distances in the embedding space such that instances belonging to the same class are closer to each other than instances from other classes. Like SCEs in their most basic form, triplet networks consider input triplets comprising a reference instance, a more similar instance (same class), and a less similar instance (different class). Here, the L2-norm (Euclidean distance) is applied in the embedding space – i.e. the focus is on learning a distance *metric* and a respective embedding into Euclidean space.

Siamese networks consider the *absolute* (Manhattan) distance between input pairs using the L1-norm as distance measure and target values of 0 and 1 for pairs belonging to the same or other classes respectively. A contrastive, energy-based loss is used that is derived from the discriminative learning framework for energy-based models such as restricted Boltzmann machines. The network is trained such that the energy is low for pairs from the same class and high for pairs of different classes. This leads to spring-like forces between the individual training examples that attract items from the same class and repel others. Additionally, there is a fixed margin around each item beyond which the repulsive force is no longer present. The size of this margin is a hyper-parameter that has to be tuned. An advantage of Siamese nets over triplet- and tuple-based approaches is the significantly lower computational cost of training. For any given dataset, there are much fewer pairs than triples (or higher order tuples) that can be derived. This, a lot less data has to be iterated over during training. Furthermore, the forward and backward propagation only needs to be computed for two input instances at a time instead of three or more. This suggests that Siamese nets should be used for more complex encoder pipelines and for exploration of suitable encoder pipeline networks where shorter training times are preferable.

Experimental results of systematically comparing SCEs, triplet nets and Siamese nets for the OpenMIIR stimulus classification using the perception trials have been presented in [65]. In the experiment, a simple linear SVC and a neural network with a single fully connected layer and Softmax output were used as classifiers. Features learned by the Siamese net disappointed in this task, whereas the triplet net was only slightly outperformed by the SCE (below significance). The confusion matrices for the SCE in combination with the two classifiers are shown in Figure 13.8. Investigating the difference between the SCE and triplet net encoder models, we found that correct predictions only overlapped by 70% which might be beneficial if they were combined. The encoder weights were much more stable throughout (cross-subject) cross-validation folds for the SCE. Further, the temporal patterns learned by a neural network classifier on top of the SCE output (visualised in Figure 13.9) turned out more crisply defined and sparse. We attribute this difference to our choice of the dot product as similarity measure. In a follow-up experiment, we also found that using

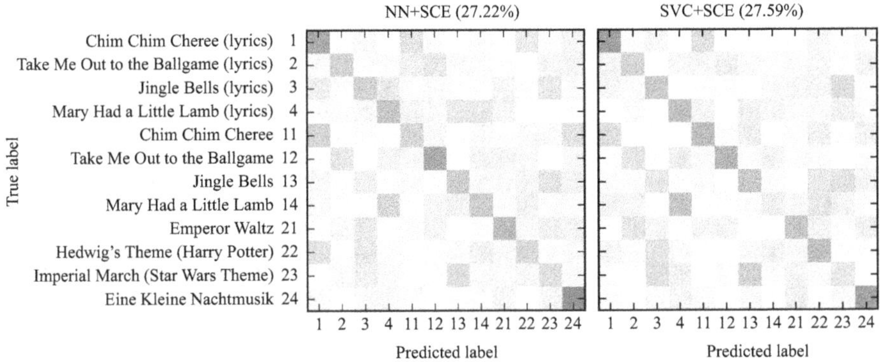

Figure 13.8 12-Class confusion matrices for the music stimuli (listed on the left) for classifiers trained on the SCE output. Middle: neural network classifier. Right: SVC. Results were aggregated over all (cross-subject) cross-validation folds

the dot product for the Siamese net drastically improved the quality of the learned features for this pre-training technique.

13.5 Interpreting trained models

Interpreting trained models and trying to understand what they have learned and where they fail is also commonly referred to as 'introspection'. Given the black-box nature of artificial neural networks, this is generally a challenge and a subject of still on-going research in deep learning. Introspection approaches for CNNs proposed so far mostly focus on visual data like images and videos [69]. Here, filters can be directly visualised and visual inspection is intuitive. Multi-channel EEG recordings can be represented as an image by using the channel axis as the second image dimension as in Figure 13.3 (left). However, the channel axis is problematic because the ordering does not imply physical proximity of neighbouring channels. Alternatively, each EEG channel can be represented as a time-frequency representation, for instance computed by a Fourier or wavelet transform. Patterns learned from such images by a CNN are generally much harder to interpret than natural image patterns. Using the transformation proposed in [52] results in images from the domain of topographic plots. In this well-studied domain, learned filters and resulting activations are potentially more meaningful. Visualisation techniques for RNNs on the other hand focus on symbolic data like characters, words or musical chords [70,71]. These approaches would need to be extended to continuous, multi-channel data to be applicable for EEG.

As a general design choice, we find that normalising input channels to $[-1,1]$ (i.e. dividing by the absolute channel maximum) in combination with using the hyperbolic

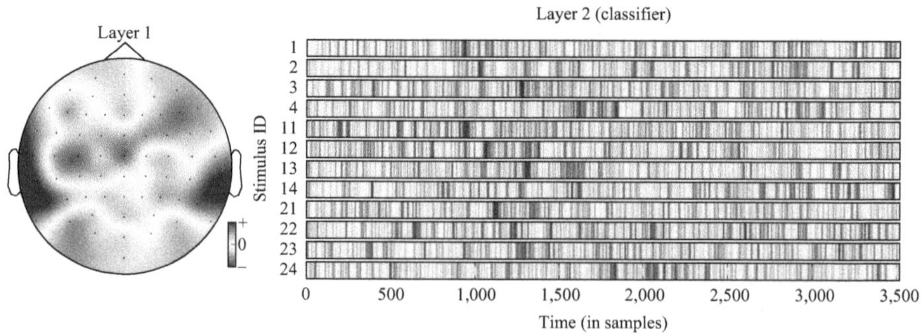

Figure 13.9 Visualisation of neural network parameters. Layer 1: a single spatial convolutional filter shown as topographic plot. Layer 2: temporal patterns for the 12 music stimuli

tangent as non-linear activation function keeps time-domain signals in a meaningful value range. If signals are converted into a time-frequency representation, activation functions commonly used in image processing like the rectified linear function can be used instead. Spatial or spatio-temporal filters that take the raw EEG channels as input can be visualised by topographic plots as shown in Figure 13.3 for filter width 1 and Figure 13.6 for filter width 3.

For the OpenMIIR dataset, we were mostly interested in spatial patterns in the filters that improve the signal-to-noise ratio as well as temporal patterns that correlate with stimulus properties. Figure 13.9 shows a visualisation example. These model parameters were averaged over the cross-validation folds in the experiment described in [65] with unsupervised pre-training of the convolutional filter as SCE and subsequent supervised training of the second (fully connected) layer for stimulus classification. The respective confusion matrix aggregated from the different cross-validation fold models is shown in Figure 13.8 (middle).

Following the advice of Haufe *et al.* [36] that model parameters for classification or decoding should not be directly interpreted in terms of the brain activity, we derived a forward model that explains how the measured signals were generated from the neural sources. To this end, we applied the regression approach proposed in [36] and trained a deconvolutional filter that reconstructs the original EEG signal from the encoder output by minimising the MSRE between the reconstructed and the actual signal over all trials. The resulting deconvolutional filter is shown in Figure 13.10.

Analysing the averaged temporal patterns visualised in Figure 13.9 shows similarities for the stimuli pairs with and without lyrics. This also explains the diagonal error patterns in the confusion matrices shown in Figure 13.8. A detailed analysis of the network layer activations as shown in Figure 13.11 reveals noticeable peaks in the encoder output and matching weights with high magnitude in the classifier layer that often coincide with downbeats – i.e. the first beat within each measure,

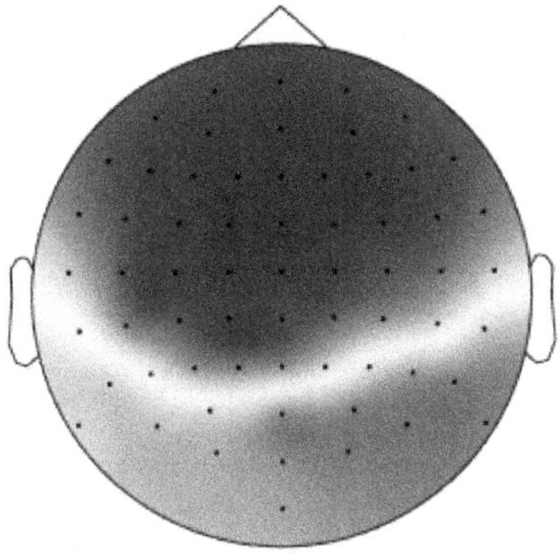

Figure 13.10 Visualisation of a forward model (deconvolutional filter) trained to reconstruct the originally recorded EEG signals from the encoder output over all perception trials

usually with special musical emphasis. These peaks are not visible in the channel-averaged EEG. This is a strong indicator that the encoder filter has successfully extracted a component from the EEG signal that contains musically meaningful information.

13.6 Conclusions

Successfully applying deep-learning techniques to decode brain activity recorded through EEG poses several challenges: data scarcity, complexity, and noisiness. Research in this application area is still in its infancy and there is still a long road ahead. However, we are certain that this road will eventually lead to success – in the form of new insights about the brain, new answers for old questions, new questions that spark further research, and maybe even completely new ways to design experiments and BCIs.

In the approaches, we described in this chapter, we focused on pre-training techniques because we are convinced that the right application of (unsupervised) pre-training is the key to successful EEG analysis with deep learning. So far, we have had success in improving the signal-to-noise ratio for the OpenMIIR music perception data – most of all through similarity-constraint encoding. There is still a lot of potential for improving the classification accuracy by using more complex encoders that possibly comprise multiple layers of neurons and recurrent connections. Investigating such options is one major direction of our on-going research efforts. Unfortunately, we have not made much progress with the imagination data. We hypothesise that tempo reproduction during imagination was not consistent enough to allow for the techniques applied so far to learn useful features. Different encoder pipelines with tolerance of temporal deviations might be the key for further progress.

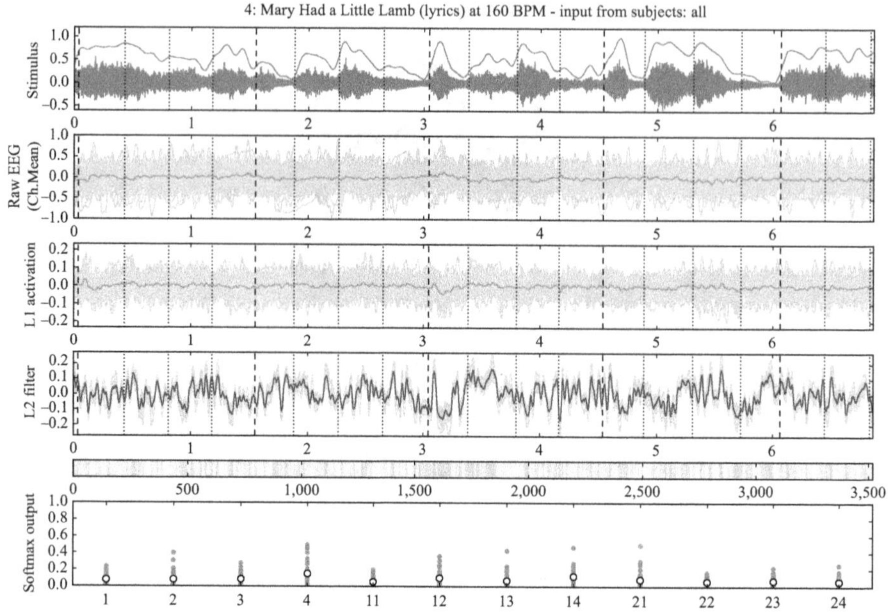

Figure 13.11 *Detailed analysis of all perception trials belonging to stimulus 4. Vertical marker lines indicate beats (dotted) and downbeats (dashed). The horizontal axis in rows 1–5 corresponds with the time in seconds or samples (row 5).* ***Top:*** *Audio stimulus (waveform) and envelope (line).* ***2nd row:*** *Raw EEG averaged over all 64 channels per trial (light grey) and overall mean (dark grey).* ***3rd row:*** *Encoder output (activation) for the individual trials (light grey) and overall mean (dark grey).* ***4th row:*** *Patterns learned by the neural network classifier for this class in the different cross-validation folds (light grey) and overall mean (dark grey).* ***5th row:*** *Alternative visualisation (as in Figure 13.9) of the averaged pattern from row 4.* ***Bottom:*** *Softmax output of the neural network classifier for the individual trials (grey) and overall mean (white) with class labels on the horizontal axis. All activations and outputs were generated using the respective test trials for each fold model in the cross-validation*

We hope that sharing our experiences so far helps others to succeed and inspires more researchers from the field of deep learning to try their methods and bring new ideas to this emerging interdisciplinary field. At the same time, we would like to encourage researchers who conduct EEG and BCI experiments to share their datasets and engage in an exchange with the deep-learning community. As a small step towards more collaboration and further scientific progress, our OpenMIIR dataset and all our code for the described techniques is available to everybody interested without any restrictions.

References

[1] J. S. Snyder and E. W. Large. Gamma-band activity reflects the metric structure of rhythmic tone sequences. *Cognitive Brain Research*, 24:117–126, 2005.

[2] L. K. Cirelli, D. Bosnyak, F. C. Manning, *et al.* Beat-induced fluctuations in auditory cortical beta-band activity: using EEG to measure age-related changes. *Frontiers in Psychology*, 5(Jul.):1–9, 2014.

[3] H. Merchant, J. Grahn, L. J. Trainor, M. Rohrmeier, and W. T. Fitch. Finding a beat: a neural perspective across humans and non-human primates. *Philosophical Transactions of the Royal Society B: Biological Sciences*, 370(1664), 20140093, 2015.

[4] J. R. Iversen, B. H. Repp, and A. D. Patel. Top-down control of rhythm perception modulates early auditory responses. *Annals of the New York Academy of Sciences*, 1169:58–73, 2009.

[5] T. Fujioka, L. J. Trainor, E. W. Large, and B. Ross. Beta and gamma rhythms in human auditory cortex during musical beat processing. *Annals of the New York Academy of Sciences*, 1169:89–92, 2009.

[6] T. Fujioka, L. J. Trainor, E. W. Large, and B. Ross. Internalized timing of isochronous sounds is represented in neuromagnetic beta oscillations. *Journal of Neuroscience*, 32(5):1791–1802, 2012.

[7] S. Nozaradan, I. Peretz, M. Missal, and A. Mouraux. Tagging the neuronal entrainment to beat and meter. *The Journal of Neuroscience*, 31(28):10234–10240, 2011.

[8] S. Nozaradan, I. Peretz, and A. Mouraux. Selective neuronal entrainment to the beat and meter embedded in a musical rhythm. *The Journal of Neuroscience*, 32(49):17572–17581, 2012.

[9] E. Geiser, E. Ziegler, L. Jancke, and M. Meyer. Early electrophysiological correlates of meter and rhythm processing in music perception. *Cortex*, 45(1):93–102, 2009.

[10] R. J. Vlek, R. S. Schaefer, C. C. A. M. Gielen, J. D. R. Farquhar, and P. Desain. Shared mechanisms in perception and imagery of auditory accents. *Clinical Neurophysiology*, 122(8):1526–1532, 2011.

[11] R. S. Schaefer, J. Farquhar, Y. Blokland, M. Sadakata, and P. Desain. Name that tune: decoding music from the listening brain. *NeuroImage*, 56(2):843–849, 2011.

[12] S. Herholz, A. Halpern, and R. Zatorre. Neuronal correlates of perception, imagery, and memory for familiar tunes. *Journal of Cognitive Neuroscience*, 24(6):1382–1397, 2012.

[13] A. R. Halpern, R. J. Zatorre, M. Bouffard, and J. A. Johnson. Behavioral and neural correlates of perceived and imagined musical timbre. *Neuropsychologia*, 42(9):1281–92, 2004.

[14] T. L. Hubbard. Auditory imagery: empirical findings. *Psychological Bulletin*, 136(2):302–329, Mar. 2010.

[15] R. S. Schaefer. *Measuring the mind's ear: EEG of music imagery*. PhD thesis, Radboud University Nijmegen, 2011.

[16] R. S. Schaefer, P. Desain, and J. Farquhar. Shared processing of perception and imagery of music in decomposed EEG. *NeuroImage*, 70:317–326, 2013.

[17] R. S. Schaefer, Y. Blokland, J. Farquhar, and P. Desain. Single trial classification of perceived and imagined music from EEG. In *Proceedings of the 2009 Berlin BCI Workshop*, 2009.

[18] S. Deng, R. Srinivasan, and M. D'Zmura. Cortical signatures of heard and imagined speech envelopes. Technical report, DTIC, 2013.

[19] J. A. O'Sullivan, A. J. Power, N. Mesgarani, *et al.* Attentional selection in a cocktail party environment can be decoded from single-trial EEG. *Cerebral Cortex*, 25(7):1697–1706, 2014.

[20] D. S. Klobassa, T. M. Vaughan, P. Brunner, N. E. Schwartz, J. R. Wolpaw, C. Neuper, and E. W. Sellers. Toward a high-throughput auditory P300-based brain–computer interface. *Clinical Neurophysiology*, 120(7):1252–1261, 2009.

[21] A. R. Murguialday, J. Hill, M. Bensch, *et al.* Transition from the locked in to the completely locked-in state: a physiological analysis. *Clinical Neurophysiology*, 122(5):925–933, 2011.

[22] A. Furdea, S. Halder, D. J. Krusienski, *et al.* An auditory oddball (P300) spelling system for brain–computer interfaces. *Psychophysiology*, 46(3):617–625, 2009.

[23] J. Höhne, M. Schreuder, B. Blankertz, and M. Tangermann. A novel 9-class auditory ERP paradigm driving a predictive text entry system. *Neuroprosthetics*, 5:99, 2011.

[24] M. Schreuder, B. Blankertz, and M. Tangermann. A new auditory multi-class brain–computer interface paradigm: spatial hearing as an informative cue. *PLoS ONE*, 5(4):e9813, 2010.

[25] S. Halder, M. Rea, R. Andreoni, *et al.* An auditory oddball brain–computer interface for binary choices. *Clinical Neurophysiology*, 121(4):516–523, 2010.

[26] N. J. Hill, T. N. Lal, K. Bierig, N. Birbaumer, and B. Schölkopf. An auditory paradigm for brain–computer interfaces. In *Advances in Neural Information Processing Systems (NIPS)*, pages 569–576, 2004.

[27] D.-W. Kim, H.-J. Hwang, J.-H. Lim, Y.-H. Lee, K.-Y. Jung, and C.-H. Im. Classification of selective attention to auditory stimuli: toward vision-free brain–computer interfacing. *Journal of Neuroscience Methods*, 197(1):180–185, 2011.

[28] N. J. Hill and B. Schölkopf. An online brain–computer interface based on shifting attention to concurrent streams of auditory stimuli. *Journal of Neural Engineering*, 9(2):026011, Apr. 2012.

[29] J. Höhne, K. Krenzlin, S. Dähne, and M. Tangermann. Natural stimuli improve auditory BCIs with respect to ergonomics and performance. *Journal of Neural Engineering*, 9(4):045003, 2012.

[30] M. S. Treder, H. Purwins, D. Miklody, I. Sturm, and B. Blankertz. The musical BCI: control by selective listening. In *Asilomar International BCI Meeting*, 2013.

[31] M. S. Treder, H. Purwins, D. Miklody, I. Sturm, and B. Blankertz. Decoding auditory attention to instruments in polyphonic music using single-trial EEG classification. *Journal of Neural Engineering*, 11(2):026009, Apr. 2014.

[32] E. R. Miranda. Brain–computer music interfacing: interdisciplinary research at the crossroads of music, science and biomedical engineering. In *Guide to Brain–Computer Music Interfacing*, pages 1–27. Springer, London, 2014.

[33] J. Deng, W. Dong, R. Socher, L.-J. Li, K. Li, and L. Fei-Fei. ImageNet: a large-scale hierarchical image database. In *IEEE Conference on Computer Vision and Pattern Recognition (CVPR'09)*, pages 248–255, 2009.

[34] P. Mirowski, D. Madhavan, Y. LeCun, and R. Kuzniecky. Classification of patterns of EEG synchronization for seizure prediction. *Clinical Neurophysiology*, 120(11):1927–1940, 2009.

[35] Y. LeCun, L. Bottou, Y. Bengio, and P. Haffner. Gradient-based learning applied to document recognition. *Proceedings of the IEEE*, 86(11):2278–2324, 1998.

[36] S. Haufe, F. Meinecke, K. Görgen, *et al.* On the interpretation of weight vectors of linear models in multivariate neuroimaging. *NeuroImage*, 87:96–110, 2014.

[37] D. Wulsin, J. Gupta, R. Mani, J. Blanco, and B. Litt. Modeling electroencephalography waveforms with semi-supervised deep belief nets: fast classification and anomaly measurement. *Journal of Neural Engineering*, 8(3), 2011.

[38] M. Längkvist, L. Karlsson, and M. Loutfi. Sleep stage classification using unsupervised feature learning. *Advances in Artificial Neural Systems*, 2012, Article ID 107046, 2012.

[39] Q. Lin, S.-q. Ye, X.-m. Huang, *et al.* Classification of epileptic EEG signals with stacked sparse autoencoder based on deep learning. In *Intelligent Computing Methodologies*, Lecture Notes in Computer Science, pages 802–810. Springer, Cham, Aug. 2016.

[40] S. Stober, D. Cameron, and J. Grahn. Classifying EEG recordings of rhythm perception. In *15th International Society for Music Information Retrieval Conference (ISMIR)*, pages 649–654, 2014.

[41] S. Stober, D. J. Cameron, and J. A. Grahn. Using convolutional neural networks to recognize rhythm stimuli from electroencephalography recordings. In *Advances in Neural Information Processing Systems 27 (NIPS'14)*, pages 1449–1457, 2014.

[42] Y. R. Tabar and U. Halici. A novel deep learning approach for classification of EEG motor imagery signals. *Journal of Neural Engineering*, 14(1):016003, 2017.

[43] H. Cecotti and A. Gräser. Convolutional neural networks for P300 detection with application to brain–computer interfaces. *IEEE Transactions on Pattern Analysis and Machine Intelligence*, 33(3):433–445, 2011.

[44] H. Cecotti and A. Gräser. Neural network pruning for feature selection-application to a p300 brain–computer interface. In *ESANN*, 2009.

[45] E. S. Nurse, P. J. Karoly, D. B. Grayden, and D. R. Freestone. A generalizable brain–computer interface (BCI) using machine learning for feature discovery. *PLoS ONE*, 10(6):e0131328, Jun. 2015.

[46] E. Nurse, B. S. Mashford, A. J. Yepes, I. Kiral-Kornek, S. Harrer, and D. R. Freestone. Decoding EEG and LFP signals using deep learning: heading TrueNorth. In *Proceedings of the ACM International Conference on Computing Frontiers*, CF '16, pages 259–266. ACM, New York, NY, USA, 2016.

[47] X. An, D. Kuang, X. Guo, Y. Zhao, and L. He. A deep learning method for classification of EEG data based on motor imagery. In *International Conference on Intelligent Computing*, pages 203–210. Springer, 2014.

[48] S. Jirayucharoensak, S. Pan-Ngum, and P. Israsena. EEG-based emotion recognition using deep learning network with principal component based covariate shift adaptation. *The Scientific World Journal*, 2014, Article ID 627892, 2014.

[49] W.-L. Zheng, J.-Y. Zhu, Y. Peng, and B.-L. Lu. EEG-based emotion classification using deep belief networks. In *2014 IEEE International Conference on Multimedia and Expo (ICME)*, pages 1–6, 2014.

[50] H. Cecotti and A. Gräser. Convolutional neural network with embedded Fourier transform for EEG classification. In *19th International Conference on Pattern Recognition (ICPR)*, pages 1–4, 2008.

[51] R. J. Williams. Simple statistical gradient-following algorithms for connectionist reinforcement learning. *Machine Learning*, 8(3–4):229–256, 1992.

[52] P. Bashivan, I. Rish, M. Yeasin, and N. Codella. Learning representations from EEG with deep recurrent-convolutional neural networks. arXiv preprint arXiv:1511.06448, 2015.

[53] L. Jingwei, C. Yin, and Z. Weidong. Deep learning EEG response representation for brain computer interface. In *2015 34th Chinese Control Conference (CCC)*, pages 3518–3523, Jul. 2015.

[54] B. L. Sturm. A simple method to determine if a music information retrieval system is a "horse". *IEEE Transactions on Multimedia*, 16(6):1636–1644, Oct. 2014.

[55] A. Petrosian, D. Prokhorov, R. Homan, R. Dasheiff, and D. Wunsch. Recurrent neural network based prediction of epileptic seizures in intra- and extracranial EEG. *Neurocomputing*, 30(1):201–218, 2000.

[56] S. Stober, A. Sternin, A. M. Owen, and J. A. Grahn. Towards music imagery information retrieval: Introducing the OpenMIIR dataset of EEG recordings from music perception and imagination. In *16th International Society for Music Information Retrieval Conference (ISMIR'15)*, pages 763–769, 2015.

[57] A. Gramfort, M. Luessi, E. Larson, *et al.* MEG and EEG data analysis with MNE-Python. *Frontiers in Neuroscience*, 7:267, 2013.

[58] F. Perrin, J. Pernier, O. Bertrand, and J. F. Echallier. Spherical splines for scalp potential and current density mapping. *Electroencephalography and Clinical Neurophysiology*, 72(2):184–187, 1989.

[59] T.-W. Lee, M. Girolami, and T. J. Sejnowski. Independent component analysis using an extended infomax algorithm for mixed subgaussian and supergaussian sources. *Neural Computation*, 11(2):417–441, 1999.

[60] J. Masci, U. Meier, D. Cireşan, and J. Schmidhuber. Stacked convolutional auto-encoders for hierarchical feature extraction. In *Artificial Neural Networks and Machine Learning–ICANN 2011*, pages 52–59. Springer, 2011.

[61] H. Bourlard and Y. Kamp. Auto-association by multilayer perceptrons and singular value decomposition. *Biological Cybernetics*, 59(4–5):291–294, 1988.

[62] X. Glorot, A. Bordes, and Y. Bengio. Deep sparse rectifier networks. In *NIPS 2010 Workshop on Deep Learning and Unsupervised Feature Learning*, 2010.

[63] W. Wang, Y. Huang, Y. Wang, and L. Wang. Generalized autoencoder: a neural network framework for dimensionality reduction. In *2014 IEEE Conference on Computer Vision and Pattern Recognition Workshops (CVPRW)*, pages 496–503, 2014.

[64] S. Stober, A. Sternin, A. M. Owen, and J. A. Grahn. Deep feature learning for EEG recordings. *arXiv preprint arXiv:1511.04306*, 2015.

[65] S. Stober. Learning discriminative features from electroencephalography recordings by encoding similarity constraints. In *Proceedings of 42nd IEEE International Conference on Acoustics, Speech and Signal Processing (ICASSP'17)*, pages 6175–6179, 2017.

[66] M. Schultz and T. Joachims. Learning a distance metric from relative comparisons. In *Advances in Neural Information Processing Systems (NIPS)*, pages 41–48, 2004.

[67] S. Chopra, R. Hadsell, and Y. LeCun. Learning a similarity metric discriminatively, with application to face verification. In *Computer Vision and Pattern Recognition, 2005. CVPR 2005. IEEE Computer Society Conference on*, vol. 1, pages 539–546. IEEE, 2005.

[68] E. Hoffer and N. Ailon. Deep metric learning using triplet network. In International Workshop on Similarity-Based Pattern Recognition, pages 84–92, 2015.

[69] J. Yosinski, J. Clune, A. Nguyen, T. Fuchs, and H. Lipson. Understanding neural networks through deep visualization. arXiv preprint arXiv:1506.06579, 2015.

[70] A. Karpathy, J. Johnson, and L. Fei-Fei. Visualizing and understanding recurrent networks. arXiv preprint arXiv:1506.02078, 2015.

[71] H. Strobelt, S. Gehrmann, B. Huber, H. Pfister, and A. M. Rush. Visual analysis of hidden state dynamics in recurrent neural networks. arXiv preprint arXiv:1606.07461, 2016.

Chapter 14

Neurofeedback games using EEG-based brain–computer interface technology

A.P. Vinod[1] *and Kavitha P. Thomas*[2]

Abstract

Brain–computer interface (BCI) is relatively a new approach to communication between man and machine, which translates brain activity into commands for communication and control. As BCI is capable of detecting human intentions, it is a promising communication tool for paralyzed patients for communicating with external world. Many of the current BCI systems employ electroencephalogram (EEG) which is the most widely used noninvasive brain activity recording technique. EEG signal carries potential features to identify and decode human intentions and mental tasks. Recently, many researchers have started exploiting the possibilities of BCI in entertainment and cognitive skill enhancement. BCI-based games have been identified as a unique entertainment mechanism nowadays, "controlling a 2-D, 3-D or virtual computer game solely by player's brain waves." BCI games work based on a neurofeedback paradigm which allows an individual to self-regulate his brain signal in response to the real-time visual or auditory feedback of his brain waves/features. This neurofeedback in a gaming environment motivates and trains the players to control his brain features toward the desired stage (self-regulation).

This chapter explores the state-of-the-art BCI technology in neurofeedback games, employing EEG signal. It also provides a survey of the existing EEG-based neurofeedback games and evaluates their success rates, challenging factors and influence on players. In neurofeedback games, a number of features extracted from EEG accompanied with sustained attention, selective attention, visuospatial attention, motor imagery, eye movements, etc. have been employed as distinct control signals. We will briefly review and compare various signal processing methodologies and machine-learning techniques employed in those studies to extract and decode the brain features.

Besides the structure and algorithms used in neurofeedback games, the therapeutic effects of neurofeedback training and its capabilities for the enhancement of

[1]Department of Electrical Engineering, Indian Institute of Technology, Palakkad, India
[2]School of Computer Science and Engineering, Nanyang Technological University, Singapore

cognitive skills will also be briefly discussed in this chapter. Neurofeedback training helps to rewire brain's underlying neural circuits and to improve brain functions. Therefore, it is considered as an effective tool for boosting cognitive skills of both healthy and the disabled. Specifically, neurofeedback has been considered as an efficient treatment modality for individuals with attention-deficit hyperactive disorder (ADHD). ADHD is characterized by three behavioral symptoms: inattention, hyperactivity and impulsivity. Along with the conventional intervention strategies such as medication, behavioral treatments, etc., neurofeedback in BCI games has also been emerging as a promising modality for treating the attention deficit.

We will also discuss portable and economical EEG recording devices currently employed in BCI-based brain training modules/games. Finally, the chapter will be concluded with a brief overview of the neurofeedback developments in the context of BCI-based games until now, their potential impact on the healthy as well as on people with neurological disorders, challenges in transferring the successful protocols from laboratories into the market and hurdles in real-time BCI system design and development.

14.1 Introduction

Since Hans Berger did the first electroencephalogram (EEG) recording on humans in 1924 [1], there has been much notional theorization about the feasibility of reading human mind utilizing EEG signals [2]. However, until Vidal [3] detected EEG events in authentic time during 1973, there has not been much research endeavors to exploit EEG patterns for multidisciplinary applications rather than clinical examinations performed with avail of EEG. But later, after 1990s, substantial research activity has commenced in this field [4], aimed at EEG signal acquisition and processing for medical and nonmedical applications. Though many noninvasive brain signal acquisition techniques are available for brain signal measurements, EEG is the most popular, due to its good temporal resolution, availability, low cost and portability. Recent findings in EEG-based neuroscience research offer tremendous contribution in developing a new technology named brain–computer interface (BCI), which establishes direct brain-to-machine communication system, bypassing brain's normal communication pathway of nerves and muscles [5]. Utilizing BCI technology, an external contrivance or computer can be controlled by thoughts. It is a potential candidate to allow locked-in patients for conveying messages to caregivers and offers revolutionary changes in development of novel rehabilitation schemes and assistive devices for the disabled [6].

Apart from the applications in medical field, fortified by the fascinating EEG-predicated BCI technology, area of neuroengineering nowadays witnesses the widespread interest in "mind-controlled" games, in healthy as well as in people with special needs who require rehabilitation program for neuronal enhancement [6–10]. BCI-driven games are capable of employing brain waves or their modulated parameters for controlling game parameters, superseding the utilization of keyboard, mouse or joystick in traditional video games, partially or plenarily. While controlling game environments with brain activations, BCI games eventually work in a neurofeedback framework that sanctions players to learn and control EEG waves/associated parameters voluntarily, on account of the received visual, auditory or tactile feedback. It is an operant conditioning approach where the feedback incentivizes the player to self-regulate his brain activity to achieve the required goal set in the game [8]. Neurofeedback games help to rewire brain's underlying neural circuits and to ameliorate brain functions [9]. It is considered as an efficacious tool for enhancing cognitive skills for the healthy as well as the neurologically incapacitated. On account of the potential of neurofeedback systems to fine tune brain capabilities, it can be considered as a good remedial modality for treating people with attention-deficit hyperactive disorder (ADHD) [11–13].

This chapter explores state-of-the-art BCI technology in neurofeedback games. First, we provide the general framework employed in neurofeedback game with BCI technology in Section 14.2. Based on the BCI interaction involved in the neurofeedback framework, the games are mainly categorized into four groups, and they are discussed in Section 14.3 including a brief survey of the existing EEG-based neurofeedback games, evaluating the gaming interfaces, signal parameters and modulating strategies, prosperity rates, challenging factors, the structure and algorithms utilized. EEG recording devices used in these systems are discussed in Section 14.4. Advantages of EEG-based neurofeedback games in the healthy and the neurologically challenged are discussed in Section 14.5. A discussion on the hurdles in transferring successful gaming protocols from laboratory to market is provided in Section 14.6. Section 14.7 concludes the chapter.

14.2 Generic framework of a neurofeedback game using BCI technology

Many research groups have begun concentrating on brain-based casual and serious gaming systems, seeing their significance in either therapeutic or nonmedical applications [14–16]. All these systems share a generic closed-loop framework but vary in control paradigms and feedback modes. In this section, the general structure of a BCI game and different types of control frameworks adapted in BCI games are presented.

Much like any BCI system, neurofeedback game establishes a closed-loop structure, composed of four basic building units, responsible for "data acquisition," "data processing," "control signal generation" and "gaming interface" as shown in Figure 14.1. All these four modules work in a synchronized manner to decode player's state of mind and to translate the identified thoughts into target game controls. This continuous interaction between the user and interface goes on, maintaining a robust interlink between the four modules, till the system terminates the application. The important functions of the various modules are described here.

14.2.1 Data acquisition

Brain activities associated with human intentions are captured using EEG. It is measured by placing electrodes on the scalp according to international standard system

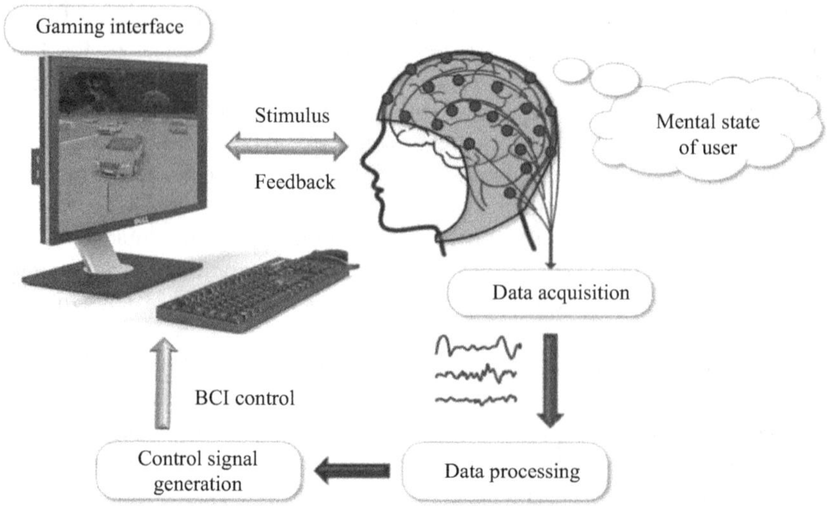

Figure 14.1 Generic closed-loop framework of the neurofeedback game

of electrode placement. It picks up the electrical activity of the brain which contains information that can be used to identify the intention of the BCI user.

14.2.2 Data processing

Artifacts due to eye movements, muscular movements, respiration, etc. affect the quality of recorded EEG. Its signal-to-noise ratio has to be improved to find out the embedded mental task, using robust signal processing methods. Besides, EEG has very low spatial resolution, due to volume conduction effect. In order to extract the most distinct subject-independent informative content from the non-stationary signal such as EEG, robust adaptive algorithms need to be employed. The extracted features can be used for generating unique command signal to control game. The data processing module performs all these tasks.

14.2.3 Control signal generation

It is the most vital stage in the design of an attractive BCI game as it has a significant impact on engaging the players in the gaming environment. In ordinary gaming interfaces, a player may enter a keyboard input, push a specific button in game console or move mouse in order to pass the informative data to the system. Instead of that, BCI-based game requires a "controller" to integrate all the modules in the neurofeedback system. For instance, the user may perform a mental task or focus on a specific part on the interface that creates unique brain event that incorporates user's expectation. In some BCI interactions, the player's purposeful intervention is not necessary; rather, the framework learns and identifies the user's mental states adaptively. These recognized information sources must be robustly classified using efficient machine-learning computations. At that point, these predicted tasks must be transformed into meaningful command signals to control gaming environment. For identification of mental state, threshold-based detection schemes, linear and non-linear discriminant methods, support vector machine tools, artificial neural network, etc. are customarily used [17–20].

14.2.4 Gaming interface

Since 2000, there have been astounding developments in gaming technology and its positive impact has been reflected in BCI applications too [14]. Evolution of virtual environments and advances in the design tools of 2-D or 3-D gaming interfaces has highly motivated the design of innovative BCI games controlled by various EEG attributes [18]. An immersive or engaging gaming environment is a key component in the successful design of any neurofeedback game. The gaming interface can be 2-D, 3-D or incorporated with virtual reality (VR). Based on the interface design and requirements, the outputs of the control generator are transformed meaningfully into game control commands. The player gets the online feedback of his brain's status from the computer screen or from the application scenario in visual or auditory mode, specific to the gaming system, and learns to self-regulate EEG parameters to gain optimum game scores.

14.3 Classification of neurofeedback games based on BCI interaction

Based on the types of BCI control paradigms adopted in the neurofeedback framework, BCI games are broadly classified into four categories, namely **active**, **reactive**, **passive** and **hybrid** [20–26]. Each of these is briefly described in this section.

14.3.1 Active BCI games

Here, the player responds to a particular stimulus to self-regulate his/her EEG signal and generates the required command signal. Brain signals have to be consciously generated by the player, in order to achieve the successful control in active BCI framework. Active games explore a number of tasks, of which EEG patterns associated with motor movements or motor imagery are the most commonly used. Specific neuronal activity occurs when a subject voluntarily imagine or make a motor movement, for example moving the right hand. The neuronal activity that occurs during motor imagery is spatiotemporally similar to the activity generated during a real movement. The characteristics used to quantify these EEG activities are extracted and subsequently classified to discriminate between various types of real or imagined movements, mapping each activity to a particular control signal. Usually, motor execution or imagery is associated with a contralateral power decrease and ipsilateral power increase in mu band (8–12 Hz) named event-related desynchronization and synchronization (ERD/ERS) over the motor cortex part of brain. Often, it is accompanied with a power increase in beta band (12–30 Hz) also, over the motor cortex.

Based on the body part associated with, the temporal and spectral characteristics of these patterns vary. However, clear distinction of EEG signals in specific frequency bands and channels occurring during various motor-related tasks enhances their suitability in generation of distinct game controls [4,5].

The following games are built up on the control parameters associated with motor intentions.

14.3.1.1 Shooter game [19]

A first-person shooter, controlled by EEG patterns associated with imagined motor movements, is presented in [19]. Player has to selectively perform left or right movement in the 3-D video game displayed in front of him using brain waves. To perform left movement in the gaming scenario, person has to focus on the screen and imagine rotating the environment in counter clockwise direction. For rightward movement, clockwise rotation imagination is required. EEG is recorded using a headband with sensors located at C3 and C4 according the 10–20 international system of electrode placement. These electrodes are over the most reactive brain regions during motor imagery. From the bipolar voltage C3–C4, a relative "mu" power is computed using a moving fast Fourier transform method. This differential mu power acts as the control parameter. If both electrodes carry similar power levels, the difference is low and the system considers it as "low mu," whereas if the differential power is higher, it is "high mu." These low and high mu values are mapped into game control commands to move a scenery toward leftward or rightward direction, respectively. Ten healthy subjects learned to control their mu rhythms according to the required paradigm after training for a few weeks and played the game successfully.

14.3.1.2 Pacman game [20]

A movement controlled BCI Pacman game is presented in [20] where the game detects intention of player's motion based on a lateralized readiness potential (LRP). LRP is a gradual negative shift within the EEG that appears over the activated motor cortex beginning some time before the real movement onset. The reported gaming interface contains a random labyrinth, displaying a single shortest path from entry (on left wall) to exit (in right wall), indicated with gray track marks. At every 1.5–2 s, Pacman makes one step forward till it finds a wall. Thus, speed of the Pacman is fixed, but direction of its motion controlled LRP brain oscillations. When Pacman faces a wall, it has to receive a command to turn either left or right. The LRP detection classifier identifies the player's intended direction from EEG and subsequently makes Pacman move. If the player moves forward with correct step in required direction, he will be able to collect fruits (rewards). But if Pacman bashes the wall due to incorrect step taken, player will not be able to collect fruit. He will lose credits too. The study reports that though a healthy subject is able to navigate "Pacman" within 40 s using traditional keyboard inputs, the brain-control integrated in the Pacman interface provides an extra fun factor in the game and helps players to seriously engage in the game. It offers a great feeling of brain's dominance in the game, even before the player becomes consciously aware of his decisions. The LRP classifier correctly identifies the readiness of players and acts accordingly. In this work, multichannel EEG recordings over primary motor

14.3.1.7 Spaceship game [31]

The game reported in [31], a 2-D interface controlled by left/right motor imagery, makes use of cognitive skills of players too during the game play. Using left or right motor imagery, player should safely position a spaceship, avoiding asteroids wisely. Besides the presence of asteroids, other distractor objects like stars will also be present during the game play, which will prompt the players to take wrong decisions. They will look like objects to be avoided, but the player has to do a careful selection on the items to be eliminated and to perform motor imagery. The work compares a number of signal processing and machine-learning techniques that can be used to tackle the main challenges in achieving continuous control of synchronous and asynchronous BCIs. Continuous game control in the designed interface is reported using features from multiband EEG (8–28 Hz) signals recorded from a set of three EEG channels (C3, Cz and C4) and they are classified with pattern recognition tools. Three subjects learned to play the game successfully and it is reported that the addition of cognitive attribute in the game makes it more challenging and engaging for the players.

14.3.1.8 Car racing game [32]

The game presented in [32] is a 3-D car racing game, named TORCS, controlled by players' motor imagery patterns associated with left hand, right hand and both hands, employing a consumer-grade EEG equipment named Emotiv EPOC. The racing simulator in this work is designed such that steering wheel, acceleration and brake values of car can be controlled smoothly by parameters obtained from autoregressive transformation of EEG signals. The game simulator also shows a shared vehicle control system which can provide quick intention recognition from EEG than from more distinct electromyography signals to apply brake on a moving car in a simulated environment. Fast recognition of intention is preferred in racing and shooter games to achieve best gaming experience, and further research may provide better insights on how to improve intention detection in games.

The ERD/S patterns associated with motor imagery in active BCIs need long training time and offer only low information transfer rate (ITR). ITR, expressed in bits per minute, is a popular metric for evaluating the performance of any BCI-based communication system [10]. It is computed as

$$\mathrm{ITR} = \frac{1}{T}\left[\log_2 N + P\log_2 P + (1-P)\log_2\left[\frac{1-P}{N-1}\right]\right] \tag{14.1}$$

where N is the number of classes/targets in the closed-loop system, P is the probability of correct target selection (also refers to target identification accuracy or classification accuracy of the closed-loop system) and T is the time taken in minutes for a single target detection. The higher the ITR, the better the performance of any BCI system. However, in active BCIs, on account of the limited availability of control commands and the challenges in extracting accurate and distinct information to distinguish various mental tasks, the achieved ITR can be low. As active BCI interaction allows output control independent of stimulus onset, asynchronous mode of BCI operation is also possible. The player can change his intentions on the go and can have a better game play experience. However, in order to achieve proper communication and control in the self-paced BCI, robust algorithms are essential to detect the intentions of players and to classify them accurately. Neurofeedback games equipped with fast control strategies and high ITRs are expected to be more popular among BCI users. Algorithms and techniques to address the subject-specificity of BCI interaction, non-stationarity of ERD/ERS patterns during motor imagery, difficulty in detecting the onset motor imagery intention in the absence of stimulus, etc. are essential to create asynchronous active BCIs with high ITR.

14.3.2 Reactive BCI games

This type of BCI uses brain patterns in response to the specific kind of stimulus presented in the gaming interface. In this paradigm, the player must focus on the interface and attend to the stimuli involved in the game to elicit stimulus-dependent brain patterns. Reactive BCIs mostly rely on visually evoked potential (VEP) in EEG that occurs based on external visual stimuli.

Steady-state visually evoked potential (SSVEP), the most commonly deployed VEP, contains exogenous responses of the brain toward visual stimulations at specific frequencies [21,23]. Repetitive stimulation by flashing of light creates constant amplitude and frequency modulations in EEG. If a stimulus flickers at 5 Hz, EEG responses are visible at frequencies of 5 Hz and its harmonics, i.e. 10, 15 Hz, etc. Another VEP used in BCI is P300, which compared to SSVEP, is an "endogenous" signal because it generates VEP component that depends on stimulus context and of player's attention. It is generated in oddball paradigms, during which a target stimulus is detected in a series of frequent irrelevant stimuli. Such oddball paradigms reliably yield P300 responses with a parieto-central scalp distribution to target compared to standard stimuli irrespective of stimulus (visual, auditory, somatosensory). It is reflected as a positive peak in EEG after 300 ms of stimulus onset. A number of games based on VEPs are discussed in this section.

14.3.2.1 Mind balance game [21]

It is an SSVEP-based game with the objective to attain one-dimensional control of a character's balance (located on a tightrope) based on EEG features associated with visual attention. The game places two checkerboards both sides of the character, and they are phase reversed at 17 and 20 Hz. The animated character is designed such that it makes random movements while walking on the rope. In order to keep the avatar safe on the rope, player must focus selectively on the required checkerboard to maintain the balance. If the user fails to focus correctly, the character will fall. Player's focused direction is identified using the subject-specific model developed during calibration stage. During that phase, player has to focus on the two checker boards for 15 s each. The system uses this recorded data to calibrate the BCI model and classify signals during online mode. Six subjects have demonstrated reliable control in mind-balance game achieving an average accuracy of 89.5%, with a bit rate of 10.3 bits/min.

14.3.2.2 Car racing game [22]

An SSVEP-based BCI game that allows users to navigate a car around a race track has been reported in [22]. In this game, BCI facilitates the real-time navigation of a car in four directions by fixing player's attention on the checkerboard placed either UP, DOWN, RIGHT and LEFT side of the car. The users are asked to attend to different directions in the required manner. The system is validated on data from five players and for two different ranges of flicker frequencies: low-frequency range: 5–8 Hz (up—5 Hz, left—6 Hz, down—7 Hz and right—8 Hz) and medium-frequency range: 12–17 Hz (up—12 Hz, left—13.3 Hz, down—15 Hz and right—17 Hz). A fully adaptive BCI system is developed and the data are recorded from six electrodes CPz, Pz, POz, P1, P2 and Fz, at a sampling rate of 256 Hz. The data processing involves a real-time blind source separation technique that performs artifact removal, a number of bandpass filters with narrow band width to extract the SSVEP and its harmonics, and an adaptive fuzzy classifier. For each subject, the optimal parameters are identified in the initial training mode and the parameters selected are used in the unsupervised free racing mode. The player's response for each event is found to reach a stable and steady state during the interval 3–6 s after the command request. Comparing the performance of SSVEP-BCI for both medium-frequency and low-frequency range, the latter offers better controllability and faster execution.

14.3.2.3 Tower protection game [33]

An SSVEP-BCI-based game using consumer grade EEG hardware, Emotiv EPOC, is proposed in [33]. The goal is to implement a BCI that generates explicit commands to control and operate a complicated, tactical game, using a cheaper EEG acquisition modality. The game is another version in the "tower defense" genre. The job of the player is to build a set of defensive structures to prevent a tower from the reach of enemies. Players have to focus, cleverly decide the defenses and select the relevant stimuli (flashing at 12.8 Hz) to protect the tower. Recorded EEG signals are bandpass filtered between 2 and 45 Hz and independent component analysis is used to extract the SSVEP features. Twenty-five healthy subjects have participated in the gaming and from the user's gaming experiences, the authors claim that it is possible to take the game out of the lab and even possible to present it during a public event.

14.3.2.4 Spacecraft game [34]

A novel SSVEP-BCI-based game interface is reported in [34]. The player has to control the left–right movement of spacecraft using SSVEP and to reach a target position in sky, avoiding obstacles. When the game engine makes the spacecraft start, the obstacles appear randomly. When an obstacle appears at distance, the left and right arrows are shown to the player without flashing. During this time, player has to choose the desired direction of spacecraft's movement to escape from obstacles. A few seconds later, the arrows begin to flash and player has to focus based on his intended direction. A stimuli frequency of 3–5 Hz is examined, which reportedly provides high level of comfort for players. The game design is intended to suit clinical applications, specifically as a neurotherapy tool for attention-deficit children as the game embeds player's selective attention too. Taking the level of comfort demanded in such an application into account, a single-electrode EEG from occipital area (Oz) recording is investigated.

14.3.2.5 Virtual claw game [35]

Another SSVEP-BCI implementation demonstrated in [35] uses a virtual claw machine, employing subjects' SSVEP response to two flickering stimuli at 15 and 17 Hz. The actions namely, movement of the claw and drop/clamp the objects by the claw are controlled by the response evoked in user as the player gazed on the 15- and 17-Hz stimulus, respectively. The data acquired from EEG channel Oz and the peak power values at 15 and 17 Hz are computed to detect the targets from SSVEP response correctly from three volunteers. The performance of the system is reported as follows: average accuracy of 96% and mean ITR of 18.23 bits/min. The study also reports the positive correlations between BCI accuracy and subject's attention level and suggests that gaming can be an efficient tool to improve attention levels of children.

14.3.2.6 Checker game [36]

A multiplayer BCI game using SSVEP is reported in [36] with the classic game of checkers. In the game, the squares of checkerboard are illuminated with flickering LEDs on the gaming screen. SSVEP response is elicited as the subject attends to a specific square, which in turn is detected by the BCI system, and is translated into a robotic arm control that moves the piece toward the selected square. EEG is recorded from electrodes PO3, PO4, PO7, PO8, O1, O2 and Oz and data are analyzed using open source BCI2000 software. Along with the game performance analysis in two-player mode, the study also investigates how the various game parameters such as interstimulus distance, interstimulus angle, distance between target stimulus and subject, number of competing stimuli and visual occlusions of the stimuli affect gaming performance and suggests a few preferred strategies too.

14.3.2.7 Memory game [37]

The study in [37] employs a BCI speller which uses neural responses to accurately predict character selection. The speller interface consists of a 6×6 matrix of characters, and the task is to perform copy spelling. The characters flashed at random intervals and the participants are instructed to spell a word, by focusing on each of the desired character and counting the number of times it flashed. The target word will be displayed on the screen, with the target letter indicated within brackets next to the word. The character detected by the BCI will be displayed after a delay of 2 s, following which the user attends to the next character. EEG was recorded from eight channels (C3, Cz, C4, T5, P3, Pz, P4, T6), at a sampling rate of 200 Hz. The data were bandpass filtered between 1 and 35 Hz and classified using stepwise linear discriminant algorithm. Five healthy subjects have participated in the experiments to write 35–70 words and successfully controlled the speller with an average accuracy of 88.47%.

14.3.2.8 Game with virtual class room [38]

This work presents a series of feedback games driven by P300 potential, to improve the subjects' attention. Classification technique using temporal and template-based features has been employed to detect P300 responses. Five subjects participated in the study and report an average error below 30% while playing games. The system uses

EEG features related to sustained visual attention in an immersive 3D VR classroom with simulated distractions and tests visual discrimination of players.

14.3.2.9 Space Invaders [39]

Based on the vintage video game "Space Invaders," [39] presents a prototype of a BCI-based video game using P300. The P300 odd-ball paradigm is used to "destroy" (select) a particular alien (target) among a number of aliens that appeared on a grid, in the game interface. The user task is to concentrate on the target alien, and the corresponding P300 response is processed by BCI. For each of the aliens appearing on the grid, the machine-learning technique assigns a probability value. The target selection is based on these probability scores. A prototype that used a game display of 36 aliens on the screen demonstrates higher transfer rate, and accurate target selection has been noted after 1–3 repetitions. This video game and the developer tools are available open source, and the researchers intend to attract larger groups of gamers by investigating BCI ergonomy.

The SSVEP/P300-based BCI paradigms offer various advantages and hence are widely used to develop real-time games. The SSVEP-BCI system can perform with reasonable accuracy, even with a very short training time, which is one of the most desirable features in BCI. These systems are relatively immune to artifacts. Multi-dimensional commands can be implemented using SSVEP and are often easier to append more commands. Short-training time, accurate and defined control commands and higher ITR are the attractive features of SSVEP-BCIs. Researchers have also identified few limitations of SSVEP, which include the visual fatigue as a result of flickering visual objects. In the situation of the control of multiple targets (usually more than tens), P300 shows its unique virtue, such as in the speller system. Though fatigue could be alleviated to some degree by the P300 protocol compared to SSVEP, the development of new paradigms with less fatigue even with the long utilization of BCI is still an open challenge. Further, for subjects with neurological disorders related to photosensitivity or severe muscular disorders, SSVEP/P300 is not a suitable option [21,23].

Interestingly, in this type of paradigm employed in reactive BCIs, player does not have to consciously generate the brain patterns, but the stimulus will trigger the generation of modulated brain waves. Though it appears easier to implement and to perform by the players, sometimes it may cause user fatigue, especially due to flickering effects from the screen. However, this paradigm is ideal for generating flexible and large number of control inputs, by changing the flashing frequency. With the goal of developing dynamic, challenging and engaging platforms, the designers of commercial video games often use complex shapes and movements, fast transitions, etc., and they may add more fatiguing factors through flashing objects. Studies reported in [40,41] present the potential of a less tiring new visual stimuli, motion onset visually evoked potentials (mVEPs). An mVEP occurs in the EEG response after the sudden motion of a moving stimulus. The perception of motion initially occurs in the primary visual cortex and extends to the medial temporal and medial superior temporal areas of the visual cortex. mVEP responses are time and phase locked to the onset of motion and consist of three main components, separable in time and positivity. The P100 positive peak occurs at around 100 ms poststimulus, N200 negative peak at around 160–200 ms and finally, the P200. P200 amplitude can be increased with more complex moving stimuli and has a latency of between 240 and 500 ms. Using mVEP-BCI-controlled game, the variation of mVEP-BCI performance with the visual complexity and more basic properties such as primary colors, image transitions, speed, dynamic movement of an object, etc. of the game are investigated [40]. The study reports various design requirements of mVEP-BCI game, to make it accurate and visually less fatiguing. Another mVEP-BCI game has been reported in [41], which used a commercial grade 3D game of car racing. The task involves navigating a car in a racing circuit and the position of car has been selected by attending to five different motion-based stimuli. The system has been validated by single-session recordings from 15 teenagers. The performance parameters, the classification accuracy and ITR are reported as 65%–70% and 10–12 bits/min, respectively. Further investigation is necessary to unveil the potential of this new candidate in BCI games.

14.3.3 Passive BCI games

Passive BCI does not incorporate any particular stimulus to elicit brain response but continuously monitors user's brain status to generate control signals. In order to establish the communication link between game and player's brain, passive BCI evaluates the attention level, relaxation level or emotional state, etc., while playing the game [42–45]. Related features can be estimated using temporal and spectral parameters of EEG. Usually, band power values in alpha (8–12 Hz), beta (13–30 Hz) and theta (4–8 Hz) [1,4] EEG rhythms are employed for quantifying attention, relaxation and emotional levels. The levels estimated from the EEG can be employed to control a cursor, the movement speed, direction or acceleration of an object, etc. depending on the setting/requirement in the gaming environment.

Many systems employ attention as the main neurofeedback parameter because of its significance in assessing a person's cognitive capabilities. In EEG-driven neurofeedback BCI games based on attention, frequency training is the most commonly used method because of its simplicity and proven efficacy of band power parameters in assessing the attention level [13]. It is reported that player has to decrease his slow wave EEG activity (delta/theta) and/or increase his fast wave activity (alpha/beta) in brain signals for enhancing attention [3,11]. The ratio of beta to theta power is considered as a potential index to assess the level of attention using EEG. Usually, a few electrodes from the frontal, parietal or occipital regions of brain will be sufficient to assess the attention/relaxation levels of the players during the game. Here, we discuss a number of passive BCI games reported in literature.

14.3.3.1 Brain ball

A typical example of a passive multiplayer game is the "Brain ball" [42,43]. An EEG head band is used to measure the EEG of players from frontal cortex. A "relaxation score" is computed using the ratio of the alpha and beta frequencies generated by the subjects' brains. This relaxation score enables the more relaxed player to roll a steel ball faster across the table. However, as the ball shifts toward the opponent, the player's excitement rises which reduces the relaxation index and causes the player to eventually lose the game. After a few gaming sessions, players learn to control his brain states more effectively to win points.

14.3.3.2 Fruit picking game [45]

Reference [45] shows BCI control of a 3-D hand, to grab a strawberry placed on a plate. The level of concentration decides the position/grasping strength of player's hand. EEG is taken by the Bio Radio 150 (created by Cleveland Medical Devices), which allows the transmission of 8 bipolar EEG channels to the host PC which plays the game. The higher the attention level, the faster the collection of the fruit. Ten subjects have played the game and controlled the attention-driven picking action successfully.

14.3.3.3 Mind-Ninja [46]

This game is similar to that of the popular Fruit-Ninja where the objective of the game is to slice as many fruits, avoiding the undesired occasional bombs that appear over a span of 1 min. Slicing the wrong object will cause the player to lose points. The player should concentrate on the game at all times maintaining the attention score above a threshold value, to proceed with the game successfully. If a deviation is found in the brainwaves, the screen becomes foggy and therefore it becomes hard for the player to move on. On the other hand, if the player is able to maintain the highest attention level, the movement of other objects in the screen will slow down, offering players better chances to slice as many fruits as possible. The Neuro Sky Mind wave Mobile EEG headset is used in this system to record brainwaves for successful game control with single electrode.

14.3.3.4 Drawing game [47]

The neurofeedback drawing game presented in [47], aimed at attention improvement, is composed of several modules including eye gaze tracking module, brain signal recording module, the attention state detection module and the drawing module. Gaze tracking module tracks the eye position to find out the point on the screen where the subject is focusing. Attention detection module computes the ratio of alpha to beta band power of EEG signals recorded from channels Fp1 and Fp2 to assess the person's level of focus in the drawing pattern. If the attention level is high, the neurofeedback system makes the cursor green and subject can freely draw the pattern displayed on the monitor. If attention is low, the cursor becomes red (as a warning to the drawer) and player has to improve his attention to proceed with the drawing. It has been found that the neurofeedback helps the players focus on drawing more effectively and enhances the attentional capability of players too.

14.3.3.5 Harvest challenge [48]

The study presents a set of mini games which employs player's attention level as the main control parameter. The attention score is measured as theta/beta value from central part of brain's frontal lobe, picked up by Mind wave neuroheadset. The mini games in harvest challenge are designed such that they test player's planning ability,

ability to wait, ability to follow instructions and capability to achieve objectives. During the journey to harvest fruits, player has to focus on the screen and select objects and relevant game elements carefully to achieve game points. Finally, in order to collect fruits in his basket, user has to sustain his attention above the threshold value for a predefined period of time. If he is not able to sustain his attention, the fruits will disappear and cannot be collected. Hence, the game requires the selective and sustained attention controls to eventually win points. The developers claim the proven benefits of this game in children with ADHD, to reinforce fundamental abilities of brain, providing notable improvement in control of inattention and impulsiveness. This study also suggests that the neurofeedback video game has the potential to be open for the public or private educational institutions, to act as a booster in learning process of the attention deficit.

14.3.3.6 Shooting game [49]

This game allows gamer to shoot enemies (robots appearing on the gaming interface) based on the subject-specific neurofeedback parameter set in the game. The shooting process is activated based on the neurofeedback parameter set for a particular subject. If the player's brain activity is in the desired state (e.g. the individual alpha peak power is greater than threshold), then the robot changes the color to red and stands still. At that time, player can shoot the robot. When the subject's brain activity is in the undesired state, robots turn red and run around. A preliminary study has been conducted on the usefulness of the proposed neurofeedback system with three subjects taking six sessions each. The neurofeedback protocols based on the power of individual upper alpha or beta/theta ratio training are reported in the investigation. The hypothesis is that the individual alpha peak frequency becomes greater after the neurofeedback training provided by the "shooting game." The results hence prove that the training is effective in increasing the alpha peak frequency, which in turn reflect the impact of game to enhance cognitive abilities for the subject to do multitasking.

14.3.3.7 Brain Cogo-Land [50]

In Cogo-Land presented in [50], the gamer can choose an animated avatar to run around an island. Speed of the selected avatar is controlled by player's ability to focus. A mobile EEG headband assesses the level of player's attention and controller generates an index in the range of 0–100. This attention index is continuously displayed in the monitor for user's information that encourages him to maintain the index as high as possible. They have to sharply focus on the screen and improve score as much as possible so that the avatar runs fast. There are three difficulty levels of the game based on the mode of fruit collection performed by the bird in its island travel. The beginner level requires only the movement control of the avatar and the points are counted based on the laps completed within a time frame. In the second level, player has to press certain button to jump and collect fruits whenever required and final game credits will be based on the avatar speed and fruits collected. Finally in the third difficult level, player has to collect fruits in a specific order. So, the attention and memory are tested during the game play. The game is specifically designed for treating the inattention symptoms of ADHD and system has been validated by testing the game in children with ADHD.

14.3.3.8 Ball game [51]

This game introduces an EEG driven gaming interface where the player's concentration is used to control the game by pushing a ball from one end of game's graphical user interface to the other end. Attention of the player is assessed using the ratio of theta to beta band power in EEG signals. EEG from channels AF3, AF4, O1 and O2 is considered to compute the level of player's attention. It is observed from experiments that the players are able to focus well and sustain their attention for longer period of time, after playing the game for a few sessions.

14.3.3.9 Multiplayer monster game [52]

It proposes a multiplayer video game in 3D environment, controlled by EEG features related to different levels of player's attention. Three different levels of attention have been detected from players based on their entropy and band power values of alpha, beta and theta bands in EEG. Visual attention of the player has direct control over the speed of the player also while it moves around the play area. In addition to these control elements by EEG features, a number of conventional control mechanisms such as keyboard inputs have also been incorporated in the game to facilitate the player's actions. The game is designed such that player is able to kill more number of enemies (monsters) in the gaming environment if he is able to acquire more lasers using his high level of attention. Each kill by every player will be recorded and tabulated in the score table during the game. The game lasts for 180 s and player with the highest score wins the game at the end.

14.3.3.10 Matrix game [53]

In this, the GUI protocol is designed such that player has to focus on a set of numbers displayed in the form of a 3 × 3-matrix textbox, memorize them and to correctly refill the matrix. Subject is able to refill the matrix correctly only if his/her attention level crosses a specific threshold. This attention level, the control parameter of game, is continuously provided in the form of a progress bar in the GUI which forms the core neurofeedback element in the game. At first, a set of vacant textboxes in the form of a 3 × 3 matrix are presented to the player in the GUI. Sample entropy of EEG signals recorded from ten channels of EEG using Emotiv EPOC neuroheadset has been used to estimate the attention level of the player. The ten channels are F3, F7, F3, P7, O1, O2, P8, F4, F8 and AF4. The schematic of the neurofeedback system is shown in Figure 14.3.

Using the above interface, our group has done an extensive analysis on whether we are able to improve attention scores by playing the game. Eight healthy subjects participated in the experiments and played the neurofeedback game continuously for 5 days. On each day, the subjects played one single game of highest difficult level. We have computed the attention scores obtained and plotted in Figure 14.4. Hence, it is evident that neurofeedback games not only provide a novel entertainment mode but also a training tool to enhance attention and related cognitive properties of brain.

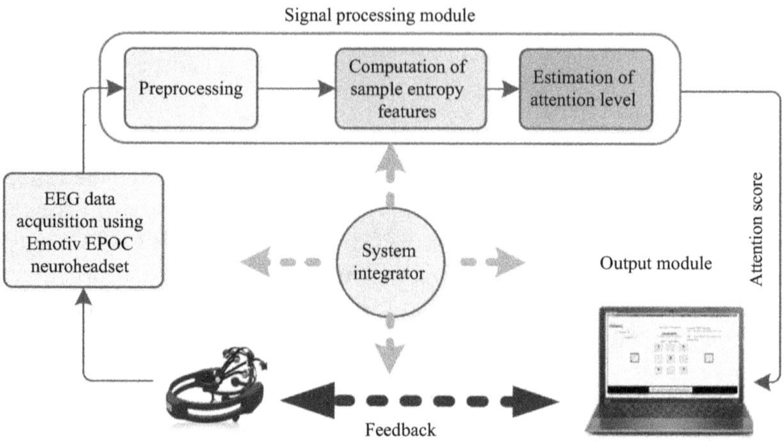

Figure 14.3 Framework of the attention-driven neurofeedback game

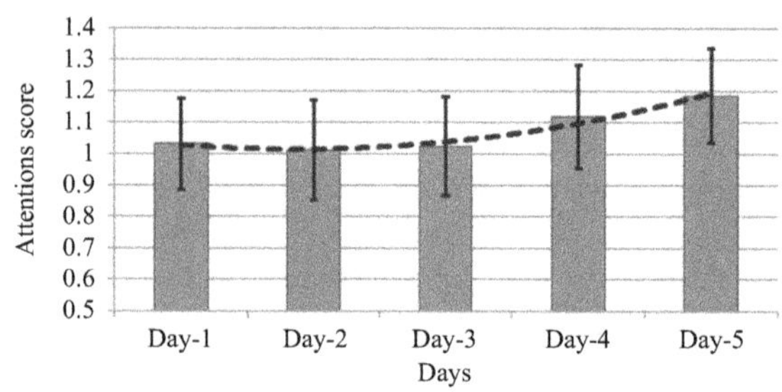

Figure 14.4 Average attention score values of eight subjects over 5 days

As discussed here, many studies indicate possibilities of passive BCI games in the healthy as well as the attention deficit. Further research is essential to quantify the effect of neurofeedback on attention enhancement and the elements in the game to be fine-tuned to improve the player's attention level/performance.

14.3.4 Hybrid games

As stated in the previous sections, most of the BCI games reported in literature generally rely on controls from a single mode. A hybrid version of any two of these modes can have multiple advantages but is rarely explored in literature. As it is possible to extend the number of brain events in hybrid BCI compared to other three paradigms, it can offer more engaging gaming environment to the players with more sophisticated control system.

There are a few studies which integrate the control paradigms in multiple modes, active, reactive or passive. A noninvasive BCI system that combines multiple neural activations to control an adapted version of the popular Tetris game is presented in [54]. The system is proposed as a treatment for children with ADHD. P300 ERP which is a response to odd-ball paradigm is used as a neurophysiologic marker of ADHD and to monitor and assess the improvement of patients. In this study, P300 ERP along with sensorimotor rhythms mu (8–12 Hz) and beta (18–24 Hz) are used to develop three different versions of BCI-Tetris game. ERD and ERS are motor responses denoted by a fall and rise in sensorimotor rhythms, respectively. In the BCI-Tetris game, the controls required are selection of orientation and position of a tile. In the system, the rotation of tile to attain the required orientation was controlled by detecting P300 from the user, and movement of tile toward left or right was controlled by motor imagery toward left or right. Thus, the paper reports an interesting approach of hybrid-BCI system that can be used in neurofeedback training and is reported to be clinically relevant.

A few more hybrid systems combining motor imagery and evoked potential are available in BCI literature [55]. However, the exact potential of combining various brain activities in a single BCI game is yet to be explored. The optimized hybrid systems will be able to provide additional benefits of BCI interaction, improving the communication speed and user's experience. When we increase the percentage of "brainy control" in the game, game can be more challenging and delivers a better feel of "mind control." However, necessary training has to be given to self-regulate specific EEG rhythms according the gaming environment.

14.4 EEG devices for neurofeedback development

The inherent complexity of designing and implementing EEG-based neurofeedback games is on account of the difficulties incurred while recording brain signals using the encephalogram technique. Traditionally, most of the research grade EEG devices are costly and gel based, which gives slight discomfort for its users over long time usage. Usually, in traditional neurofeedback games, mental actions/reactions are classified and mapped as command signals to control the gaming environment. If the recording device gives discomfort or the mapped control signals go wrong frequently, the player may gradually lose his interest and involvement in the gaming interface, even though the gaming environment is so appealing. Motivated by these facts, invention of more comfortable and economical recording devices is a hot topic now in BCI hardware research community. Nowadays, the neuroscientists and engineers collaborate together to establish smooth, fast and robust communication between man and computer, to avoid any involvement of muscular activation in BCI-based communication, with quality ensured EEG recording devices.

Table 14.1 shows the list and details of EEG devices available in market. The first five rows show the most commonly used EEG amplifiers in core-stream BCI applications. They provide research grade EEG at the expense of several thousands of dollars. But, low cost, dry and portable EEG devices are more appreciated in gaming applications. Rows 6–13 show the available wireless EEG headsets which can contribute in the progress of neurofeedback games. BCI community has explored

Table 14.1 EEG devices in market, examples are on the right side

No.	Name	Channels	Wired/wireless
1	Neuroscan	16/32/64/128	Wired
2	Brain amp	32/64/128	Wired
3	MOBIlab+	8/16/32	Wired
4	Biosemi "Pin Type"	16/32/128	Wired
5	G.tech	16/32/64	Wired
6	Enobio	8/20/32	Wireless
7	Emotiv EPOC	14	Wireless ($799)
8	Emotiv Insight	5	Wireless ($299)
9	Neurosky mindset	1	Wireless ($199)
10	InteraXon Muse	4	Wireless ($249)
11	Melon EEG headband	3	Wireless
12	Mind play Brainband XL	1	Wireless ($232)
13	iWin's Aurora	1	Wireless ($299)

Neuroscan

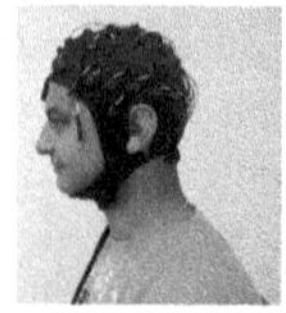

Emotiv Insight

most of these devices in game control, especially the eighth item, named Emotiv EPOC [51–53], which can be bought at $799. It has 14 EEG electrodes namely AF3, F7, F3, FC5, T7, P7, O1, O2, P8, T8, FC6, F4, F8, AF4. A smaller version of it with five channels, Epoc insight, is also available in market since 2016, which is of $299. Neurosky mindset, a single channel device available at $199, is one of the cheapest devices, up to our knowledge. Advances in design technology will help to develop new hardware devices to help the transfer of BCI games outside the laboratory and hopefully the next decade will witness the fully fledged brain controlled entertainment modalities. However, validation investigations are yet to be performed to examine the quality of the brain waves captured by the upcoming portable/dry EEG devices.

14.5 Benefits of neurofeedback games

Literature clearly shows numerous benefits of neurofeedback-based BCI games, especially in entertainment, and cognitive skill enhancement. An additional benefit of neurofeedback training is reported recently which discusses the advantages of playing a neurofeedback game before starting a general BCI experiment. All of these have been discussed in the following subsections.

14.5.1 Novel entertainment modality

In the exchanges of BCI gatherings discussed in [14,16], it is highlighted that BCI for gaming will stand out amongst the most encouraging applications of BCI, after the assistive technology innovation. Video games are of great acceptance nowadays and integration of brain control will attract gamers extensively, provided if BCI is equipped at its best performance possible. Brain is the controller of all our actions and it is possible to detect the intention/preparation of the subsequent actions going to be performed by the player. The players in EEG-based game reported in [32] mention that they are able to control the steering wheel before they started consciously doing so. This type of intention prediction, focus/attention recognition, understanding the patterns of EEG toward various kinds of stimuli, etc. will give direct mind control feeling in BCI games, if the system achieves high classification accuracies and ITRs in the closed-loop real-time paradigm. The best thing with BCI control in games is its unique way of interaction with the gaming world, which cannot be offered by any of the existing/emerging technologies.

14.5.2 Cognitive enhancement tool in the neurologically challenged as well as healthy

Based on the impact of neurofeedback training to rewire the neuronal circuits of brain, it has been accepted as one of the relevant treatment modalities for people with neurological disorders, especially for ADHD. ADHD is characterized by three behavioral symptoms: inattention, hyperactivity and impulsivity. The available intervention strategies such as medication or behavioral treatments have a number of disadvantages, especially in terms of generalization and long-term effects [44,50]. Therefore, neurofeedback in BCI games has been emerging as one of the most promising modalities for treating the attention deficient. Particularly, there are few studies that have examined the impact of the video games in ADHD and its capacity to reduce impulsivity and inattention. The pilot study in [50] also reports the benefits of BCI game in enhancing attention/cognitive skills of the ADHD after a few weeks of neurofeedback game play. However, based on the neurofeedback game's benefits in brain skill enhancement, they can be used to improve cognitive abilities of healthy people too. Neurofeedback strategies based on EEG parameters associated with cognitive attributes, such as attention, memory, working memory, can provide alternative pathways to enhance brain's power. It has been reported that neurofeedback paradigms utilizing alpha, beta and theta control have significant impact on a person's cognitive skills. The work presented in [56] reports enhancement in cognitive skills followed by a few days of subject-specific alpha band training even in stroke patients. Our previous work also shows the capability of attention score enhancement for a group of eight healthy persons attending a simple neurofeedback game for a few days [53]. The study also indicates the possibilities in cognitive skill enhancement followed by EEG-based attention training. If we provide the relevant neurofeedback control in highly engaging and immersive games, it will motivate players to play games with better focus, which in turn regulate the brain rhythms positively and enhance associated skills.

14.5.3 BCI performance booster

BCIs are communication systems that translate brain activity into control commands to operate a computer application or some external devices. For any BCI to be tested in real time, it is necessary to perform a calibration or training procedure, using which a subject-specific model is developed. In order to develop this model, long and repetitive trials may be required, eventually resulting in user fatigue and low performance. Hence, the motivation of subjects to take part in a BCI experiments is of great importance on achieving final performance goal. Most of the BCI studies are oriented on optimizing signal processing and machine-learning layers in the BCI loop, neglecting this human aspect. Motivated by the advances on usage of BCI for video games and virtual environments, a recent study in [57] investigates the effect of prior gaming experience on brain pattern modulation in BCI communication. The study uses an SMR-based neurofeedback game prior to actual BCI experiment aiming to systematically identify specific elements that contribute to excellent BCI control. In the reports, it is argued that experienced gamers could have better performance in BCI training due to enhanced sensorimotor learning derived from gaming. To achieve this, a pilot study with 12 participants is conducted. Results show that a strong gaming profile not only could possibly enhance the performance in BCI training through motor imagery but it can also increase EEG rhythm activity. Another study in [58] also indicates the impact of gaming before BCI experiments. However, these studies give valuable insights to the novel application of BCI games in general BCI purposes. Further research is essential to validate the methodologies, to prove its impact on BCI performance and to quantify the contribution of prior gaming experience on various types of BCI applications.

14.6 Challenges in practical implementation

Neurofeedback-based BCI game falls under the category of "serious games" which repurpose videogame's core technology of entertainment to further applications. Looking at state-of-the-art designs, BCI games carry high potential to be explored for the healthy as well as the neurologically challenged, allowing players to control a game partially or fully with brain signals. This interaction between brain and computer can be in an active way: to send control commands continuously, in reactive way: to send control commands in response to a stimulus, in a passive way: to adapt the game content based on the user's brain activity, or with a multi-modal fusion of these. A number of games have been reported using these modalities, designed in 2-D, 3-D and virtual environments. Unfortunately, detailed scientific reports on the neurofeedback experience provided by these games and impact of various gaming factors are limited. The mode of giving feedback is a significant element in any BCI game because of its high influence on player's EEG, game performance and user's interest/motivation. Literature shows that providing feedback in VR mode shows superior performance to 2-D feedback mode [17]. Also, the players will have better immersive and engaging gaming experience if the games are equipped with VR elements. Hence, new games exploring VR environment have to be developed to attract more users to attempt BCI gaming experience.

Another problem associated with BCI technique is the misunderstanding in the public that BCI has a capability to read one's mind, and it prevents people from using this technology. However, with the help of several technological breakthroughs, such as Internet, usage of BCI games has to be promoted, removing the over-simplified beliefs regarding this fascinating technology. Making the interaction protocol transparent to the user will also help in increasing the public interest in neurofeedback games. Due to these reasons, there is a lack of commercialization and acceptance by the general public toward neurofeedback games.

Additionally, the recognition performance and ITR achieved in a BCI-based game are lower compared to traditional interfaces. Current BCI systems can offer transfer rates of up to 60 bits/min—a significantly lower bandwidth compared to the conventional human–computer communications media [59,60]. In order to attain real-time control using BCI, intense training is also required. It is necessary to develop novel protocols and methodologies to reduce the training effort. To date, there is no single brain feature that serves all purposes of a robust BCI interaction. P300 and SSVEP are good in selection of targets without much conscious efforts, but they do not provide an immersive control that can be achieved by self-paced motor imagery protocols. The detection of motor imagery takes time compared to VEP, making the loop of information exchange in BCI slower. Hence, new interaction paradigms that encourage and engage player in a stronger and faster way are appreciated.

The recording device is also an important component in realizing a BCI game. Though there are a number of popular BCI devices such as Emotiv, Neurosky, etc., for EEG recording using wireless and dry electrodes, applicable to games, there are a number of demerits too. Emotiv EPOC is a 14-channel device, having frontal, parietal or occipital electrodes. No electrodes are present in motor cortex, which limits this device's application in motor imagery-based BCIs. Also, Neurosky is a single electrode device, and hence not a robust candidate that can be applied in hybrid paradigms. Though it is difficult to develop a prototype that suits all types of gaming interactions, it will be worth to have a portable and economical device which covers at least one electrode from all the brain lobes.

Though portable EEG devices have positively helped the emerging BCI games, most of the state-of-the-art games have been tested only in a small population of subjects, played in a restricted environment, away from real-life scenario. Investigation of large-scale BCI studies, close to real-life scenario, is yet to be carried out.

Using the existing devices, a few companies such as Emotiv, Neurosky, Uncle Milton, Mind Games, Mind Technologies, Brain-Master and OCZ technology show interest in developing games [14,18]. However, the neural parameters and data processing algorithms adopted by them are proprietary and hence inaccessible [17]. These games offer potential entertainment in its limited setup, but further control of brain signal and brain-driven games can be achieved only by bridging the gap between the game design and control activated by brain. Subject-specific signal parameters and game performance need to be further investigated. Good research from multidisciplinary fields will help to fine tune the optimum scenarios in BCI games to achieve their ultimate purposes, entertainment or rehabilitation.

One main reason for slow transfer from research labs to market can be the coordination between multidisciplinary groups, especially the BCI researchers and game developers [7]. The aim of a BCI researcher may be on integrating brain events with the gaming interface, with less importance on the game content. In order to make a BCI game more engaging and interesting, than just to be feasible with BCI inputs, game control paradigms should be developed with consistent interactions and discussions with multidisciplinary groups. Game developers and BCI scientists have to try concurrently to alleviate the challenges in existing control paradigms, using motor imagery, attention/relaxation, VEPs and P300. New control parameters have to be identified and integrated in the emerging BCI games. It is also necessary to find out the optimum neurofeedback strategies applicable to various types of game genres and significant to evaluate gamer's performance by game developers. The evaluation will help to decide what type of control signals and feedback mode will be suitable in various games depending on its genre, namely action, strategy, role-playing, adventure, sports and puzzle solution. Advances in neuroscience and gaming technology will eventually help scientists and engineers to address all these hurdles and transfer the benefits of BCI games to the general public.

14.7 Conclusion

In recent years, many proof-of-concept investigations have shown that BCI games are of great importance, in medical and entertainment applications. This chapter presented a variety of examples showcasing the latest developments in BCI-based neurofeedback game designs. Neurofeedback games utilize a number of features extracted from EEG, associated with motor imagery, motor movements, selective attention, sustained attention, visuospatial attention, VEPs, etc. Efficient signal processing schemes and machine-learning algorithms employed in those studies extract informative EEG pattern and accurately decode player's intentions. The classified signals are then successfully mapped into game control commands which provide a new mode of playing the game only by "thoughts." The feedback about brain state motivates the player to self-regulate his/her brain waves and thus enhances game performance gradually. In order to make the subject actively engage and involve in this new mode of gaming system, it would be better to give a more interesting feedback rather than simple feedback (like graphically representing the modulated brain indices). More realistic, interactive and motivationally engaging gaming environments accompanied with attractive visual or auditory feedback, emulated with real-life situations like race cars, will enhance player's interest and motivation in playing game. Further, visually appealing gaming interface may significantly reduce the training period when compared to the traditional feedback systems too. The promising benefits of neurofeedback game offered to the healthy as well as the cognitively challenged persons ensure that it is going to be one of the most important applications of BCI in future. However, in order to transfer the game play experience from laboratory to external world, it is essential to optimize existing neurofeedback systems. Suitable headset selection, proper gaming environments, robust algorithms, effective mapping of control signals, control actions, feedback modalities, etc. will ultimately help to enhance use's neurofeedback-based gaming experience.

References

[1] Niedermeyer E., "Historical aspects," In E. Niedermeyer and F. Lopes da Silva (Eds.), *Electroencephalography: Basic Principles, Clinical Applications, and Related Fields*, 5th ed., Lippincott Williams & Wilkins, Baltimore, pp. 1–16, 2004.

[2] Vidal J. J., "Toward direct brain-computer communication," *Annual Review of Biophysics and Bioengineering*, vol. 2, no. 1, pp. 157–180, 1973.

[3] Vidal J. J., "Real-time detection of brain events in EEG," *Proceedings of the IEEE*, vol. 65, no. 5, pp. 633–641, 1977.

[4] Gaggioli A., Keshner E. A., Weiss P. L. and Riva G. (Eds.), *Advanced Technologies in Rehabilitation: Empowering Cognitive, Physical, Social and Communicative Skills Through Virtual Reality, Robots, Wearable Systems and Brain–Computer Interfaces*, IOS Press, Amsterdam, 2009.

[5] Wolpaw J. R., Birbaumer N., McFarland D. J., Pfurtscheller G. and Vaughan T. M., "Brain–computer interfaces for communication and control," *Clinical Neurophysiology*, vol. 113, pp. 767–791, June 2002.

[6] Nijholt A., Bos D. P.-O. and Reuderink B., "Turning shortcomings into challenges: brain–computer interfaces for games", *Entertainment Computing*, vol. 1, pp. 85–94, 2009.

[7] Bos D. P.-O., Reuderink B., Laar B., *et al.*, "Brain-computer interfacing and games," In D. S. Tan and A. Nijholt (Eds.), *Brain–Computer Interfaces*, Springer, London, pp. 149–178, Chapter. 10, 2010.

[8] Neuper C. and Pfurtscheller G., "Neurofeedback training for BCI control," In B. Graimann, G. Pfurtscheller and B. Allison (Eds.), *Brain–Computer Interfaces*, ser. *The Frontiers Collection*, Springer, Berlin, Heidelberg, pp. 65–78, 2010.

[9] Vernon D. J., "Can neurofeedback training enhance performance? An evaluation of the evidence with implications for future research," *Applied Psychophysiology and Biofeedback*, vol. 30, pp. 347–364, 2005.

[10] Yuan P., Gao X., Allison B., Wang Y., Bin G. and Gao S., "A study of the existing problems of estimating the information transfer rate in online brain–computer interfaces," *Journal of Neural Engineering*, vol. 10, no. 2, p. 026014, 2013.

[11] Loo S. K. and Barkley R. A., "Clinical utility of EEG in attention deficit hyper active disorder," *Applied Neuropsychology*, vol. 12, no. 2, pp. 64–76, 2005.

[12] Fox D. J., Tharp D. F. and Fox L. C., "Neurofeedback: an alternative and efficacious treatment for attention deficit hyperactivity disorder," *Applied Psychophysiology and Biofeedback*, vol. 30, no. 4, pp. 365–373, December 2005.

[13] Lim C. G., Lee T. S., Guan C., *et al.*, "A brain–computer interface based attention training program for treating attention deficit hyperactivity disorder," *PLoS ONE*, vol. 7, no. 10, p. e46692, doi:10.1371/journal.pone.0046692, 2012.

[14] Van Erp J., Lotte F. and Tangermann M., "Brain–computer interfaces: beyond medical applications," *Computer*, vol. 45, no. 4, pp. 26–34, 2012.

[15] Lotte F., Faller J., Guger C., *et al.*, "Combining BCI with virtual reality: towards new applications and improved BCI," *Towards Practical Brain–Computer Interfaces*, Springer, Berlin, Heidelberg, pp. 197–220, 2012.

[16] Future BNCI, A Roadmap for Future Directions in Brain/Neuronal Computer Interaction, Available online: http://bnci-horizon-2020.eu/images/bncih2020/FBNCI_Roadmap.pdf. 2015.

[17] Gurkok, H., Nijholt, A. and Poel, M., "Brain-computer interface games: towards a framework" *In International Conference on Entertainment Computing*, pp. 373–380. Berlin, Heidelberg: Springer, 2012.

[18] Nijboer F., Allison B. Z., Dunne S., *et al.*, "A preliminary survey on the perception of marketability of brain–computer interfaces and initial development of a repository of BCI companies," In G. R. Müller-Putz, R. Scherer, M. Billinger, *et al.* (Eds.), *Proceedings of the 5th Int. Brain–Computer Interface Conference*. Graz: Verlag der Technischen Universitaet Graz, p. 4, 2011.

[19] Pineda J. A., Silverman D. S., Vankov A. and Hestenes J., "Learning to control brain rhythms: making a brain–computer interface possible," *IEEE Transactions on Neural Systems and Rehabilitation Engineering*, vol. 11, no. 2, pp. 181–184. doi:10.1109/TNSRE.2003.814445, 2003.

[20] Krepki R, Blankertz B, Curio G, Müller K. R., "The Berlin brain–computer interface (BBCI)—towards a new communication channel for online control in gaming applications," *Multimedia Tools and Applications*, vol. 33, pp. 73–90, 2007.

[21] Lalor E. C., Kelly S. P., Finucane C., *et al.* "Steady-state VEP-based brain–computer interface control in an immersive 3D gaming environment," *EURASIP Journal on Signal Processing*, vol. 19, pp. 3156–3164, 2005.

[22] Martinez P., Bakardjian H. and Cichocki, A., "Fully online multi-command brain–computer interface with visual neurofeedback using SSVEP paradigm," *Computational Intelligence and Neuroscience*, doi:10.1155/2007/94561, 2007.

[23] Bos D. P.-O., Reuderink B., De Laar B. V., *et al.*, "Brain-computer interfacing and games,". In D. S. Tan and A. Nijholt (Eds.), *Brain–Computer Interfaces. Applying our Minds to Human-Computer Interaction*, Springer London, London, pp. 149–178. doi:10.1007/978-1-84996-272-8, 2010.

[24] Scherer R., Friedrich E. C. V., Allison B. Z., *et al.*, "Non-invasive brain–computer interfaces: enhanced gaming and robotic control," In *Lecture Notes in Computer Science*, vol. 6691, pp. 362–269, Springer-Verlag, Berlin, 2011.

[25] Tangermann M., Krauledat M., Grzeska K., *et al.*, "Playing pinball with non-Invasive BCI," In *Advances in Neural Information Processing Systems*, MIT Press, Cambridge, MA, USA, vol. 21, pp. 1641–1648, 2009.

[26] Congedo M., Goyat M., Tarrin N., *et al.*, ""Brain Invaders": a prototype of an open-source P300-based video game working with the OpenViBE platform," In G. R. Müller-Putz, R. Scherer, M. Billinger, A. Kreilinger, V. Kaiser, and C. Neuper (Eds.), *Proceedings of the 5th International Brain–Computer Interface Conference*. Graz: Verlag der Technischen Universitaet Graz, pp. 1–6, 2011.

[27] Wolpaw J. R., McFarland D. J., Neat G. W., and Forneris C. A., "An EEG-based brain–computer interface for cursor control," *Electroencephalography and Clinical Neurophysiology*, vol. 78, pp. 252–259, 1991.

[28] Bonnet L., Lotte F. and Lécuyer A., "Two brains, one game: design and evaluation of a multiuser BCI video game based on motor imagery," *IEEE Transactions on Computational Intelligence and AI in Games*, vol. 5, no. 2, pp. 185–198, 2013.

[29] An H.-S., Kim J.-W. and Lee S.-W., "Design of an asynchronous brain–computer interface for control of a virtual Avatar," *Brain–Computer Interface (BCI), 2016 4th International Winter Conference on IEEE*, 2016.

[30] Ang K. K., Chin Z. Y., Zhang H. and Guan C., "Filter bank common spatial pattern (FBCSP) in brain–computer interface," *IEEE International Joint Conference on in Neural Networks*, pp. 2390–2397, 2008.

[31] Coyle D., Garcia J., Satti A. R. and McGinnity T. M., "EEG-based continuous control of a game using a 3 channel motor imagery BCI: BCI game," *2011 IEEE Symposium on Computational Intelligence, Cognitive Algorithms, Mind, and Brain (CCMB)*, pp. 1–7, 2011.

[32] Kim D. and Cho S. B., "A brain–computer interface for shared vehicle control on TORCS car racing game," *10th IEEE International Conference on Natural Computation (ICNC)*, pp. 550–555, 2014.

[33] Van Vliet M., Robben A., Chumerin N., Manyakov N. V., Combaz A. and Van Hulle M. M., "Designing a brain–computer interface controlled video-game using consumer grade EEG hardware," 2012 *Biosignals and Biorobotics Conference (BRC)*, pp. 1–6, 2012.

[34] Parafita R., Pires G., Nunes U. and Castelo-Branco M., "A spacecraft game controlled with a brain–computer interface using SSVEP with phase tagging," *IEEE Serious Games and Applications for Health (SeGAH), 2013 IEEE 2nd International Conference on*, pp. 1–6, 2013.

[35] Po-Lei L., Hao-Teng S. and Hsiang-Chih C., "Design a brain–computer interface gaming system using steady-state visual evoked potential," *IEEE International Conference on Consumer Electronics-Taiwan (ICCE-TW)*, pp. 5–6, 2014.

[36] Akhtar A., Norton J. J., Kasraie M. and Bretl T., "Playing checkers with your mind: an interactive multiplayer hardware game platform for brain–computer interfaces," *36th Annual International Conference of the IEEE Engineering in Medicine and Biology Society (EMBC)*, pp. 1650–1653, 2014.

[37] Angeloni C., Salter D., Corbit V., Lorence T., Yu Y. C. and Gabel, L. A., "P300-based brain–computer interface memory game to improve motivation and performance," *38th IEEE Annual Northeast Bioengineering Conference (NEBEC)*, pp. 35–36, 2012.

[38] Rohani D. A., Sorensen H. B. D. and Puthusserypady S, "Brain–computer interface using P300 and virtual reality: a gaming approach for treating ADHD," *36th Annual International Conference of the IEEE Engineering in Medicine and Biology Society (EMBC)*, pp. 3606–3609, 2014.

[39] Congedo M., Goyat M., Tarrin N., *et al.*, "'Brain Invaders': a prototype of an open-source P300-based video game working with the OpenViBE platform," *5th International Brain–Computer Interface Conference 2011 (BCI 2011)*, pp. 280–283, September 2011.

[40] Beveridge R., Marshall D., Wilson S. and Coyle D., "Classification effects on Motion-Onset Visual Evoked Potentials using commercially available video games," *IEEE Computer Games: AI, Animation, Mobile, Multimedia, Educational and Serious Games (CGAMES)*, pp. 28–37, 2015.

[41] Beveridge R., Wilson S. and Coyle D., "Can teenagers control a 3D racing game using motion-onset visual evoked potentials?," *Brain–Computer Interfaces*, vol. 4, no. 1–2, pp. 102–113, 2017.

[42] Hjelm S. I., "Research+ design: the making of Brain ball," *Interaction*, vol. 10, pp. 26–34, 2003.

[43] Hjelm S. I., Eriksson E. and Browall C, "Brainball—using brain activity for cool competition," *Proceedings of the First Nordic Conference on Human-Computer Interaction*, p. 59, 2000.

[44] Pope A. T. and Palsson O. S., "Helping video games 'rewire our minds'," *Tech. Rep.*, NASA Langley Research Center, 2001.

[45] Jiang L., Guan C., Zhang H., Wang C. and Jiang B., "Brain–computer interface based 3D game for attention training and rehabilitation," *6th IEEE Conference on Industrial Electronics and Applications*, Singapore, pp. 124–127, June 2011.

[46] Joselli M., Binder F., Clua E. and Soluri E., "MindNinja: concept, development and evaluation of a mind action game based on EEGs," *2014 Brazilian Symposium on Computer Games and Digital Entertainment (SBGAMES)*, pp. 123–132, 2014.

[47] Moon J.-H., Park K.-H. and Lee S.-W., "Neurodrawing: neurofeedback for enhancing attention by drawing," *4th IEEE International Winter Conference on Brain–Computer Interface (BCI)*, pp. 1–2, 2016.

[48] Munoz J. E., Lopez D. S., Lopez J. F. and Lopez A., "Design and creation of a BCI videogame to train sustained attention in children with ADHD," *10th IEEE Computing Colombian Conference*, pp. 194–199, 2015.

[49] Liu Y., Sourina O. and Hou X., "Neurofeedback games to improve cognitive abilities," *International IEEE Conference on Cyberworlds (CW)*, pp. 161–168, 2014.

[50] Lee X. Y., Koukouna E., Lim C. G., *et al.*, "Can we play with ADHD? An alternative game-based treatment for inattentive symptoms in attention-deficit/hyperactivity disorder," *Subconscious Learning via Games and Social Media*, Singapore: Springer, pp. 69–86, 2015.

[51] Shenjie S., Thomas K. P., and Vinod A. P., "Two player EEG-based neurofeedback ball game for attention enhancement," *IEEE International Conference on Systems, Man and Cybernetics (SMC)*, pp. 3150–3155, 2014.

[52] Khong A., Jiangnan L., Thomas K. P. and Vinod A. P., "BCI based multi-player 3-D game control using EEG for enhancing attention and memory," *IEEE International Conference on Systems, Man and Cybernetics (SMC)*, pp. 1847–1852, 2014.

[53] Thomas K. P, Vinod A. P. and Guan C., "Enhancement of attention and cognitive skills using EEG based neurofeedback game," *6th IEEE International IEEE/EMBS Conference on Neural Engineering (NER)*, pp. 21–24, 2013.

[54] Pires G., Torres M., Casaleiro N., Nunes U. and Castelo-Branco M., "Playing Tetris with non-invasive BCI," *IEEE 1st International Conference on Serious Games and Applications for Health (SeGAH)*, pp. 1–6, 2011.

[55] Ma T., Li H., Deng L., *et al.*, "The hybrid BCI system for movement control by combining motor imagery and moving onset visual evoked potential," *Journal of Neural Engineering*, vol. 14, p. 026015, 2017.

[56] Kober S. E., Schweiger D., Reichert J. L., Neuper C. and Wood G., "Upper alpha based neurofeedback training in chronic stroke: brain plasticity processes and cognitive effects," *Applied Psychophysiology and Biofeedback*, vol. 42, pp. 69–83, 2017.

[57] Vourvopoulos A., Liarokapis F. and Chen M.-C., "The effect of prior gaming experience in motor imagery training for brain–computer interfaces: a pilot study," *7th International Conference on* Games and Virtual Worlds for Serious Applications (VS-Games), 2015.

[58] Vourvopoulos A., Badia S. B. and Liarokapis F., "EEG correlates of video game experience and user profile in motor-imagery-based brain–computer interaction," *The Visual Computer*, vol. 33, no. 4, pp. 533–546, 2017.

[59] Ahn M, Lee M, Choi J, Jun S. C., "A review of brain–computer interface games and an opinion survey from researchers, developers and users," *Sensors*, vol. 14, pp. 14601–14633, 2014.

[60] Lecuyer A., Lotte F., Reilly R. B., Leeb R., Hirose M. and Slater M., "Brain–computer interfaces, virtual reality, and videogames," *Computer*, vol. 41, pp. 66–72, 2008.

Index

www.ingramcontent.com/pod-product-compliance
Lightning Source LLC
LaVergne TN
LVHW011909200926
848484LV00002B/578